Communications in Computer and Information Science 2063

Rationale

The CCIS series is devoted to the publication of proceedings of computer science conferences. Its aim is to efficiently disseminate original research results in informatics in printed and electronic form. While the focus is on publication of peer-reviewed full papers presenting mature work, inclusion of reviewed short papers reporting on work in progress is welcome, too. Besides globally relevant meetings with internationally representative program committees guaranteeing a strict peer-reviewing and paper selection process, conferences run by societies or of high regional or national relevance are also considered for publication.

Topics

The topical scope of CCIS spans the entire spectrum of informatics ranging from foundational topics in the theory of computing to information and communications science and technology and a broad variety of interdisciplinary application fields.

Information for Volume Editors and Authors

Publication in CCIS is free of charge. No royalties are paid, however, we offer registered conference participants temporary free access to the online version of the conference proceedings on SpringerLink (http://link.springer.com) by means of an http referrer from the conference website and/or a number of complimentary printed copies, as specified in the official acceptance email of the event.

CCIS proceedings can be published in time for distribution at conferences or as post-proceedings, and delivered in the form of printed books and/or electronically as USBs and/or e-content licenses for accessing proceedings at SpringerLink. Furthermore, CCIS proceedings are included in the CCIS electronic book series hosted in the SpringerLink digital library at http://link.springer.com/bookseries/7899. Conferences publishing in CCIS are allowed to use Online Conference Service (OCS) for managing the whole proceedings lifecycle (from submission and reviewing to preparing for publication) free of charge.

Publication process

The language of publication is exclusively English. Authors publishing in CCIS have to sign the Springer CCIS copyright transfer form, however, they are free to use their material published in CCIS for substantially changed, more elaborate subsequent publications elsewhere. For the preparation of the camera-ready papers/files, authors have to strictly adhere to the Springer CCIS Authors' Instructions and are strongly encouraged to use the CCIS LaTeX style files or templates.

Abstracting/Indexing

CCIS is abstracted/indexed in DBLP, Google Scholar, EI-Compendex, Mathematical Reviews, SCImago, Scopus. CCIS volumes are also submitted for the inclusion in ISI Proceedings.

How to start

To start the evaluation of your proposal for inclusion in the CCIS series, please send an e-mail to ccis@springer.com.

Zhaoquan Gu · Wanlei Zhou · Jiawei Zhang ·
Guandong Xu · Yan Jia
Editors

Network Simulation and Evaluation

Second International Conference, NSE 2023
Shenzhen, China, November 22–24, 2023
Proceedings, Part I

Springer

Editors
Zhaoquan Gu
Harbin Institute of Technology
Shenzhen, China

Wanlei Zhou
City University of Macau
Macau, China

Jiawei Zhang
Peng Cheng Laboratory
Shenzhen, China

Guandong Xu
Peng Cheng Laboratory
Shenzhen, China

Yan Jia
Peng Cheng Laboratory
Shenzhen, China

ISSN 1865-0929 ISSN 1865-0937 (electronic)
Communications in Computer and Information Science
ISBN 978-981-97-4518-0 ISBN 978-981-97-4519-7 (eBook)
https://doi.org/10.1007/978-981-97-4519-7

This Springer imprint is published by the registered company Springer Nature Singapore Pte Ltd.
The registered company address is: 152 Beach Road, #21-01/04 Gateway East, Singapore 189721, Singapore

Preface

Cyberspace technology has already penetrated various areas such as science, economy, politics as well as defense and military. Cyberspace has now become one of the most important areas which draws attention from both academics, professionals, and industrial communities. With the emergence of new technologies and applications in cyberspace, cyberspace security has been also facing great challenges. On the one hand, new technologies emerged with the prevalence of data science and artificial intelligence can largely advance and exceed the traditional security technologies, which largely improve the security defense capability of cyberspace. On the other hand, the malicious usage of new technologies, or exploiting the defects in new technologies and new applications have further exacerbated the security threat of cyberspace. To better handle the security risks caused by the new technologies and new applications and develop effective approaches and techniques for defending such potential security concerns, the network simulation technology can realize the security evaluation of these technologies by simulating in a controllable network environment, which has significant research value and great social benefit.

The purpose of the NSE Conference is to bring together world-class researchers, practitioner and engineers in the theoretical and practical areas related to network simulation and security evaluation to share their latest research results and discuss about the future development directions.

The NSE Conference was held in Shenzhen, China, from November 22–24, 2023. The conference received 73 paper submissions, of which 57 were accepted. A single-blind review process was employed, with each paper receiving an average of 2.15 reviews. We extend our gratitude to all the authors who submitted their work, whether accepted or not, for their willingness to share their insights and research findings with us.

We hope this conference will facilitate improving the cyberspace security capabilities and further promote cyberspace technology collaborations in various application domains.

January 2024

Wanlei Zhou
Zhaoquan Gu

Organization

General Chairs

Yan Jia	Pengcheng Laboratory, China
Guandong Xu	Pengcheng Laboratory, Australia

Program Committee Chairs

Wanlei Zhou	Macau City University, China
Zhaoquan Gu	Harbin Institute of Technology (Shenzhen), China

Program Committee

Aiping Li	National University of Defense Technology, China
Chao Zhang	Tsinghua University, China
Cheng Huang	Sichuan University, China
Cheng Qiao	Guangzhou University, China
Cuiyun Gao	Harbin Institute of Technology (Shenzhen), China
Denghui Zhang	Guangzhou University, China
Fengwei Zhang	Southern University of Science and Technology, China
Guangxia Xu	Guangzhou University, China
Haiyan Wang	Peng Cheng Laboratory, China
Hao Peng	Zhejiang Normal University, China
He Debiao	Wuhan University, China
Hong Chen	University of Electronic Science and Technology of China, China
Hua Wang	Victoria University, Australia
Huan Huo	University of Technology Sydney, Australia
Hui Lu	Guangzhou University, China
Huixiang Zhang	Northwestern Polytechnical University, China
Jianchao Luo	University of Electronic Science and technology of China, China
Jianye Yang	Guangzhou University, China
Jiawei Zhang	Peng Cheng Laboratory, China

Kun Xiao	University of Electronic Science and Technology of China, China
Le Wang	Guangzhou University, China
Lei Luo	University of Science and Technology of China, China
Lijie Wang	University of Electronic Science and Technology of China, China
Lijing Ren	Guangzhou University, China
Lirong Chen	University of Electronic Science and Technology of China, China
Liyi Zeng	Peng Cheng Laboratory, China
Mohan Li	Guangzhou University, China
Qing Liao	Harbin Institute of Technology (Shenzhen), China
Rong Jiang	National University of Defense Technology, China
Shanqing Guo	Shandong University, China
Shen Su	Guangzhou University, China
Shudong Li	Guangzhou University, China
Sun Jianguo	Harbin Engineering University, China
Tao Li	Nankai University, China
Wen Xia	Harbin Institute of Technology (Shenzhen), China
Wenying Feng	Peng Cheng Laboratory, China
Xiangyu Song	Peng Cheng Laboratory, China
Xiaofeng Wang	Jiangnan University, China
Xun Yi	Royal Melbourne Institute of Technology, Australia
Ye Wang	National University of Defense Technology, China
Yi Jin	Beijing Jiaotong University, China
Yonghao Gu	Beijing University of Posts and Telecommunications, China
Yun Li	University of Electronic Science and Technology of China, China
Zhenyu Li	Guilin University of Electronic Technology, China
Zhi Wang	Nankai University, China
Zhijie Qiu	University of Electronic Science and Technology of China, China
Zhijuan Li	Harbin Engineering University, China

Additional Reviewers

Aiping Li
Chao Zhang
Cheng Huang
Cheng Qiao
Denghui Zhang
Fengwei Zhang
Guangxia Xu
Haiyan Wang
Hao Peng
Hao Tan
Hao Yan
Haoliang Tan
He Debiao
Hong Chen
Hua Wang
Huan Huo
Huan Zhang
Hui Lu
Huixiang Zhang
Jianchao Luo
Jianming Li
Jiawei Zhang
Kun Xiao
Le Wang
Lei Luo
Lijie Wang
Lijing Ren
Lirong Chen
Liyi Zeng
Mohan Li
Rong Jiang
Shanqing Guo
Shen Su
Shudong Li
Sun Jianguo
Tao Li
Wenying Feng
Xiangyu Song
Xiaofeng Wang
Xun Yi
Ye Wang
Yi Jin
Yonghao Gu
Yun Li
Zhang Zhiqiang
Zhaoquan Gu
Zhenyu Li
Zhi Wang
Zhijie Qiu
Zhijuan Li

Contents – Part I

Cybersecurity Systems and Applications

Contents – Part II

Test and Evaluation for Cybersecurity

Threat Detection and Defense

Cybersecurity Attack and Defense

SecretKeeper: Robust Vertical Federated Learning Against Label Inference Attacks

Chong Fu[1], Yuwen Pu[1], Qiao Zhang[1], Jiayu Pan[2], Jing Qiu[3], Xuhong Zhang[1], Yiming Wu[4](✉), and Shouling Ji[1]

[1] Zhejiang University, Hangzhou 310027, China
{fuchong,yw.pu,j0k1ng,zhangxuhong,sji}@zju.edu.cn
[2] The Ohio State University, Columbus 43210, USA
pan.743@osu.edu
[3] Guangzhou University, Guangzhou 510006, China
qiujing@gzhu.edu.cn
[4] Zhejiang University of Technology, Hangzhou 310023, China
wyiming@zjut.edu.cn

Abstract. Federated learning (FL) is a privacy-preserving machine learning paradigm that allows multiple participants to collaboratively train one model without revealing their local datasets. As one variant of FL, vertical federated learning (VFL) is designed for the scenario where a few participants' local datasets share the same sample space but differ in the feature space, with the training labels owned by one of the participants. The participant owning the training labels is referred to as the active party, while the remaining participants are known as passive parties. Though VFL is considered to be able to protect each participant's local data, recent studies find that VFL has potential risks of privacy leakage. A typical privacy risk of VFL is the leakage of labels. By abusing the parameters of the trained local bottom model, a malicious passive party can infer training labels privately owned by the active party and even labels of testing samples. Existing defense methods against label inference attacks are limited to a specific type of VFL framework. Specifically, existing defenses are designed for VFL without model splitting and cannot be applied to VFL with model splitting. To bridge this gap, we propose the first defense for split-learning-based VFL against label inference attacks. The experimental results across multiple datasets and model architectures demonstrate the efficacy of our defense in reducing the label inference accuracy of the malicious party, while incurring only minimal performance degradation in the federated models' original task.

Keyword: Vertical federated learning · Split learning · Label inference attack

1 Introduction

Federated learning (FL) is a privacy-preserving machine learning paradigm that allows multiple participants to collaboratively train one model without revealing their local datasets. With the promulgation of privacy protection laws such as General Data Protection Regulation (GDPR) [1, 11, 22] and California Consumer Privacy Act (CCPA)

Z. Gu et al. (Eds.): NSE 2023, CCIS 2063, pp. 3–18, 2024.
https://doi.org/10.1007/978-981-97-4519-7_1

[9, 20], world-wide IT companies and researchers put much effort into developing FL frameworks with the purpose of mining the value of distributed data while protecting user privacy, including Federated Artificial Intelligence Technology Enabler (FATE) developed by Tencent [23], TensorFlow Federated developed by Google [8], and PySyft developed by OpenMined [16].

As one variant of FL, vertical federated learning (VFL) [14, 17, 24] is designed for the scenario where a few participants' local datasets share the same sample space but differ in the feature space, with the training labels owned by one of the participants. For example, a car rental company and a bank may share the same group of users but have different user features. Besides, the labels of user credits are owned by the bank. The participant who owns training labels is referred to as the active party, while other participants are known as passive parties [26]. Generally, the active party determines the training task of VFL and the passive parties are employed by the active party to offer additional features for the training task. Both the active party and the passive party utilize a local bottom model to transform local features into embeddings.

Once the local features have been converted into embeddings, in one scenario, the active party employs a top model to aggregate the embeddings from all participants and output the predicted label. This framework is known as VFL with model splitting. In another scenario, the active party simply combines embeddings from all participants by summing them up and utilizes the softmax function to derive the final predicted label. This framework is referred to as VFL without model splitting [7]. In real-world VFL platforms, VFL with model splitting is more practical because it facilitates superior aggregation and in-depth analysis of embeddings submitted by all participants [21].

Though VFL is considered to be able to protect each participant's local data, recent studies find that VFL has potential risks of privacy leakage. A typical privacy risk of VFL is the leakage of labels [7, 26]. By abusing the parameters of trained local bottom model, the malicious passive party can infer training labels privately owned by the active party and even labels of samples beyond the training dataset, e.g., labels of testing samples. Existing works aiming to defend VFL against label inference attacks only consider the VFL framework without model splitting and cannot be applied to VFL with model splitting. However, as demonstrated above, VFL with model splitting is the more widely used framework. Thus, to bridge this gap, we propose SecretKeeper, a robust VFL framework against label inference attacks, which is compatible with the VFL framework with model splitting. Based on the intuition of limiting the feature extraction ability of the passive party's bottom model, we design four modules for SecretKeeper, including defensive pre-training, periodic suppression, secure architecture and gradient protection. Specifically, SecretKeeper strengthens VFL's robustness against label inference attacks from three perspectives. Firstly, in the perspective of training process, SecretKeeper uses the novel strategies of defensive pre-training and periodic suppression. With the active party's bottom model pretrained and the malicious passive party's local training suppressed periodically, the VFL framework tends to rely more on local features of the active party rather than the malicious passive party. By this means, the leakage of labels from the malicious passive party's trained bottom model is mitigated. Secondly, in the perspective of VFL architecture, SecretKeeper uses a specially designed top model, which has more layers to process local outputs of the malicious passive party, with the purpose of reducing the information leakage caused by the passive party's model parameters. Thirdly, in the perspective of gradient protection, SecretKeeper adds random noise to the gradients sent to passive parties, further reducing the privacy leaked to potential malicious parties. Experimental results on multiple datasets and model architectures show that SecretKeeper can significantly decrease the label inference accuracy of the malicious passive party.

In summary, our work makes the following contributions.

- We present SecretKeeper, a novel VFL framework that is robust to label inference attacks. To our best knowledge, this is the first robust VFL framework that is compatible with multiple designs of VFL architectures, including VFL with model splitting and VFL without model splitting.
- SecretKeeper is made up of four modules, including defensive pre-training, periodic suppression, secure architecture and gradient protection. Besides, our ablation studies show that every module of SecretKeeper is useful for preventing label leakage.
- Experimental results over multiple datasets show that SecretKeeper can effectively decrease the label inference accuracy with only minimal degradation of the federated model's performance on the original task. Furthermore, we use visualization techniques like TSNE and GradCAM to derive a deeper understanding towards the underlying mechanism of SecretKeeper.

2 Background and Related Work

2.1 Vertical Federated Learning

Vertical federated learning (VFL) is designed for the scenario where a few datasets (usually two) share the same sample space but differ in the feature space, as shown in Fig. 1a. For example, a car rental company and a bank located in one city may share the same group of users. Thus, part of their datasets share the same sample space. However, their datasets have different feature space, e.g., the bank's user features are about payment records while the car rental company's user features are about car rental history.

In this scenario, assuming that the bank has the labels of users' credit ratings and wants to train a machine learning model to predict a new user's credit rating. Obviously, if the bank can extends its dataset by incorporating user features stored by the car rental company, the larger dataset can improve the model's performance of predicting users' credit ratings. However, directly transporting private user data among companies is strictly prohibited by laws and regulations like GDPR and CCPA. As a result, the bank has to use the privacy-preserving machine learning paradigm, VFL, to achieve the goal of incorporating the car rental company's user features. In real-world cases, the bank may pay the car rental company for providing valuable data.

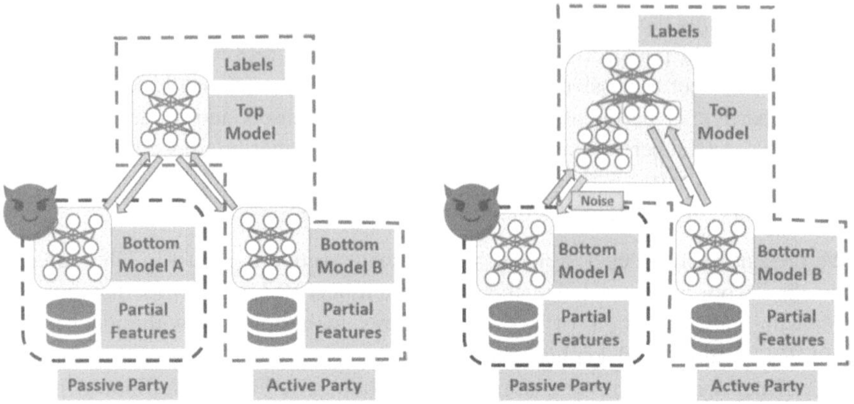

(a) Normal VFL architecture (vulnerable to label inference attacks).

(b) Secure VFL architecture built against label inference attacks.

Fig. 1. Comparison between the normal VFL architecture and our secure VFL architecture designed against label inference attacks.

In VFL, the participant who owns training labels is referred to as the active party, while other participants are the passive parties. In this paper, we consider the typical case of two participants. In this two-party scenario, either the active party or the passive party has one bottom model to project their local features into embeddings. During the inference stage of VFL, the passive party sends the embedding encoded by its bottom model to the active party. The active party then has two ways to aggregate and process the embeddings, including the received embedding and its own embedding. On one occasion, the active party may have one top model to aggregate the processed features from all participants and derive the final predicted label. Such a framework is named as VFL with model splitting. For another occasion, the active party may simply aggregate all processed features by summing them up and use the softmax function to derive the final predicted label. In this case, the framework is named as VFL without model splitting. In real-world VFL platforms, VFL with model splitting is more widely used because the top model, as a parameterized classifier, has better classification ability than the rule-based summing-up method.

2.2 Label Inference Attacks Against VFL

Recent studies find that VFL is vulnerable to label inference attacks [7]. The passive party is the potential adversary, whose goal is to infer the labels owned by the active party. The passive label inference attack is a fundamental variant of label inference attacks. Specifically, after the training process is finished, the adversary gets a trained local bottom model. Then the adversary can abuse the feature extraction ability of the trained bottom model to infer labels. Assuming that the adversary has a small amount of labeled auxiliary data, he/she can add a classification layer to the trained bottom model, then use the labeled auxiliary data and the local unlabeled data to fine-tune the local model in a semi-supervised manner. The adversary can choose classic semi-supervised learning methods, e.g., Mix-Match [3] or Mix-Text [4]. Then the adversary gets a trained local model with classification ability. With this model, the adversary can infer labels of training samples by feeding local features to it. The adversary can even infer labels of samples beyond the training dataset, e.g., the labels of testing samples. Such an attack is called the passive label inference attack, as the adversary passively follows normal procedures at the training stage, without conducting any malicious behavior.

Another attack is the active label inference attack, where the adversary actively triggers the VFL model to leak more information in the training stage. Specifically, the adversary maliciously increases the local learning rates so as to make its bottom model optimized faster than that of the benign active party. Then the VFL model tends to rely more on the adversary's bottom model, leaving the adversary a bottom model with better feature extraction ability. Finally, the adversary leverages the method of the passive label inference attack to infer labels. As demonstrated above, the passive label inference attack is based on the feature extraction ability of the adversary's local trained bottom model. Thus, the better trained bottom model helps the adversary achieve better label inference accuracy. Such an attack procedure is named as the active label inference attack because the adversary actively interferes the training process.

The passive and active label inference attacks can exploit the privacy risks of both VFL with model splitting and VFL without model splitting. The reason is that the adversary, i.e., the malicious passive party, is in charge of the training of its bottom model in either VFL architecture, which satisfies the requirement of passive/active label inference attack. There is another variant of label inference attack named as the direct label inference attack, which is only applicable for the VFL without model splitting. As demonstrate in Sect. 1, the VFL without model splitting simply sums local embeddings up and sends the sum to the softmax function to obtain the predicted result. Thus, the passive party can access the gradients of the classification layer in this VFL architecture, which causes more severe information of the labels. Specifically, the adversary can directly infer the training labels of samples by analyzing the signs of gradients [25].

A variant of directly label inference attack is the batch-level label inference attack [26]. The direct label inference attack requires the adversary to access the gradient of an individual sample, i.e., the batch size is 1. When the batch size is greater than 1, the adversary can only get the overall gradient of multiple samples in one batch, making the gradient sign of each individual training sample unavailable. Zou et al. proposed the batch-level label inference attack to address this problem. Given an overall gradient of one batch and its corresponding local features, the batch-level label inference attack uses a gradient descend-based optimization method to inverse gradient signs of all samples within this batch. After the gradient sign of each sample is recovered, the adversary follows steps of the direct label inference attack to infer labels. Like the naive direct label inference attack, the batch-level label inference attack is only applicable for the VFL without model splitting, as it requires the adversary to access the gradients of the output layer.

2.3 Existing Defenses Against Label Inference Attacks

Fu et al. illustrate four possible defense methods against label inference attacks [7], including noisy gradients, gradient compression, discrete SGD (a customized version of SignSGD [2]). Another defense is the privacy-preserving deep learning [19]. The four defenses follow the idea of decreasing the information leakage within the gradient sent to the passive party. Specifically, the noisy gradients defense adds random noise to the gradients. The gradient compression defense filters out gradients of small absolute values. The discrete SGD defense discretizes the gradient. The privacy-preserving deep learning defense utilizes differential privacy and gradient compression to prevent privacy leakage from gradients. These four defenses are proved to be not effective against passive and active label inference attacks [7]. As for the direct label inference attack, noisy gradient and privacy-preserving deep learning have relatively good defense performance than the other two defense methods.

Zou et al. proposed CAE to defend VFL without model splitting against direct label inference attack and the batch-level label inference attack [26]. Their idea is to obfuscate the true training labels to soft fake labels during the training stage. During the inference stage, the VFL model produces soft fake labels, then the soft fake labels are de-obfuscated into real predicted labels. The obfuscation and de-obfuscation processes are implemented with a encoder-decoder model. CAE is effective against direct/batch-level label inference attacks but fails to defend against passive/active label inference attacks. Zou et al. further proposed DCAE, which combines CAE with the defense method of discrete SGD, to make the defense more effective against passive/active label inference attack [26]. Note that both CAE and DCAE are based on the assumption that the aggregation rule is simple summing up embeddings of all participants' bottom models, which only holds for VFL without model splitting. Thus, CAE and DCAE is not suitable for VFL with model splitting.

3 Threat Model

Two participants collaboratively train a VFL model. One participant is the active party, who owns training labels and partial features. The active party is in charge of training the top model and its own bottom model. Another participant is the passive party, who also owns partial features and controls the training of its bottom model. Note that the partial features owned by the two parties do not overlap. The passive party is the adversary and its attack goal is to infer the training labels held by the active party. Besides, the adversary also intends to infer labels of samples beyond the training dataset, i.e., labels of testing samples.

4 Methodology

As demonstrated in Sect. 1, we design SecretKeeper from three perspectives, including secure training process, secure VFL architecture and secure gradient transmission. Algorithm 1 shows an overall illustration of SecretKeeper. In this section, we present the detailed methodology of each module within SecretKeeper.

4.1 Secure VFL Architecture

As mentioned by Fu et al. [7], shallower top model causes more severe privacy leakage. This is because shallower top model makes the VFL model rely more on the feature extraction ability of bottom models, which empowers the passive party's potential of inferring labels. Based on this observation, we design a special top model architecture, which uses relatively more layers to process the passive party's embeddings, i.e., outputs of the passive party's bottom model. On the contrary, the top model uses fewer layers to process the outputs of the active party's bottom model. With more layers to suppress the VFL model's reliance on the passive party's bottom model, the passive/active attack is also mitigated. The illustration of the specially designed VFL architecture is shown in Fig. 1b. The secure VFL architecture module corresponds to line 2 in Algorithm 1. For more details of the secure VFL architecture, please refer to Sect. 5.2.

4.2 Secure Training Process

From the illustrations of the passive and active label inference attacks in Sect. 2.2, we can draw a conclusion that both attacks are based on the feature extraction ability of the local trained bottom model. Specifically, the passive label inference attack directly fine-tunes a classification model from the trained bottom model, while the active label inference attack uses a malicious local optimizer to further empower the bottom model's feature extraction ability. Thus, to prevent label leakage in VFL, we can slow down the optimization of the passive party's bottom model during the training stage, with the purpose of limiting the feature extraction ability of the adversary's bottom model. Following this intuition, we design two sub modules of SecretKeeper, i.e., defensive pre-training and periodic suppression.

Defensive Pre-training. SecretKeeper firstly trains the trusted active party's bottom model and top model for several iterations, with the potentially malicious passive party's bottom model frozen. Specifically, in the first few training iterations of VFL, the active party mutes the gradients sent to the passive party, i.e., back-propagates zero gradients to the passive party. By this means, the active party initially has a better trained bottom model than the passive party, preventing the top model from relying much on the passive party's bottom model. The defensive pre-training module corresponds to lines 13–15 in Algorithm 1. For more details of the parameter settings for defensive pre-training, please refer to Sect. 5.3.

Periodic Suppression. After the defensive pre-training stage is finished, SecretKeeper further uses periodic suppression to limit the passive party's local optimization. Specifically, the active party mutes the gradients sent to the passive party for one iteration every

N iterations. The periodic suppression module corresponds to lines 13–15 in Algorithm 1. For more details of the parameter settings for periodic suppression, please refer to Sect. 5.3.

4.3 Secure Gradient Transmission

Considering that the gradient sent to the passive party may cause label leakage, we incorporate the operation of adding noise to gradients as one sub module of SecretKeeper. In the back-propagation process of each iteration, the active party firstly adds random Gaussian noisy to the gradients, then sends the noisy gradients to the passive party. Note that the gradients sent to the active party's bottom model are kept as original, without adding any noise. By this means, the passive party will find it harder to exploit privacy leakage from the gradients, e.g., the information about training labels hidden in the signs of the gradients. The secure gradient transmission module corresponds to line 16 in Algorithm 1.

5 Experiment Setup

5.1 Datasets

Following the common practice of fitting vision datasets in the VFL setting, we cut an image sample by half and each half represents the partial features hold by one participant [7, 13, 14]. The details of datasets for evaluation are as follows.

CIFAR-10. CIFAR-10 [12] is a 10-class dataset built for image classification, which is made up of 60,000 3-channel images in the size of 32 $\times$ 32 pixels. There is no overlap among 10 classes, including airplane, automobile, bird, cat, deer, dog, frog, horse, ship and truck. Following the common practice of using CIFAR-10, we take 50,000 samples as the training dataset and other 10,000 as the test dataset. CIFAR-10 is a balanced dataset, i.e., for each class, there are 5,000 training samples and 1,000 training samples.

Algorithm 1 SecretKeeper

Require: Top model parameters θ_{top}, bottom model parameters of the passive party θ_p, bottom model parameters of the active party θ_a, learning rate η, ground-truth label y, number of iterations for defensive pre-training N_{pre}, interval of periodic suppression N_{Int}.

Server executes:
Change the model architecture of the top model and initialize θ_{top}
Initialize θ_p, θ_a

for Id_{iter} in total number of training iterations **do for**
each batch b of sample Ids **do**
$o_a \leftarrow$ **ParticipantForwardProp**(θ_a, b)
$o_p \leftarrow$ **ParticipantForwardProp**(θ_p, b)
$o_{concate} \leftarrow Concat(o_a, o_p)$ //concatenating bottom model outputs
$o_{final} \leftarrow \theta_{top}(o_{concate})$
$L \leftarrow LossFunc(o_{final}, y)$

$g_{top} \leftarrow \frac{\partial L}{\partial \theta_{top}}$ //updating top model

$\theta_{top} \leftarrow \theta_{top} - \eta \cdot g_{top}$
$g \leftarrow \frac{\partial L}{\partial o_a}$
$g \leftarrow \frac{\partial L}{\partial o_p}$
if $Id_{iter} \leq N_{pre}$ or $Id_{iter} \% N_{Int} == 0$ **then**
$g_p \leftarrow zeros\_like(g_p)$ //defensive pre-training
end if
$g_p \leftarrow g_p + random\_like(g_p)$ //noisy gradient
ParticipantBackProp(θ_a, g_a, o_a)
ParticipantBackProp(θ_p, g_p, o_p)
end for end
for

ParticipantForwardProp(θ, b):
return bottom model forward outputs $\theta(b)$

ParticipantBackProp(θ, g_o, o): g
$\leftarrow g_o \cdot \frac{\partial o}{\partial \theta}$
$\theta \leftarrow \theta - \eta \cdot g$ //updating bottom model

CIFAR-100. The dataset size and image size of CIFAR-100 [12] are the same as these of CIFAR-10, i.e., 60,000 images in the size of 32 × 32 pixels. The difference is that CIFAR-100 has 100 classes, including beaver, bed, bicycle and so on. Following the common practice of using CIFAR-100, we take 50,000 samples as the training set and other 10,000 samples as the test dataset. CIFAR-100 is also balanced so that there are 500 training samples and 100 testing samples per class. We choose CIFAR-100 for the purpose of evaluating the performance of SecretKeeper when there is a large number of classes.

CINIC-10. CINIC-10 [5] extends CIFAR-10 by adding down sampled images from ImageNet [6], which is made up of 270,000 images (4.5 times of CIFAR-10) in the size of 32×32 pixels. Following the common practice of using CINIC-10, we take 180,000

samples as the training dataset and other 90,000 as the test dataset. Like CIFAR-10, CINIC-10 is also a balanced 10-class dataset. Thus, there are 18,000 training samples and 9,000 testing samples per class. We choose CINIC-10 to evaluate the performance of SecretKeeper in the situation where the dataset size is large.

5.2 Model Architectures

Following the implementations in [7], for either participant of VFL, we use ResNet-18 [10] as the bottom model's architecture. As for the top model, we use a fully connected network composed of 4 linear layers as its model architecture. As for the special top model's architecture demonstrated in Sect. 4.1, it is composed of six fully connected layers. Specifically, during the forward process, we first use three fully connected layers to encode the output embeddings of the passive party's bottom model, then concatenate the encoded embeddings with the output embeddings of the active party's bottom model. After that, the concatenated embeddings is sent to the remaining three fully connected layers to derive the final prediction result.

5.3 Parameter Settings and Evaluation Metrics

We follow the parameter settings and evaluation metrics used in previous work [7, 26]. Besides, we set the number of iterations for defensive pre-training (N_{pre}) as 30% of the total iterations. The interval of periodic suppression (N_{Int}) is set to 3. For CIFAR-10 and CINIC-10, the top-1 accuracy is the evaluation metric for the VFL model's performance on the original task, as well as the attack performance of label inference attacks. On the CIFAR-100 dataset, the evaluation metric is the top-5 accuracy, following the practice in previous works.

6 Evaluation

6.1 Defense Performance

The defense performance of SecretKeeper against the passive label inference attack is shown in Table 1. It can be seen that SecretKeeper significantly decreases the label inference accuracy on the three datasets. For example, on the training dataset of CIFAR-10, SecretKeeper successfully decreases the passive attack's label inference accuracy from 82.01% to 25.39%. As demonstrated in previous work [7], with 40 labeled auxiliary samples and a randomly initialized bottom model with inference head, the passive attack can achieve 16.98% inference accuracy on the training dataset of CIFAR-10. In other words, no defense can reduce the inference accuracy to be lower than 16.98%. Therefore, it's reasonable to consider a reduction of the inference accuracy to 25.39% as excellent defense performance.

SecretKeeper can also prevent the adversary from inferring labels of samples beyond the training dataset. For example, as shown in Table 1, we can see that SecretKeeper decreases the label inference accuracy from 65.16% to 25.61% on testing dataset of CIFAR-10. This is because SecretKeeper limits the feature extraction ability of the

Table 1. Defending Against Passive Label Inference Attack

Dataset			CIFAR-10	CIFAR-100	CINIC-10
Metric(Accuracy)			Top-1	Top-5	Top-1
Quantity of Auxiliary Labeled Samples Known by The Adversary			40	400	40
VFL Model's Original Task Accuracy	W/O Defense		79.93%	73.63%	74.59%
	With Defense		76.6%	72.81%	73.83%
	Only Benign		74.21%	70.45%	72.10%
Inference Accuracy of The Passive Label Inference Attack	Training Dataset	W/O Defense	82.01%	70.17%	74.12%
		With Defense	25.39%	22.62%	31.91%
	Testing Dataset	W/O Defense	65.16%	47.01%	57.79%
		With Defense	25.61%	22.97%	30.91%

adversary's bottom model, making the adversary's classifier lose the ability to precisely classify both training samples and testing samples. We will further analyze the underlying mechanism of SecretKeeper and show the evidence that supports this point of view in Sect. 7.

Beside testing the degradation of label inference accuracy, we should also test the side effect of SecretKeeper, i.e., the degradation on the VFL model's performance on the original task. We use the term "only benign" to denote the situation where the active party independently trains its bottom model and top model on local partial features. Obviously, any defense against label inference attacks should not decrease the original task performance to the same accuracy as the "only benign" situation, or the active party can simply refuse to collaborate with the passive party. From Table 1, we can see that though SecretKeeper causes the original task accuracy to degrade to some extent, e.g., from 73.63% to 72.81% on CIFAR-100 (measured by top-5 accuracy), the decreased accuracy is higher than that of the "only benign" situation (70.45%). Thus, with SecretKeeper, the active party still benefits from collaborating with the passive party. SecretKeeper is also effective against the active label inference attack. As shown in Table 2, SecretKeeper also significantly decreases the label inference accuracy of the active attack, e.g., from 77.23% to 37.91% on the training dataset of CINIC-10. Besides, the degradation of the original task accuracy is minor than that of cancelling the collaboration with the passive party. For example, the original task accuracy on CINIC-10 is 73.17% with SecretKeeper, which is higher than that of the "only benign" situation (72.10%).

6.2 Comparison with Other Defenses

We compare SecretKeeper with four defense methods, including discrete stochastic gradient descent (DiscreteSGD), privacy preserving deep learning (PPDL), gradient compression (GC) and DiscreteSGD-enhanced confusional auto encoder (DCAE). Introductions of the four methods are listed as follows.

Table 2. Defending Against Active Label Inference Attack

Dataset			CIFAR-10	CIFAR-100	CINIC-10
Metric(Accuracy)			Top-1	Top-5	Top-1
Quantity of Auxiliary Labeled Samples Known by The Adversary			40	400	40
VFL Model's	W/O Defense		79.92%	72.24%	73.52%
Original Task	With Defense		76.20%	72.82%	73.17%
Accuracy	Only Benign		74.21%	70.45%	72.10%
Inference Accuracy of The Active Label Inference Attack	Training	W/O Defense	82.23%	67.95%	77.23%
	Dataset	With Defense	20.55%	22.11%	37.91%
	Testing	W/O Defense	59.68%	45.14%	56.52%
	Dataset	With Defense	20.59%	21.15%	33.43%

Gradient Compression. Gradient compression is originally designed for reducing the communication cost in distributed learning, which prunes gradients of relatively small absolute values to zeros. Existing work shows that distributed learning can converge even if 90% of the gradient values are pruned. Considering that the only information available for the malicious passive party during the training stage of VFL is the back-propagated gradients, gradient compression may mitigate the problem of label leakage in VFL because it reduces the information contained in the gradients.

DiscreteSGD. Following the same idea of Gradient Compression, DiscreteSGD also reduces the information contained in the back-propagated gradients. DiscreteSGD converts the form of gradients from consecutive numbers to discrete numbers. Firstly, DiscreteSGD sets several intervals, e.g., $[-1.0, -0.5]$, $[-0.5, 0.0]$, $[0.0, 0.5]$, and $[0.5, 1.0]$. Then the gradient of each parameter is set to its closest interval boundary. For example, the gradient value 0.46 will be set to 0.5.

Table 3. Results of the ablation study. "DPT", "PS", "SA" and "NG" represents "Defensive Pre-Training", "Periodic Suppression", "Secure VFL Architecture" and "Noisy Gradient", respectively.

Muted Module			DPT	PS	SA	NG	None
Original Task Accuracy	W/O Def		79.92%	79.92%	79.92%	79.92%	79.92%
	With Def		76.63%	76.41%	76.94%	77.37%	76.20%
Inference Accuracy of The Active Label Inference Attack	Training	W/O Def	82.23%	82.23%	82.23%	82.23%	82.23%
	Dataset	With Def	24.80%	25.18%	29.70%	33.11%	20.55%
	Testing	W/O Def	59.68%	59.68%	59.68%	59.68%	59.68%
	Dataset	With Def	25.05%	24.28%	28.31%	31.28%	20.59%

Privacy Preserving Deep Learning. Privacy preserving deep learning [19] composes gradient compression and differential privacy, which adds noise to gradients and prunes gradient of small values at the same time. By composing these two defensive strategy, privacy preserving deep learning makes a better defense.

DiscreteSGD-Enhanced Confusional Auto Encoder. Zou et al. [26] proposed the defense method "confusional auto encoder" to mitigate label leakage in VFL, which makes the VFL model to output confused fake labels rather than normal labels. By this means, the malicious passive party gets confused gradients in the back-propagation process during the training stage. Then the performance of inference attacks based on gradients degrades. To make the defense further capable of defending against passive/active label inference attacks, the authors combine the confusional auto encoder with DiscreteSGD and name the composed method as the DiscreteSGD-enhanced confusional auto encoder.

For the four defenses, the parameter settings are set following previous work [7, 26] to achieve the balance between defense performance and VFL model's original task accuracy. Specifically, for gradient compression, the compression rate is set to 0.1, i.e., pruning 90% smallest gradient values to zeros. For DiscreteSGD, we set 24 intervals in the region of the highest and lowest gradient values in the first iteration. For PPDL, the parameter θ_u is set to 0.5. As for DCAE, the parameter λ is set to 1.0 and the number of intervals are 12.

We compare the defense performance of SecretKeeper with the four existing defense methods against the active label inference attack. As shown in Fig. 2(a), in the perspective of maintaining the VFL model's accuracy on the original task, the performance of SecretKeeper is similar with or superior than existing methods. As for the effectiveness of decreasing the inference accuracy of label inference attacks, SecretKeeper outperforms all the evaluated defense methods, as shown in Fig. 2(b).

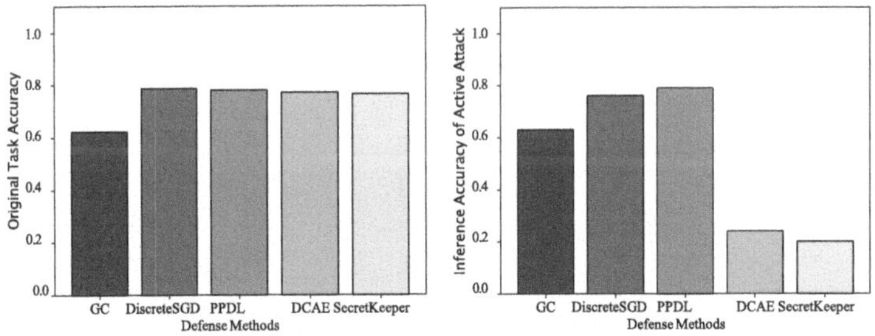

(a) VFL Model's Accuracy on The Original Task

(b) Inference Accuracy of The Active Attack

Fig. 2. Comparison of defense performance across several baseline methods.

6.3 Ablation Study

SecretKeeper includes 4 modules, including defensive pre-training, periodic suppression, secure VFL architecture and noisy gradient. We conduct ablation studies to make sure that each module contributes to the effectiveness of SecretKeeper. The ablation studies are conducted on the CIFAR-10 dataset. As shown in Table 3, each of the four modules contributes to the defense performance, i.e., muting anyone among the four modules causes the increase of label inference accuracy. Thus, the four modules are all valuable components of SecretKeeper.

7 Analysis

In this section, we use visualization techniques to further analyze the underlying working mechanism of SecretKeeper. As demonstrated in Sect. 1, the basic idea of SecretKeeper is to limit the feature extraction ability of the adversary's local bottom model. To verify this intuition, we use TSNE [15] to visualize the outputs of the adversary's trained bottom model on the training dataset of CIFAR-10. As shown in Fig. 3, the outputs of adversary's bottom model can be easily clustered into 10 groups when there is no defense applied. However, with SecretKeeper, the outputs are more close to the state of random mixing, indicating that the adversary's bottom model's feature extraction ability evidently degrades. Such a phenomenon proves that SecretKeeper successfully limits the feature extraction ability of the adversary's bottom model.

We also use the explanation tool GradCAM [18] to analyze the entire VFL model's reliance on the adversary's local features. As shown in Table 4, when the adversary conducts the active label inference attack and there is no defense, the VFL model relies more on the adversary's local features when classifying an input. However, with SecretKeeper applied, the adversary cannot gain such an advantage anymore and the VFL model relies on both parties instead of preferring the adversary. To draw a conclusion, SecretKeeper limits the feature extraction ability of the adversary's bottom model and prevents the VFL model from relying too much on the adversary's local features. By this means, the adversary's label inference ability by abusing the trained bottom model is suppressed.

Table 4. GradCAM visualization of the passive and active label inference attacks on CIFAR-10. The left half of image is the datum of the adversary.

Original Image	
GradCAM under Active Attack & No Defense	
GradCAM under Active Attack **& SecretKeeper**	

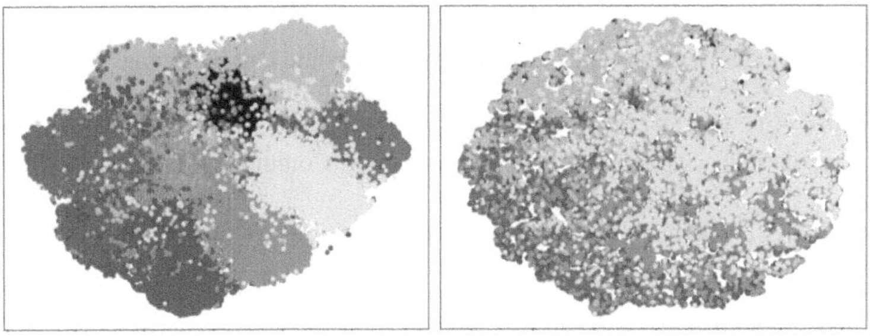

(a) w/o defense (b) with defense

Fig. 3. TSNE visualization of the adversary's bottom model's outputs.

8 Conclusion

In this paper, we propose the first split-learning-compatible robust VFL framework against label inference attacks, which consists of four modules, including defensive pre-training, periodic suppression, secure VFL architecture and noisy gradient. Experimental results across various datasets demonstrate that our framework can effectively defend against passive/label inference attacks and outperforms state-of-the-art defenses, with a negligible degradation on the VFL model's performance on the original task. Then we use visualization techniques to analyze the underlying mechanism of our framework, motivating future improvements towards more robust VFL.

References

1. Albrecht, J.P.: How the GDPR will change the world. Eur. Data Prot. L. Rev. **2**, 287 (2016)
2. Bernstein, J., Wang, Y.X., Azizzadenesheli, K., Anandkumar, A.: signSGD: compressed optimisation for non-convex problems. In: International Conference on Machine Learning, pp. 560–569. PMLR (2018)
3. Berthelot, D., Carlini, N., Goodfellow, I., Papernot, N., Oliver, A., Raffel, C.A.: Mixmatch: a holistic approach to semi-supervised learning. Advances in Neural Information Processing Systems 32 (2019)
4. Chen, J., Yang, Z., Yang, D.: Mixtext: linguistically-informed interpolation of hidden space for semi-supervised text classification. In: Proceedings of the 58th Annual Meeting of the Association for Computational Linguistics, pp. 2147–2157 (2020)
5. Darlow, L.N., Crowley, E.J., Antoniou, A., Storkey, A.J.: Cinic-10 is not imagenet or cifar-10. arXiv preprint arXiv:1810.03505 (2018)
6. Deng, J., Dong, W., Socher, R., Li, L.J., Li, K., Fei-Fei, L.: Imagenet: a large-scale hierarchical image database. In: 2009 IEEE Conference on Computer Vision and Pattern Recognition, pp. 248–255. IEEE (2009)
7. Fu, C., et al.: Label inference attacks against vertical federated learning. In: 31st USENIX Security Symposium (USENIX Security 22), pp. 1397–1414 (2022)
8. Google: TensorFlow Federated. https://www.tensorflow.org/federated
9. Harding, E.L., Vanto, J.J., Clark, R., Hannah Ji, L., Ainsworth, S.C.: Understanding the scope and impact of the California consumer privacy act of 2018. J. Data Prot. Priv. **2**(3), 234–253 (2019)

10. He, K., Zhang, X., Ren, S., Sun, J.: Deep residual learning for image recognition. In: Proceedings of the IEEE Conference on Computer Vision and Pattern Recognition, pp. 770–778 (2016)
11. Hoofnagle, C.J., Van Der Sloot, B., Borgesius, F.Z.: The European union general data protection regulation: what it is and what it means. Inf. Commun. Technol. Law **28**(1), 65–98 (2019)
12. Krizhevsky, A., Hinton, G., et al.: Learning multiple layers of features from tiny images (2009)
13. Liu, Y., et al.: A communication efficient collaborative learning framework for distributed features. arXiv preprint arXiv:1912.11187 (2019)
14. Liu, Y., Zhang, X., Wang, L.: Asymmetrical vertical federated learning. arXiv preprint arXiv: 2004.07427 (2020)
15. Van der Maaten, L., Hinton, G.: Visualizing data using t-SNE. J. Mach. Learn. Res. **9**(11), 2579–2605 (2008)
16. OpenMined: PySyft. https://github.com/OpenMined/PySyft
17. Romanini, D., et al.: Pyvertical: a vertical federated learning framework for multi-headed splitnn. arXiv preprint arXiv:2104.00489 (2021)
18. Selvaraju, R.R., Cogswell, M., Das, A., Vedantam, R., Parikh, D., Batra, D.: Grad-CAM: visual explanations from deep networks via gradient-based localization. In: Proceedings of the IEEE International Conference on Computer Vision, pp. 618–626 (2017)
19. Shokri, R., Shmatikov, V.: Privacy-preserving deep learning. In: Proceedings of the 22nd ACM SIGSAC Conference on Computer and Communications Security, pp. 1310–1321 (2015)
20. State of California Department of Justice: California Consumer Privacy Act. https://oag.ca.gov/privacy/ccpa
21. Tencent: Split learning-based vertical federated learning framework. https://github.com/FederatedAI/FATE/tree/master/python/federatedml/nn/hetero/nn_component
22. Voigt, P., Von dem Bussche, A.: The EU General Data Protection Regulation (GDPR). A Practical Guide, 1st edn., Springer International Publishing, Cham (2017). https://doi.org/10.1007/978-3-319-57959-7
23. Webank, T.: Federated Artificial Intelligence Enabler. https://github.com/FederatedAI/FATE
24. Yang, Q., Liu, Y., Chen, T., Tong, Y.: Federated machine learning: concept and applications. ACM Trans. Intell. Syst. Technol. (TIST) **10**(2), 1–19 (2019)
25. Zhu, L., Liu, Z., Han, S.: Deep leakage from gradients. Advances in Neural Information Processing Systems 32 (2019)
26. Zou, T., et al.: Defending batch-level label inference and replacement attacks in vertical federated learning. IEEE Trans. Big Data **1**, 1–12 (2022)

Quantify the Success Probability of Honest-But-Curious Relay Attack for Modulation Based Physical Layer Network Coding

Qiao Liu[1], Qi Han[2(✉)], Yuting Yang[1], Jin Cao[1], Hui Li[1], and Yong Wang[1]

[1] School of Cyber Engineering, National Key Lab of ISN, Xidian University, Xi'an, Shaanxi, China

[2] Xi'an Institute of Applied Optics, Xi'an, Shaanxi, China

hanqiwildginger@gmail.com

Abstract. Physical layer network coding (PNC) has been considered a valid way to improve transmission efficiency, especially in the distributed network. Furthermore, due to the received signal at the relay is a superposed codeword that cannot be decomposed to recover the individual information, PNC is also viewed as a promising approach for physical layer security. In this paper, we reveal that information leakage risk exists for the honest-but-curious relay. Under such an attack, the honest-but-curious relay tries to recover the information from both transmitting sides while complying with the transmission protocol honestly. Then, we quantify the success probability of the honest-but-curious relay attack for modulation-based PNC. To the best of our knowledge, such a formulation is the first numerical formulation. We analyzed the BPSK, QPSK, and 16-QAM respectively, then we formulate the general attack successful probability for M-QAM. With the information-theoretic analysis results, it has been clearly shown that the attack successful probability is decreased with the increasing modulation order. Finally, the simulation is performed to demonstrate the theoretical analysis.

Keywords: Physical Layer Network Coding · Honest-but-Curious Relay · Attack Success Probability · Physical Layer Security

In recent years, with the rapid development of science and technology, wireless communication technology plays more and more important roles in modern life. From personal communication to industrial development and then to the military defense, users want not only high-speed and low-latency transmission, but also reliability and security to be guaranteed.

Dating back to the original research by Shannon in 1948, the information has been quantified in [1]. Until 2000, the concept of information flow was introduced in [2]. It was proved that the combination of information can increase the capacity of the network. This research opened up a new field called Network Coding (NC). In the traditional network, the relay can only copy, amplify, and forward the received signal, but network coding breaks this limitation, it allows nodes in the network to perform certain encoding

Z. Gu et al. (Eds.): NSE 2023, CCIS 2063, pp. 19–32, 2024.
https://doi.org/10.1007/978-981-97-4519-7_2

operations, and then broadcast to downstream nodes. Thus, the throughput performance has been improved with implied NC. Subsequently, the physical layer network coding (PNC) proposed by S. Zhang et al. in [3] can be regarded as a branch field of network coding. The basic idea is to use the mixed signals that are naturally generated when electro-magnetic waves are superimposed. In the following decade after proposing, PNC has become a research hot-spot because of the high throughput improvement it can bring. In particular, since the received signal at the relay is a weighted sum when using physical layer network coding, the relay cannot decode the code sent by each source node individually. Thus, data leakage can be avoided especially for the Honest-but-Curious relay, which obeys the transmission protocol honestly but tries to recover the transmitted information from users.

Combining PNC with modulation is a research topic in physical layer network coding. In 2006, S. Zhang et al. proposed the concept of PNC for the first time in [3] and presented a relay demodulation mapping scheme based on BPSK and QPSK modulation schemes. Thereafter, the paper [4] analyzes the Symbol Error Rate (SER) of the PNC system with BPSK, QPSK, and 16PSK modulation when the relay adopts Amplify-and-Forward (AF) and Decode-and-Forward (DF). In the paper [5], the PNC system using M-QAM modulation is analyzed and it is proved that using M-QAM modulation has better performance than using M-PSK modulation. In paper [6], a bit-interleaved coding modulation scheme based on PNC is studied. In paper [7], the authors propose a cooperative orthogonal PNC scheme for a double-hop cooperative relay network and based on decoding and forwarding protocol propose an orthogonal physical layer code-decode-forward relay protocol. Paper [8] proposes a PNC system based on binary continuous phase shift keying, paper [9] studies a PNC scheme using MFSK modulation, and paper [10] proposes a joint scheme of hierarchical modulation and physical layer network coding (HM-PNC) for downlink asymmetric bidirectional relay channels. Paper [11] proposes a constellation shaping scheme and applies it to a PNC system based on M-QAM modulation. Paper [12] proposes a non-binary irregular repeating cumulative modulation coding PNC scheme for Gaussian TWRC.

In addition, the physical layer network coding system needs to send different signals to the same relay on the same channel at the same time, it is quite difficult for the relay to decode a single message, so the physical layer network coding can greatly improve the security of the system. A secure beamforming scheme for multiple-input multiple-output (MIMO) bidirectional relay systems based on PNC was studied in [13]. In paper [14], the authors discuss the security performance of PNC and show that PNC is a secure means against passive eavesdroppers. In paper [15], the authors consider a Gaussian channel network employing lattice-based PNC as a routing strategy. On the premise that communication is subject to adversarial attacks in the form of eavesdropping, the problem of data security is solved from the perspective of information theory. The authors focus on the security issues in smart homes, a new secure packet data exchange protocol using the physical layer method is proposed in [16]. Furthermore, the anonymous transmission problem of physical layer network coding, the physical layer security enhancement scheme based on artificial noise is studied in [17, 18]. In paper [19], the authors propose a hybrid MIMO phased- array time-modulated directional modulation (DM) scheme for physical layer security in millimeter-wave (mmWave) wireless communications. In paper [20] the authors applying physical layer security to the wireless vehicular network and consider the eavesdropping attack for vehicle-to-infrastructure communications. In paper [21], the authors introduce appropriate performance metrics for evaluating physical layer security in ultra-reliable and low latency communication (URLLC) and investigate the trade-off between latency, reliability, and security. As an important technology in 6G, non-orthogonal multiple access (NOMA) is also usually combined with PNC to enhance system security from the physical layer. In paper [22], the authors investigate the physical security of cooperative NOMA systems. In paper [23] the authors propose to design safe beamforming vectors on relay nodes and NOMA techniques are used to enhance physical layer security in visible light communication systems. In paper [24], the authors investigates the secrecy outage performance for a large-scale downlink system with full-duplex NOMA transmission aided by artificial noise. In paper [25] the authors investigate the physical layer security for a reconfigurable intelligent surface-aided NOMA 6G networks. In paper [26], the authors study the physical layer security of a vehicular network employing a reconfigurable intelligent surface. And the authors discuss the challenges, solutions, and visions of physical layer security in 6G networks. In paper [27], the authors discuss the role of physical layer security in 6G. Likewise, the authors reviewed physical layer security for 5G wireless networks in [28]. A large number of scholars focus on the security improvement brought by PNC to the system. In particular, the mobile Ad Hoc networks, which has attracted widespread attention in recent years. In paper [29], the authors propose a Sight-based Cooperative Jamming scheme to improve the physical layer security performance of mmWave Ad Hoc communication. Because it is a multi-hop, centerless, self-organizing wireless network, the traditional upper-layer security technology is difficult to implement, and its security must be considered from the bottom layer. Physical layer network coding technology brings more possibilities to such network structures.

Much research has been done on PNC systems utilizing some modulation as a means of enhancing security, but there is no research analyzing how much the security performance that PNC system can bring to communication. In this paper, we quantify the security performance of PNC by studying the probability of an attacker recovering a single message. Comparing the PNC security performance with different modulations, we conclude that the higher-order modulation in the PNC system not only improves the throughput but also greatly improves the security.

The main contributions of this paper is listed as:

- We define the success probability of Honest-but-Curious relay attack for modulation based PNC. To the best of our knowledge, such definition is the first numerical formulation.
- We analyze the attack success probability of BPSK, QPSK and 16-QAM based PNC respectively. We classify different cases by listed the mapping rules in relay. Then we calculate the attack success probability for each cases. The final probability is the sum of all cases.
- We provide a general derivation for M-QAM based PNC. By dividing MQAM into two orthogonal independent $\sqrt{M}$-ASK signals, we first analyze the attack success probability for one $\sqrt{M}$-ASK signal. Then we generalize it into M-QAM situation.

- We conduct simulations to evaluate the theoretical analysis. The simulation results demonstrate that the theoretical analysis formulation is the upper bound of attack success probability for modulation based PNC.

The rest of this paper is organized as follows. Section 2 presents the basic model for modulation based PNC in the view of transmission and security. Section 3 analyzes the attack success probability of Honest-but-Curious relays for BPSK and QPSK based PNC. Such analysis is a basic for general case. Section 4 analyzes the attack success probability of Honest-but-Curious relays for General M-QAM based PNC. Section 4 conducts numerical results and simulations to demonstrate the theoretical analysis. Finally, Sect. 4 concludes this work and points out the future work.

1 System Model

1.1 Transmission Model

Consider a network with two source nodes N_i, i = 1, 2 and one relay node *R*, where N_i's exchange information with the help of *R*. Furthermore, we assume symbol- level and carrier-phase synchronization, and the use of power control, so that the frames from N_1 and N_2 arrive at *R* with the same phase and amplitude. Unlike network coding that requires three-time slots, this model only needs two-time slots to complete information exchange, which is called PNC. The transmission schedule is given in Fig. 1. N_1 and N_2 try to exchange their messages S_1, S_2 using relay *R*. In the first time slot, N_i sends the modulated signal a_i at the same time, and the relay tries to decode the received signal $Y = a_1 + a_2 + n_0$.The combined signal received by *R* during the one-time slot is

$$S_R(t) = S_1(t) \oplus S_2(t), \tag{1}$$

where $S_R(t)$ is the signal received by *R*, and is $\oplus$ a defined mapping as a helper for relay decoding. In the second time slot, the relay broadcasts the decoded signal to the terminals, then each node uses its information to recover the message sent by the other.

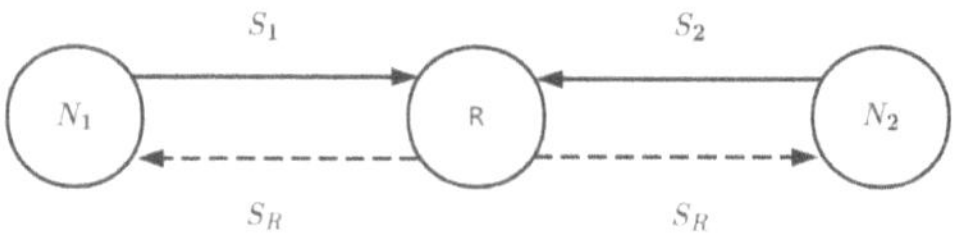

Fig. 1. Physical layer network coding

1.2 Security Model

We consider the relay is Honest-but-Curious, which obeys the transmission protocol honestly but curious about the exchanged information from both users sides.

Due to the nature of the PNC system, it is necessary for the Honest-but-Curious relay to obtain *Y*. Under such assumption, such relay is an effective attacker. Actually, the Honest-but-Curious relay can conduct more threatened than extra eavesdropper. If the attacker recovers the information sent by the source node through Y, it will threaten the information security. So we define that when Si are sent, the attacker decodes S_i's information per bit by Y as a successful attack.

Now let us describe more details about the probability of attack success. When the attacker obtains the signal Y, as the same of the relay, it will decode Y instantly. But unlike the relay, which broadcasts the decoded signal through a special mapping, the attacker does not need to map and broadcast, it just needs to decode Y. We assume that the decoded Y is $\hat{Y}$, and N_1, N_2 sent information a_1, a_2, we believe that the probability of the attacker recovered every bit of a_1 and a_2 using $\hat{Y}$ without any other information is quantifiable:

$$P_{\text{att}} = p(a_1, a_2)p(a_1, a_2 \ are\ recovered), \tag{2}$$

where $p(a_1, a_2)$ is the probability of the N_1, N_2 sending a_1, a_2, and $p(a_1, a_2 \ are\ recovered)$ is the probability of the attacker recovers a_1, a_2 accurately, it needs to determine at the same time what a_i is sent by N_i. Note that the worst possibility for an attacker is a random guess completely. For example the random guess probability is 0.5 for BPSK modulation, then the attacker recovers the information that sent by N_1 and N_2 has the probability of 0.25. We use P_w to represent the worst probability of attack success, it is obvious that if the calculated P_{att} is less than P_w, it is better to the attacker guess randomly. So we can define the P_{att} more stricter:

$$P_{att} = \begin{cases} p(a_1, a_2)p(a_1, a_2 arerecovered), & P_{att} > P_w \\ p_w, & P_{att} \le P_w \end{cases} \tag{3}$$

2 Success Probability Analysis of Honest-But-Curious Relay Attack for BPSK and QAM Based PNC

2.1 Analysis for BPSK Based PNC

Now let us assume the use of BPSK modulation in all the nodes and call it PNC_BPSK. The data bit of N_i is $S_i \in \{1, 0\}$, and $a_i \in \{-1,1\}$ is representing the BPSK modulated bit of S_i such that $a_i = 2S_i - 1$.

From the model introduced in Sect. 1, the relay should broadcast the necessary information to source nodes then the exchange of information is completed, so we need a PNC mapping scheme $\oplus$ to modulate and demodulate the message from the relay. The symbol $\oplus$ represents modulo 2 operation in this paper. Table 1 gives the mapping rules in the PNC using BPSK modulation.

Now in this paper, we assume that the received signal energy for one bit is unity, and the noise is Gaussian white whit density $N_0/2$. The bit error rate (BER) transmitted from the source node to the relay in the first time slot is consistent with the BER of traditional BPSK modulation $p_{bpsk} = Q(\sqrt{2/N_0})$, where $Q(x) = \int_x^{+\infty} \frac{1}{\sqrt{2\pi}} e^{-\frac{1}{2}t^2} dt$.

Table 1. PNC_BPSK mapping

BPSK modulation				PNC mapping		
S_1	S_2	a_1	a_2	$\hat{Y}$	$\oplus$	E_{Output}
0	0	−1	−1	−2	0	−1
0	1	−1	−1	0	1	1
1	0	+1	−1	0	1	1
1	1	+1	+1	+2	0	−1

In the broadcast phase, according to Table 1, the signal space at relay is {−2,0, 2}, and the probability of occurrence is 0.25, 0.5 and 0.25. By referring the derivation in [3], we apply the maximum posterior probability criterion (MAP):

$$\begin{aligned} 1 &= \frac{P(\hat{Y}=0|Y)}{P\big(\hat{Y}=-2|Y\big)+P(\hat{Y}=2|Y)} = \frac{0.5P(Y|\hat{Y}=0)}{0.25P(Y|\hat{Y}=-2)+0.25P(Y|\hat{Y}=2)} \\ &= 2\frac{exp(-\frac{Y^2}{N_0})/\sqrt{\pi N_0}}{exp(-\frac{(Y+2)^2}{N_0})/\sqrt{\pi N_0}+exp(-\frac{(Y-2)^2}{N_0})/\sqrt{\pi N_0}} \end{aligned} \tag{4}$$

then we can obtain the threshold:

$$\begin{cases} d_1 = -1 - \frac{N_0}{4} ln(1+\sqrt{1-e^{-8/N_0}}) \\ d_2 = 1 + \frac{N_0}{4} ln(1+\sqrt{1-e^{-8/N_0}}), \end{cases} \tag{5}$$

when the received signal $Y < d_1$ we judge that $\widehat{Y} = -2$; when the received signal $Y > d_2$ we judge that that $\widehat{Y} = 2$; when the received signal $d_1 < Y < d_2$ we judge that $\widehat{Y} = 0$. From this, we can easily get the bit error rate (BER) of PNC_BPSK as follows:

$$\begin{aligned} PNC_{BPSK} &= \frac{1}{2}\int_{-\infty}^{d_1}\frac{1}{\sqrt{\pi N_0}}e^{-\frac{Y^2}{N_0}dr} + \frac{1}{2}\int_{d_2}^{+\infty}\frac{1}{\sqrt{\pi N_0}}e^{-\frac{Y^2}{N_0}dr} \\ &+ \frac{1}{4}\int_{d_1}^{d_2}\frac{1}{\sqrt{\pi N_0}}e^{-\frac{(Y+2)^2}{N_0}dr} + \frac{1}{4}\int_{d_1}^{d_2}\frac{1}{\sqrt{\pi N_0}}e^{-\frac{(Y-2)^2}{N_0}dr} \end{aligned} \tag{6}$$

2.2 Analysis for 16-QAM Based PNC

Next, we discuss the impact of higher-order modulations on the probability of attack success. First, we focus on all nodes using QAM modulation. The amplitude and phase of the signal are simultaneously modulated as two independent parameters, and one symbol can be expressed as

$$e_k(t) = A_k \cos(\omega_c t + \theta_k), kT_B \le t \le (k+1)T_B, \tag{7}$$

Set

$$X_k = A_k \cos\theta_k, Y_k = -A_k \sin\theta_k, \tag{8}$$

then (7) is

$$e_k(t) = X_k \cos \omega_c t + Y_k \sin \omega_c t. \tag{9}$$

It is obvious that $e_k(t)$ can be regarded as the sum of two quadrature amplitude- shift keying(ASK) signals.

Then we take 16-QAM as an example. The data bit of N_i is $S_i \in \{1011, 1001, 0010, 0011, 1010, 1000, 0000, 0001, 1100, 0100, 0110, 1111, 1110, 0101, 0111\}$. There are two ways to generate the 16-QAM signal. The first way is the quadrature amplitude modulation method, i.e. two independent quadrature 4ASK signals are superimposed to form a 16-QAM signal, any M-QAM modulation can be considered as two independent quadrature $\sqrt{M}$-ASK signals. The second way to generate a 16-QAM signal is composite phase shifting. We will not describe too much here. Now all nodes in the PNC system use 16-QAM modulation, to facilitate the analysis of its performance, we decompose a 16-QAM signal into two independent orthogonal 4ASK signals, and set the modulation level to be $\{-3, -1, 1, 3\}$, then we only need to analyze the performance of one ASK signal, the other is the same.

The signal space of N_i is $S_i \in \{00, 01, 10, 11\}$, we can easily get the mapping $f(\cdot)^{-1}$

$$\{00, 01, 10, 11\} \xrightarrow{f-1} \{-3, -1, 1, 3\},$$

that is $a_i \in \{-3, -1, 1, 3\}$, then mapping of signals superimposed at the relay is

$$\{-6, -4, -2, 0, 2, 4, 6\} \xrightarrow{f} \{00, 01, \{00, 10\}, 11, \{00, 10\}, 01, 00\}.$$

We assume that the 4ASK signal $\{00, 01, 10, 11\}$ is sent with the same probability, then the probability of the signal through the addition operation at the relay is show in Table 2.

Table 2. The probability of $\widehat{Y}$

$\widehat{Y}$	−6	−4	−2	0	2	4	6
P	1/16	1/8	3/16	1/4	3/16	1/8	1/16

According to the MAP:

$$\Pr(Y|\hat{Y}_i)P_i - \Pr(Y|\hat{Y}_{i+1})P_{i+1} > 0, \tag{10}$$

where the P_i is the probability of $\hat{Y}_i$, and $\Pr(Y\ \hat{Y}_i)$ is the probability of sending $\hat{Y}_i$ and receiving Y at the relay, if the expression (11) is greater than 0, it can be determined that the transmission signal is $\hat{Y}_i$, if the expression (11) is less than 0, it can be regarded that the signal $\hat{Y}_{i+1}$ is sent, then we let the expression (11) equal to 0 and thresholds for adjacent signals can be determined:

$$d_i = \frac{1}{2}\left(\widehat{Y}_i \mid \widehat{Y}_{i+1}\right) + \frac{\sigma^2 \ln P_i/P_{i+1}}{\widehat{Y}_{i+1} - \widehat{Y}_i} \tag{11}$$

where P_i and $\hat{Y}_i$ are known, and the $\sigma^2 = \frac{No}{}$. The thresholds are shown below

d_1	d_2	d_3
$-5-(N_0 \ln 2)/4$	$-3-(N_0 \ln 3/2)/4$	$-1-(N_0 \ln 3/4)/4$
d_4	d_5	d_6
$1+(N_0 \ln 3/4)/4$	$3+(N_0 \ln 3/2)/4$	$5+(N_0 \ln 2)/4$

If $Y < d_1$ we judge that $\hat{Y} = 6$, if $d_1 < Y < d_2$ we judge that $\hat{Y} = 4$ and so on. We suppose that

$$f_i = \frac{1}{\sqrt{\pi N_0}} e^{-\frac{\left(Y-\widehat{Y_i}\right)^2}{N_0}}$$

is the probability density function (PDF) when the relay received signal is $\hat{Y}_i$. Then we can obtain the BER of PNC_4ASK:

$$\begin{aligned} PNC\_4ASK = & \frac{1}{16} \int_{d_1}^{+\infty} f_1 dY + \frac{1}{8}\left(\int_{-\infty} d_1 f_2 dY + \int_{d_2}^{+\infty} f_2 dY \right) \\ & + \frac{3}{16}\left(\int_{-\infty}^{d_2} f_3 dY + \int_{d_3}^{+\infty} f_3 dY \right) + \frac{1}{4}\left(\int_{-\infty} d_3 f_4 dY + \int_{d_4}^{+\infty} f_4 dY \right) \\ & + \frac{3}{16}\left(\int_{-\infty}^{d_4} f_5 dY + \int_{d_5}^{+\infty} f_5 dY \right) + \frac{1}{8}\left(\int_{-\infty}^{d_5} f_6 dY + \int_{d_6}^{+\infty} f_6 dY \right) \\ & + \frac{1}{16} \int_{-\infty}^{d_6} f_7 dY \end{aligned}$$

The BER of PNC_16QAM is:

$$PNC\_16QAM = 1 - (1 - PNC\_4ASK)^2 \tag{12}$$

Now we analyze the $P_{\text{att\_16QAM}}$. Table 3 shows that the scheme of PNC system in broadcast phase.

We briefly describe the calculation of $P_{\text{att\_16QAM}}$ from seven cases in Table 4. Such as Cases3, $P\left(\widehat{Y} = -2\right) \int_{d_2}^{d_3} f_3 dY$ is the probability that the sum of the signals sent by two source nodes is $\hat{Y} = 2$ and the received signal $d_2 < Y < d_3$, 1/3 is the probability that the attacker randomly guess on3 of the three possible transmitted signals. Other cases are calculated in the same way as Case3. Note that we are now talking about the P_{att} of one 4ASK signal, and the 16QAM signal is generated by two independent and orthogonal 4ASK signals, so the $P_{\text{att\_16QAM}}$ of PNC_16QAM is

$$P_{att\_16QAM} = \begin{cases} \left(\sum_{i=1}^{i=7} Pc_i\right)^2, P_{att\_16QAM} > \frac{1}{256} \\ \frac{1}{256}, P_{att\_16QAM} \le \frac{1}{256} \end{cases}$$

Table 3. PNC_QAM mapping

QAM modulation				PNC mapping		
$S1$	$S2$	$a1$	$a2$	$\hat{Y}$	$\oplus$	EOutput
00	00	−3	−3	−6	00	00
00	01	−3	−1	−4	01	01
00	10	−3	+1	−2	10	10
00	11	−3	+3	0	11	11
01	00	−1	−3	−4	01	01
01	01	−1	−1	−2	00	00
01	10	−1	+1	0	11	11
01	11	−1	+3	+2	10	10
10	00	+1	−3	−2	10	10
10	01	+1	−1	0	11	11
10	10	+1	+1	+2	00	00
10	11	+1	+3	+4	01	01
11	00	+3	−3	0	11	11
11	01	+3	−1	+2	10	10
11	10	+3	+1	+4	01	01
11	11	+ 3	+ 3	+ 6	00	00

2.3 Analysis for General M-QAM Based PNC

Now we focus on the P_{att} of M-QAM. Generally, we divide M-QAM into two orthogonal independent $\sqrt{}$M-ASK signals for processing.

Next, we analyze one $\sqrt{}$M-ASK signal. The data bit of N_{i} is

$$s_i \in \left\{c_i, c_2, \ldots, c_{\sqrt{M}}\right\},$$

and

$$a_i \in \left\{-\left(\sqrt{M}-1\right), -\left(\sqrt{M}-1\right)+2, -\left(\sqrt{M}-1\right)+4, \ldots, \sqrt{M}-1\right\},$$

let Quantify the Success Probability of Honest-but-Curious Relay Attack

$$d_i = -\frac{1}{2}[(3A_1 + A_2) + 2(i-1)] - \frac{N_0 \ln\frac{i+1}{i}}{4}$$

where $i \in \{1, 2, \ldots, \sqrt{M}-1\}$. We define $d_0 = -\infty$ and $d_{2\sqrt{M}-1} = +\infty$, then the Patt of M-QAM in PNC system is

$$P_{att} = \begin{cases} \dfrac{1}{M^2}\left(\sum_{i=0}^{2\sqrt{M}-1}\int_{d_i}^{d_{i+1}} f_{i+1} dY\right)^2, & P_{att} > \frac{1}{M^2} \\ \frac{1}{M^2}, & P_{att} \le \frac{1}{M^2} \end{cases}$$

Table 4. The Cases of Attack Successful

Cases	$(a1, a2)$	PC
$\hat{Y} = -6$	$(-3, -3)$	$P(\hat{Y} = -6) \int_{-\infty}^{d_1} f_1 dY$
$\hat{Y} = -4$	$(-3, -1)$ $(-1, -3)$	$\frac{1}{2} P(\hat{Y} = -4) \int_{d_1}^{d_2} f_2 dY$
$\hat{Y} = -2$	$(-3, +1)$ $(+1, -3)$ $(-1, -1)$	$\frac{1}{3} P(\hat{Y} = -2) \int_{d_2}^{d_3} f_3 dY$
$\hat{Y} = 0$	$(-3, +3)$ $(+3, -3)$ $(-1, +)$ $(+1, -1)$	$\frac{1}{4} P(\hat{Y} = 0) \int_{d3}^{d_4} f_4 dY$
$\hat{Y} = +2$	$(-1, +3)$ $(+3, -1)$ $(+1, +1)$	$\frac{1}{3} P(\hat{Y} = +2) \int_{d4}^{d_5} f_5 dY$
$\hat{Y} = +4$	$(+1, +3)$ $(+3, +1)$	$\frac{1}{2} P(\hat{Y} = +4) \int_{d5}^{d_6} f_6 dY$
$\hat{Y} = +6$	$(+3, +3)$	$P(\hat{Y} = +6) \int_{d6}^{+\infty} f_7 dY$

3 Numerical Results and Simulation

In the previous sections, we analyze security performance of PNC based on different modulations in theoretical way. Now we plot the $P_{\text{att\_bpsk}}$, $P_{\text{att\_qpsk}}$ and $P_{\text{att\_16qam}}$ in Fig. 2.

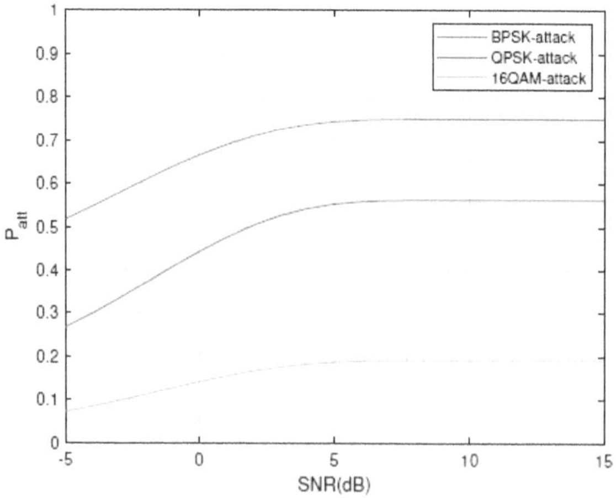

Fig. 2. P_{att} with Different Modulations

When the SNR is relatively low, the probability of attack success is smaller than when the SNR is high. That is because when the SNR is low, the attacker can obtain very little useful information, but as the SNR increases, the impact of noise on the attacker becomes smaller and smaller, and the information can be recovered more easily. In addition, we can find in Fig. 2 that with the increase of modulation order, the probability of a successful attack decreases significantly. This shows that the application of high-order modulation in the PNC system will not only improve the transmission efficiency but also enhance the security performance of the system.

We have proposed the theoretical value of P_{att}. Now we focus on the simulated P_{att} based on the BPSK, QPSK, and 16-QAM modulations in Fig. 3. In the three modulations, their theoretical and simulated values eventually converge. Note that when the SNR is poor, although the two are relatively close, there is also a certain gap. It should be emphasized here that the premise of an attacker's successful attack is that its decoding is correct. That is to say, if the superimposed signal $\hat{Y}$ falls within the threshold, we decode it according to the corresponding rules, but due to the influence of noise, the value of the superimposed signal may fall within the wrong threshold range, and the attacker recovers the signal according to the wrong $\hat{Y}$ value is also wrong. Because there are factors of decoding errors, its simulation results will be lower than the theoretical results, but when the SNR is good, the probability of decoding errors is greatly reduced, and the theoretical and simulated values tend to be consistent. Therefore, the theoretical value can be used as the upper limit of the attack success.

4 Conclusion and Future Work

It was originally proposed that the PNC system greatly improves the network throughput. Due to its superimposed nature at the relay, we found that the PNC system also guarantees the security of transmission. In this paper, by quantifying the P_{att} in the PNC system with different modulations, the contribution of PNC system to secure transmission is confirmed, and it is demonstrated that utilizing higher-order modulations in PNC systems will greatly improve its security.

In early 2022, the OPPO Research Institute released the "Zero Power Communication" white paper [30]. Point out that the zero-power terminal has a simple RF and baseband structure, while it needs to transmit data in an ultralow-power mode. Therefore, these two aspects will bring strong constraints to the signal modulation and coding methods for zero-power terminals, it is difficult to support QPSK and QAM modulation. Therefore, low-order ASK, FSK, and PSK modulations are better choices. In addition, trusted access and secure transmission remain important for zero-power communication. However, due to the extremely limited hardware and software capabilities, very small memory capacity, ultra-low power consumption, and the dependence on external power supply, it is difficult for zero-power terminals to use complex security mechanisms as in LTE or NR. Therefore, a lightweight security mechanism suitable for zero-power communication needs to be studied, so it should be researched on the enhanced security scheme that combines the physical layer and the transport layer. However, in this paper, we found that the security of using low-order modulation in PNC system is very low, so it is necessary to study the combination of low-complexity modulation technology and PNC system in the future.

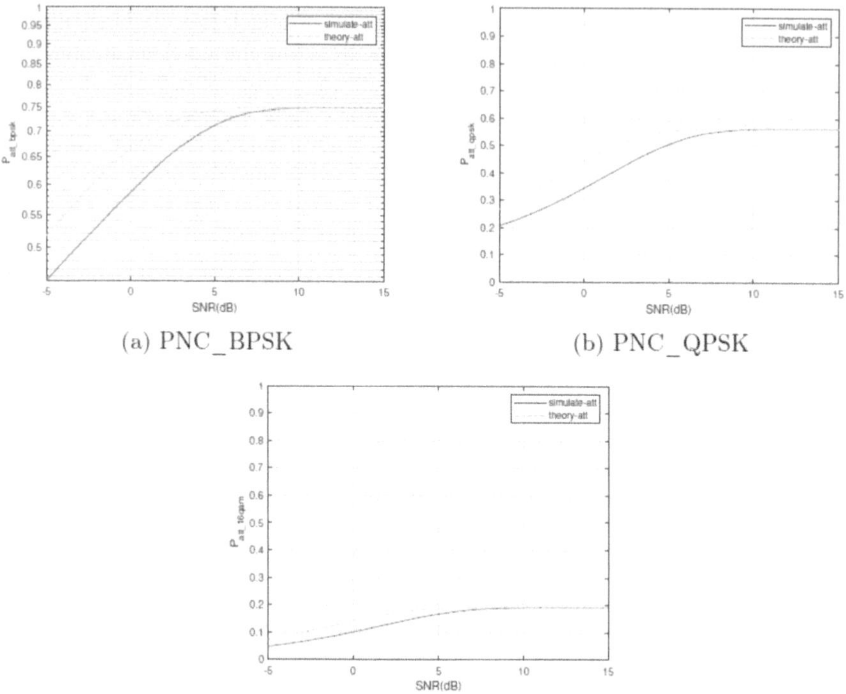

(a) PNC_BPSK (b) PNC_QPSK

(c) PNC_16QAM

Fig. 3. Simulate results

References

1. Shannon, C.E.: A mathematical theory of communication. Bell Syst. Tech. J. **27**(3), 379–423 (1948)
2. Ahlswede, R., Cai, N., Li, S.Y., Yeung, R.W.: Network information flow. IEEE Trans. Inf. Theory **46**(4), 1204–1216 (2000)
3. Zhang, S., Liew, S.C., Lam, P.P.: Hot topic: physical-layer network coding. In: Proceedings of the 12th Annual International Conference on Mobile Computing and Networking, pp. 358–365 (2006)
4. Cui, T., Gao, F., Tellambura, C.: Differential modulation for two-way wireless communications: a perspective of differential network coding at the physical layer. IEEE Trans. Commun. **57**(10), 2977–2987 (2009)
5. Namboodiri, V., Venugopal, K., Rajan, B.S.: Physical layer network coding for two-way relaying with QAM. IEEE Trans. Wirel. Commun. **12**(10), 5074–5086 (2013)
6. Li, X., Zhang, S., Qian, G.: Mapping and coding design for channel coded physical-layer network coding. In: 2013 International Workshop on High Mobility Wireless Communications (HMWC), pp. 120–125. IEEE (2013)
7. Yang, H., Chong, P.H., Meng, W., Li, B., Guan, Y.L.: Cooperative quadrature physical layer network coding in wireless relay networks. Int. J. Commun. Syst. **28**(1), 112–126 (2015)
8. Dang, X., Liu, Z., Li, B., Yu, X.: Noncoherent multiple-symbol detector of binary CPFSK in physical-layer network coding. IEEE Commun. Lett. **20**(1), 81–84 (2015)

9. Yu, Q.Y., Zhang, D.Y., Chen, H.H., Meng, W.X.: Physical-layer network coding systems with MFSK modulation. IEEE Trans. Veh. Technol. **65**(1), 204–213 (2015)
10. Tang, M., Chen, J., Zhang, Y., Zhang, Y.: Performance analysis for physical-layer network coding with hierarchical modulation. In: 2016 8th IEEE International Conference on Communication Software and Networks (ICCSN). pp. 33–37. IEEE (2016)
11. Donati, D., Flanagan, M.F.: Physical-layer network coded QAM with trellis shaping for the two-way relay channel. In: 2016 50th Asilomar Conference on Signals, Systems and Computers, pp. 250–254. IEEE (2016)
12. Yang, L., Yang, T., Yuan, J., An, J.: Achieving the near-capacity of two-way relay channels with modulation-coded physical-layer network coding. IEEE Trans. Wirel. Commun. **14**(9), 5225–5239 (2015)
13. Jayasinghe, K., Jayasinghe, P., Rajatheva, N., Latva-Aho, M.: Secure beamforming design for physical layer network coding based MIMO two-way relaying. IEEE Commun. Lett. **18**(7), 1270–1273 (2014)
14. Lu, K., Fu, S., Qian, Y., Zhang, T.: On the security performance of physical-layer network coding. In: 2009 IEEE International Conference on Communications, pp. 1–5. IEEE (2009)
15. Forutan, V., Fischer, R.F.: On the security of lattice-based physical-layer network coding against wiretap attacks. In: SCC 2015, 10th International ITG Conference on Systems, Communications and Coding, pp. 1–6. VDE (2015)
16. Liu, Q., Zhang, W., Ding, S., Li, H., Wang, Y.: Novel secure group data exchange protocol in smart home with physical layer network coding. Sensors **20**(4), 1138 (2020)
17. Trushina, O.: On the anonymity of physical-layer network coding against wiretapping. In: 2016 XV International Symposium Problems of Redundancy in Information and Control Systems (REDUNDANCY), pp. 158–161. IEEE (2016)
18. Deng, D., Yang, Z.l., Zhao, M.: PHY security enhancement in analog network coding based on artificial noise. In: 2014 Sixth International Conference on Wireless Communications and Signal Processing (WCSP), pp. 1–6. IEEE (2014)
19. Wang, W.Q., Zheng, Z.: Hybrid MIMO and phased-array directional modulation for physical layer security in mmWave wireless communications. IEEE J. Sel. Areas Commun. **36**(7), 1383–1396 (2018)
20. Mensi, N., Rawat, D.B., Balti, E.: Physical layer security for V2I communications: reflecting surfaces vs. relaying. In: 2021 IEEE Global Communications Conference (GLOBECOM), pp. 01–06. IEEE (2021)
21. Chen, R., Li, C., Yan, S., Malaney, R., Yuan, J.: Physical layer security for ultra-reliable and low-latency communications. IEEE Wirel. Commun. **26**(5), 6–11 (2019)
22. Chen, J., Yang, L., Alouini, M.S.: Physical layer security for cooperative NOMA systems. IEEE Trans. Veh. Technol. **67**(5), 4645–4649 (2018)
23. Yesilkaya, A., et al.: Physical-layer security in visible light communications. In: 2020 2nd 6G Wireless Summit (6G SUMMIT), pp. 1–5. IEEE (2020)
24. Gong, C., Yue, X., Zhang, Z., Wang, X., Dai, X.: Enhancing physical layer security with artificial noise in large-scale NOMA networks. IEEE Trans. Veh. Technol. **70**(3), 2349–2361 (2021)
25. Zhang, Z., Zhang, C., Jiang, C., Jia, F., Ge, J., Gong, F.: Improving physical layer security for reconfigurable intelligent surface aided NOMA 6G networks. IEEE Trans. Veh. Technol. **70**(5), 4451–4463 (2021)
26. Makarfi, A.U., Rabie, K.M., Kaiwartya, O., Li, X., Kharel, R.: Physical layer security in vehicular networks with reconfigurable intelligent surfaces. In: 2020 IEEE 91st Vehicular Technology Conference (VTC2020-Spring), pp. 1–6. IEEE (2020)
27. Chorti, A., et al.: Context-aware security for 6g wireless: the role of physical layer security. IEEE Commun. Stand. Mag. **6**(1), 102–108 (2022)

28. Sánchez, J.D.V., Urquiza-Aguiar, L., Paredes, M.C.P., Osorio, D.P.M.: Survey on physical layer security for 5G wireless networks. Ann. Telecommun. **76**(3), 155–174 (2021)
29. Zhang, Y., Shen, Y., Jiang, X., Kasahara, S.: Secure millimeter-wave ad hoc communications using physical layer security. IEEE Trans. Inf. Forensics Secur.nsics Secur. **17**, 99–114 (2021)
30. Zero power communication (2022). https://www.oppo.com/content/dam/oppo/en/mkt/newsroom/press/oppo-releases-zero-power-communication-whitepaper/white-paper_en.pdf

Detecting SQL Injection and XSS Attacks Using ASCII Code and CNN

Na Wei and Bailin Xie(✉)

School of Information Science and Technology, Guangdong University of Foreign Studies, Guangzhou, China
bailinxie@gdufs.edu.cn

Abstract. Nowadays, SQL injection attacks and XSS attacks are the two most popular attacks. Because of the shortcomings of traditional detection methods, this study proposes to use the ASCII code corresponding to the character as the feature vector, and then adopt the convolutional neural network (CNN) to detect SQL injection and XSS attacks. In the experiments, the evaluation indicators are accuracy, recall, and F1 score. On the SQL and XSS datasets, the accuracy of the method achieves 98.5% and 98.0%, the recall achieves 98.3% and 97.7%, and the F1 achieves 98.5% and 97.9%, respectively. Considering that the attack data in real life is difficult to obtain, we extracted small-scale samples and verified the effectiveness of our method on small-scale samples. Meanwhile, we compared our method with three machine learning methods. Experimental results show that the accuracy, recall and F1 values of our method are better than other methods. This method does not require manual extraction of data features and can detect new types of attacks.

Keywords: SQL injection attack · XSS attack · ASCII code · Convolutional neural network

1 Introduction

In recent years, SQL injection and XSS attacks have ranked in the top 10 of all cyber-attacks in OWASP [1], as shown in Fig. 1. The principle of SQL injection attacks is similar to that of XSS attacks. Attackers send untrusted data as part of a command to the parser for execution, which ultimately leads to sensitive data being stolen, arbitrarily manipulating the background, and obtaining the highest privileges.

Existing methods for detecting Web attacks typically rely on rules, blacklists, or whitelists [15]. However, with the continuous improvement of attack technology, Web attacks have become more sophisticated and covert. Many new types of Web attacks can not be effectively detected using existing methods. Currently, some researchers have turned to using artificial intelligence algorithms to detect Web attacks.

This study uses ASCII code and convolutional neural network (CNN) to detect SQL injection and XSS attacks. Experimental results show that the method has higher detection accuracy than other methods and is easier to implement.

Z. Gu et al. (Eds.): NSE 2023, CCIS 2063, pp. 33–45, 2024.
https://doi.org/10.1007/978-981-97-4519-7_3

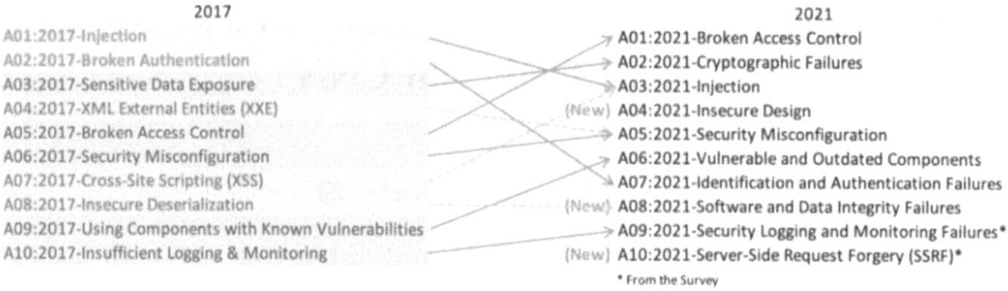

Fig. 1. OWASP top 10 2017–2021.

The remainder of this paper is organized as follows. Section 2 introduces recent works on data preprocessing and detection models for SQL injection and XSS attacks. Section 3 presents the detection method for SQL injection and XSS attacks. In Sect. 4, experiments are presented and the results are analyzed. Finally, Section 5 concludes the paper.

2 Related Works

Web attacks are based on vulnerabilities of Web applications. Among them, SQL injection and XSS attacks are the two popular attacks in our life, and their datasets consist of string statements. Most researchers utilized feature extraction methods to process the data of SQL injection and XSS attacks. Zhou et al. [19] extracted the output length, the number of keywords (alert, script, image, etc.), and special characters (', ", +, etc.) as features of XSS attacks. Uwagbole [16] extracted SQL tags, single quotes, semicolons and comments at the injection point, and marked these features for regular matching. Due to the fact that feature extraction methods rely on researchers' prior knowledge, the selected features may result in poor detection results. Nowadays, natural language processing (NLP) is widely used as a data preprocessing method for Web attack detection. Li et al. [10] used Word2vec and Bag-of-Word (BoW) to convert the data of SQL injection and XSS attacks into feature vectors, and the experimental results show that the Word2vec method is better than BoW. Lu et al. [11] proposed a detection model based on semantic learning. It utilized SQL syntax trees to achieve semantic position embedding. This method has good generalization performance and can detect unknown attacks. In addition, Kar et al. [8] converted blank characters to spaces and removed backticks and redundant parentheses. As shown in Fig. 2, the SQL queries are normalized to specific tokens after the conversion rules.

Traditional detection methods are based on rules and regular expressions that match the input with existing rules to detect Web attacks. Atoum [3] proposed a hybrid technique combining static analysis and dynamic analysis. It implemented tracking in the dynamic analysis phase and compared strings in the static analysis phase to identify SQL injection attacks. Choras et al. [4] used a genetic algorithm to generate different regular expressions between populations and the same regular expression within a population and finally calculated the fitness function to identify SQL injection attacks. The main drawback of methods based on rules and regular expressions is the difficulty of maintaining rules. With

```
seLEcT * FroM users whERE age > 25
SELECT STAR FROM USRTBL WHERE USRCOL GT INT

SeLeCt * fRom users WHeRe user_name = ' Jerry ' AnD user_sex = 1
SELECT STAR FROM USRTBL WHERE USRCOL ET SQUT CHAR SQUT AND USRCOL ET INT
```

Fig. 2. Examples of SQL query tokenization.

the development of machine learning technology, machine learning algorithms have been widely applied in Web attack detection. Shar and Tan [14] proposed a hybrid approach based on k-means clustering and combining static and dynamic attributes to detect SQL and XSS vulnerabilities in PHP. This method does not require labeled data labels, so it is less expensive. Wang et al. [17] extracted similarity-based features and difference-based features from webpages by comparing online social networks (OSNs) with general networks and fed them into ADTree and AdaBoost to detect XSS attacks in OSNs. Hasan [5] compared 23 different machine learning algorithms and summarized 5 classifiers with better performance, Boosted Trees, Bagged Trees, Linear Discriminant algorithm, Cubic SVM, and Fine Gaussian SVM. Joshi [6] implemented SQL injection detection using Naive Bayes and Gradient Boosting classifier, and experimental results showed that the Gradient Boosting classifier had higher accuracy. Ross et al. [13] designed a data generation mechanism to generate normal and malicious traffic, and used the J48 decision tree, random forest, SVM, neural network, and JRip rule-based algorithms to classify SQL injection. The experimental results showed that the neural network algorithm achieved the same accuracy but with much less time cost for modeling and detection. However, these methods also have some drawbacks. When the samples are unbalanced, the performance of the KNN algorithm is poor, the logistic regression is prone to underfitting, and the decision tree is prone to overfitting. Nowadays, deep learning is widely used in Web attack detection. Jothi et al. [7] used the index as the only representation of the word, and input it into a neural network model containing the global average pooling and MLP to detect SQL injection attacks. In [2], Abaimov and Bianchi proposed a method to detect injection attacks using the convolutional deep neural network and supplemented it with domain-specific semantic knowledge. Zhang et al. [18] designed a graph intrusion detection (GID) framework combining topology with graph neural network (GNN). This method can learn the network structure and node embedding information and detect the attack in the industrial Internet of Things. Lei et al. [9] proposed an LSTM-attention-based model for XSS attack detection, which adds an attention layer after the LSTM layer to focus attention on more important units.

3 The Proposed Approach

Traditional detection methods cannot detect new attacks, and machine learning based methods are difficult to extract key data features. The convolutional neural network (CNN) is a computational model that imitates the structure and function of biological brains, and the convolutional layers can automatically extract features. In this study, we adopt CNN to extract features of attack data and detect SQL injection and XSS attacks. Since CNN cannot process the attack data directly, we use ASCII code to convert the strings of the attack data to numeric vectors.

3.1 SQL Injection Attacks and XSS Attacks

SQL Injection Attacks. When the input parameters are not filtered (or not completely filtered), SQL injection attacks will occur. The attacker splices malicious requests into SQL queries, which the server parses for its illegitimate purposes. There are two conditions for triggering SQL injection attacks: the parameters passed from the front end to the back end are user-controllable; the controlled parameters must be brought into the database for querying.

According to the type of injection point, SQL injection is divided into numeric injection and character injection.

Numeric Injection. Fig. 3 shows a normal SQL query and two numeric injection samples. The type of the injection point "id" is numeric, and the information corresponding to the "id" is returned to the front end page. The attacker enters "and $1 = 1$" (sample①) or "and $1 = 2$" (sample②) to judge the numeric injection.

```
normal SQL query :
SELECT * FROM users WHERE id = 1

SQL injection - numeric injection samples:
① SELECT * FROM users WHERE id = 1 and 1 = 1
② SELECT * FROM users WHERE id = 1 and 1 = 2
```

Fig. 3. Examples of numeric injection.

Character Injection. The normal SQL query and the character injection samples are shown in Fig. 4. The attacker enters "user'#" (sample①) or "user'–" (sample②) to close the single quote to the left of "user". "#" and "–" are comment characters in SQL syntax, so the attacker can successfully log into the account without entering the password.

XSS Attacks. XSS (Cross-Site Scripting) attacks occur during the rendering of HTML documents on the target website, where the attacker injects malicious code into the HTML page to execute abnormal script commands. XSS attacks can be divided into Reflected XSS, Stored XSS, and DOM-based XSS attacks. Reflected XSS is a non-persistent XSS that executes malicious commands only when the user clicks on a URL containing malicious scripts. Stored XSS is persistent XSS. The attacker stores malicious scripts in the application's database and executes them every time data is retrieved from the database. Reflected XSS and Stored XSS inevitably generate data interactions, while DOM-based XSS directly launches the attack on the front end.

normal SQL query :

SELECT * FROM users WHERE username = 'user' AND password = '123'

SQL injection - character injection samples:

① SELECT * FROM users WHERE username = 'user'#' AND password = 'xxx'

② SELECT * FROM users WHERE username = 'user'-- ' AND password = 'xxx'

Fig. 4. Examples of character injection.

3.2 Data Processing

Data vectorization is a popular data processing method in deep learning. The processing operations for SQL injection and XSS attack data include word segmentation and vectorization, that is, unstructured data is converted into digital data.

ASCII is a unified set of encoding rules that use 128 decimal numbers to represent different symbols. The conversion rules are shown in Fig. 5, where the blue part represents non-printing control characters and the rest represents printable characters. SQL injection and XSS attack data consist of English letters, numbers, and special symbols. These characters have one-to-one corresponding values in ASCII code, so we use ASCII code to represent the feature vectors of the data.

ASCII															
DEC	**Symbol**	**DEC**	**Symbol**	**DEC**	**Symbol**	**DEC**	**Symbol**	**DEC**	**Symbol**	**DEC**	**Symbol**	**DEC**	**Symbol**	**DEC**	**Symbol**
0	NUL	16	DLE	32	(space)	48	0	64	@	80	P	96	`	112	p
1	SOH	17	DC1	33	!	49	1	65	A	81	Q	97	a	113	q
2	STX	18	DC2	34	"	50	2	66	B	82	R	98	b	114	r
3	ETX	19	DC3	35	#	51	3	67	C	83	S	99	c	115	s
4	EOT	20	DC4	36	$	52	4	68	D	84	T	100	d	116	t
5	ENQ	21	NAK	37	%	53	5	69	E	85	U	101	e	117	u
6	ACK	22	SYN	38	&	54	6	70	F	86	V	102	f	118	v
7	BEL	23	ETB	39	'	55	7	71	G	87	W	103	g	119	w
8	BS	24	CAN	40	(	56	8	72	H	88	X	104	h	120	x
9	TAB	25	EM	41	)	57	9	73	I	89	Y	105	i	121	y
10	LF	26	SUB	42	*	58	:	74	J	90	Z	106	j	122	z
11	VT	27	ESC	43	+	59	;	75	K	91	[	107	k	123	{
12	FF	28	FS	44	,	60	<	76	L	92	\	108	l	124	|
13	CR	29	GS	45	-	61	=	77	M	93	]	109	m	125	}
14	SO	30	RS	46	.	62	>	78	N	94	^	110	n	126	~
15	SI	31	US	47	/	63	?	79	O	95	_	111	o	127	DEL

Fig. 5. Conversion rules corresponding to ASCII code.

After analyzing the datasets, we counted the percentage of samples with lengths in the range of 100, 200, 300, and 400. It can be seen from Fig. 6 that most of the SQL and XSS attack samples in the datasets are within 400 characters, so we fixed the sample length at 400. If the length is less than 400, add 0 at the end; otherwise, only the first 400 characters will be retained, and the remaining characters will be discarded.

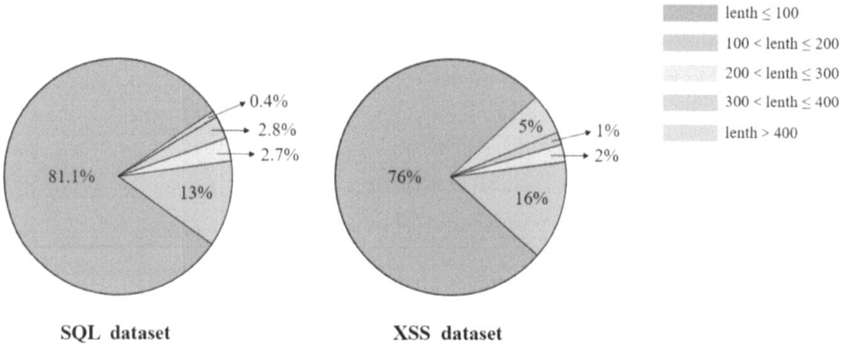

Fig. 6. The sample length distribution in the SQL datasets and XSS datasets.

We convert each character to the corresponding ASCII value to form a vector of numbers with a length of 400. Next, we reshape each vector into a 20 * 20 matrix to extract the attack features more accurately. Take the following SQL injection sample as an example, its conversion to an ASCII matrix is shown in Fig. 7.

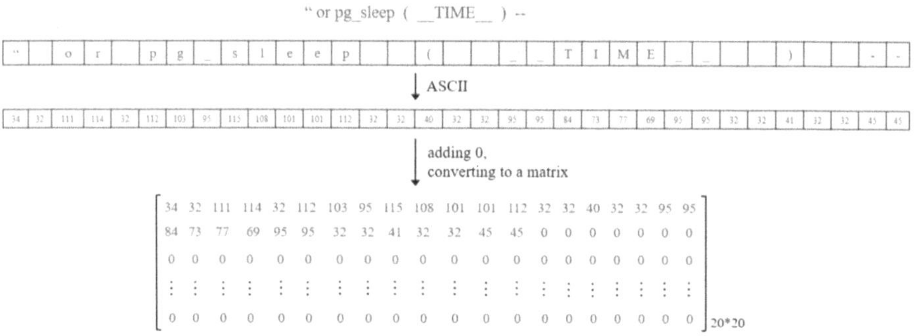

Fig. 7. The process of constructing an ASCII matrix.

3.3 Model Training and Testing

The basic structure of CNN includes convolutional layers, pooling layers, and fully connected layers. The convolutional layer can automatically extract the features of the input data. The fully connected layer completes the classification of the data based on the extracted features. However, if we only use fully connected layers, it will lead to the loss of spatial information and generate too many parameters, resulting in overfitting. Based on the above two points, we use CNN to detect SQL injection and XSS attacks. The CNN we constructed is shown in Fig. 8. After two convolution and pooling operations, the flattened data is input to the fully connected layer, and the classification results of the model are given.

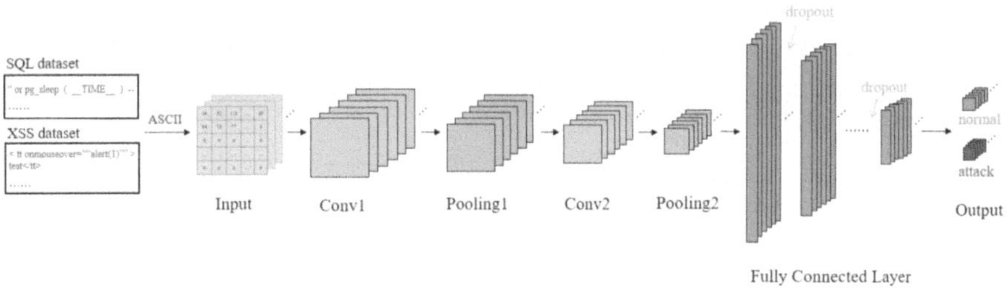

Fig. 8. The structure of CNN.

Convolutional Layer. This layer is the core of CNN, and its function is to automatically extract the key features of the data through convolution calculation. The convolution kernel performs multiplication and addition operations on each matrix from left to right and from top to bottom according to the step size. The convolution is calculated by (1), where m represents the size of the input matrix, $x \in X$ represents the input ASCII matrix, $k \in K$ is the convolution kernel, T represents the transpose of the matrix, b is the bias unit, and $z_{u,v}$ represents the element at (u, v) in the output feature map.

$$z_{u,v} = \sum_{i=0}^{m} \sum_{j=0}^{m} x_{i+u,j+v} \cdot k_{i,j}^{T} + b \tag{1}$$

The size of the feature map is calculated by (2), where m is the same variable as in formula (1), p is the number of layers filled with 0, n is the size of the kernel, and s represents the step size.

$$f = \frac{m + 2p - n}{s} + 1 \tag{2}$$

In our model, the size of the convolution kernel is 3 * 3 and the step size is 1. The number of convolution kernels in the first and second layers is 256 and 128, respectively. We use the activation function ReLU to remove redundant data and maximize the representativeness of the data features. The convolution operation of ASCII data is shown in Fig. 9.

Pooling Layer. It is added after the convolutional layer to reduce the amount of data in the network and effectively prevent overfitting. The pooling functions include max pooling, average pooling, and L2-norm pooling. The size of the output matrix after pooling can also be calculated by (2). We use max pooling to select the largest number to retain the features of the region, which is calculated by (3), where $y_{i,j}$ represents the output of max pooling and $x_{p,q}$ represents the element (p, q) in the rectangular area R_{ij}. In this study, the size of the pooling kernel is 2 * 2.

$$y_{i,j} = \max_{p,q \subset R_{ij}} x_{p,q} \tag{3}$$

Fully Connected Layer. It maps the learned feature representations to the label space for sample classification. Multiple fully connected layers can be connected in a network,

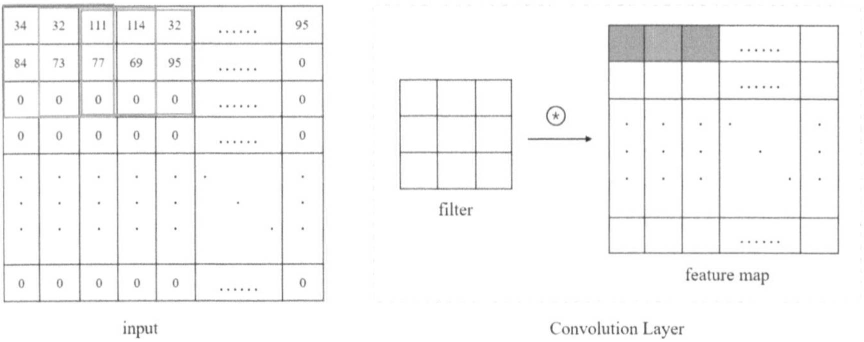

Fig. 9. Convolution process of ASCII data.

with neurons in each layer connected to all neurons in the previous layer. This fully connected structure leads to a large number of parameters in the network, so dropout regularization can be added to the model to alleviate overfitting. In our CNN model, there are 6 fully connected layers with a number of neurons 128, 64, 32, 16, 8, and 1. We add dropout in the first and penultimate layers to randomly inactivate neurons with a probability of 0.2 during model training.

4 Evaluation

4.1 Dataset

We collect the SQL dataset and XSS dataset both from the GitHub website [12]. These two datasets contain normal samples and attack samples with labels, where 0 represents normal samples and 1 represents attack samples. The SQL dataset contains 30,918 samples, 19,536 normal samples, and 11,382 SQL injection attack samples. The XSS dataset contains 13,686 samples, 6313 normal samples, and 7373 XSS attack samples, as shown in Table 1. The training sets and testing sets are divided in the ratio of 7:3.

Table 1. SQL dataset and XSS dataset.

Datasets	Total	Class	Numbers
SQL dataset	30918	Normal (0)	19536
		Attack (1)	11382
XSS dataset	13686	Normal (0)	6313
		Attack (1)	7373

4.2 Experimental Results and Analysis

To validate the method proposed in this study, we conducted two comparison experiments on the SQL and XSS datasets. The first experiment demonstrates the effectiveness of the ASCII method, and the second experiment proves the CNN model constructed in this study has better performance. We evaluate the experimental results using accuracy, recall, and F1 score.

In the first experiment, we select two different data processing methods, Word2vec and the ASCII code method proposed in this study, and both choose the same detection model CNN to compare the experimental results. The optimizer for the experiment is Adam, the loss function is binary cross-entropy, and the maximum value of epoch is 5. The experimental results are shown in Table 2.

Table 2. Comparison with Word2vec on the SQL and XSS datasets.

Datasets	Methods	Accuracy	Recall	F1 score
SQL dataset	Word2vec + CNN	0.977	0.971	0.975
	ASCII + CNN	**0.985**	**0.983**	**0.985**
XSS dataset	Word2vec + CNN	0.971	0.970	0.971
	ASCII + CNN	**0.980**	**0.977**	**0.979**

The experimental results show that the detection performance of our method is superior to Word2vec. On the SQL and XSS datasets, the accuracy of our method achieves 98.5% and 98.0%, the recall achieves 98.3% and 97.7%, and the F1 achieves 98.5% and 97.9%, respectively.

In the second experiment, we compare the CNN model constructed in this study with Support Vector Machine (SVM), K-Nearest Neighbor (KNN), and Random Forest (RF). The experimental results are shown in Table 3.

Table 3. Comparison with machine learning methods on the SQL and XSS datasets

Datasets	Methods	Accuracy	Recall	F1 score
SQL dataset	SVM	0.965	0.962	0.963
	KNN	0.939	0.948	0.945
	RF	0.970	0.971	0.971
	CNN	**0.981**	**0.977**	**0.979**
XSS dataset	SVM	0.960	0.957	0.958
	KNN	0.938	0.944	0.942
	RF	0.970	0.965	0.967
	CNN	**0.977**	**0.976**	**0.976**

From the experimental results in Table 3, it can be seen that our method has higher accuracy, recall, and F1 score than other methods. Machine learning algorithms only perform shallow feature data analysis, resulting in poor performance. Deep learning models can analyze the semantic information structure of the data. CNN is one of the classic models of deep learning. It extracts deep features from SQL injection and XSS attack data and also considers the influence of surrounding features to obtain better performance.

The slightly different scores achieved in the first and second experiments using CNN are due to the training sets and test sets were selected randomly. Our method performs better on the SQL dataset than on the XSS dataset. The above two experiments indicate that the method of combining ASCII code with CNN can achieve better detection performance than other methods and can effectively detect Web attacks.

Small-Scale Samples. Due to the attack samples being difficult to obtain in real life. We randomly selected 1000 samples from the SQL datasets and XSS datasets as small-scale samples, including 500 normal samples and 500 attack samples, as shown in Figs. 10 and 11.

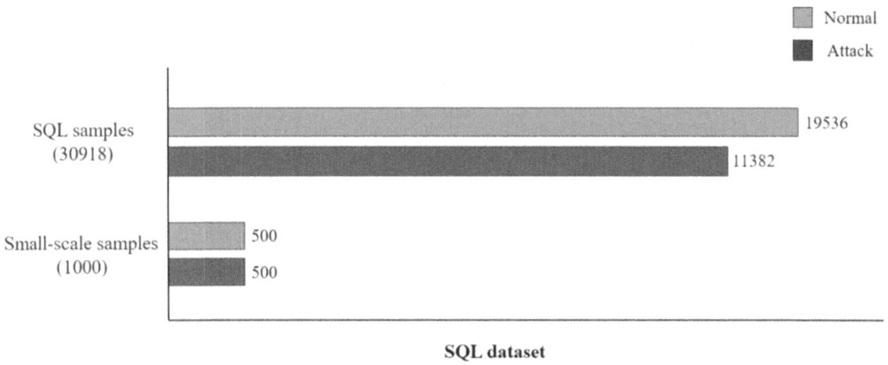

Fig. 10. Small-scale samples of SQL dataset.

To verify the effectiveness of our method on small-scale samples, we conducted two sets of comparison experiments, Word2vec, and machine learning methods. The experimental results are shown in Tables 4 and 5.

On small-scale SQL and XSS datasets, the accuracy of our method achieves 95.3% and 94.8%, the recall achieves 94.2% and 94.3%, and the F1 achieves 95.0% and 94.5%, respectively. Compared with other methods, our method still obtains the highest score, which shows the method of combining ASCII code with CNN can also show excellent detection ability in the case of scarce samples.

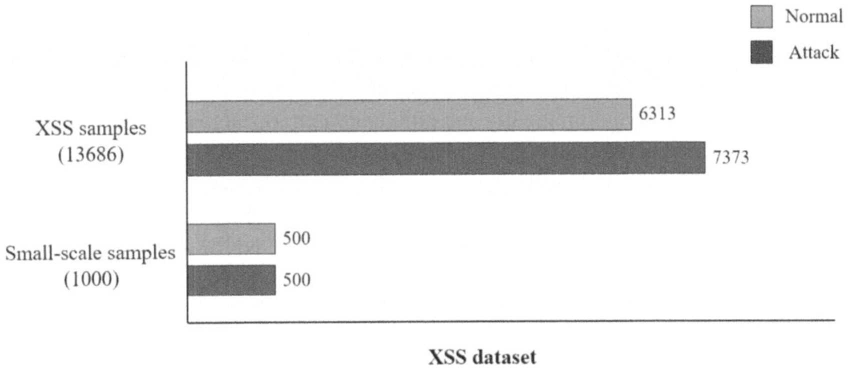

Fig. 11. Small-scale samples of XSS dataset.

Table 4. Comparison with Word2vec on the small-scale sample.

Datasets	Methods	Accuracy	Recall	F1 score
SQL dataset (small-scale)	Word2vec + CNN	0.930	0.932	0.932
	ASCII + CNN	**0.953**	**0.942**	**0.950**
XSS dataset (small-scale)	Word2vec + CNN	0.928	0.935	0.938
	ASCII + CNN	**0.948**	**0.943**	**0.945**

Table 5. Comparison with machine learning methods on the small-scale sample.

Datasets	Methods	Accuracy	Recall	F1 score
SQL dataset (small-scale)	SVM	0.917	0.920	0.920
	KNN	0.889	0.882	0.887
	RF	0.914	0.916	0.916
	CNN	**0.926**	**0.930**	**0.928**
XSS dataset (small-scale)	SVM	0.915	0.917	0.920
	KNN	0.880	0.875	0.878
	RF	0.909	0.911	0.912
	CNN	**0.931**	**0.929**	**0.929**

5 Conclusion

In this study, we propose to use ASCII code to represent the feature vectors of the data and adopt the CNN model to detect SQL injection and XSS attacks. We conducted experiments comparing our method with Word2vec and three machine learning methods using SQL and XSS samples. The results show that our method outperforms the other methods for detecting Web attacks effectively. Considering that attack samples are difficult to obtain in real life, we further conduct comparison experiments with small-scale samples to demonstrate the effectiveness of our approach in the case of sample scarcity.

The CNN model can automatically and efficiently extract data characteristics, which brings great convenience to detect SQL injection and XSS attacks.

However, as attack methods become increasingly sophisticated and covert, simple CNN models may not be effective in detecting complex Web attacks. In future work, we will improve the data preprocessing mechanism and CNN model to improve the detection performance.

References

1. A01:2021 - broken access control. [EB/OL]. https://owasp.org/Top10/A01_2021-Broken_Access_Control/. Accessed 17 Dec 2022
2. Abaimov, S., Bianchi, G.: Coddle: code-injection detection with deep learning. IEEE Access **7**, 128617–128627 (2019)
3. Atoum, J.O., Qaralleh, A.J.: A hybrid technique for SQL injection attacks detection and prevention. Int. J. Database Manag. Syst. **6**(1), 21 (2014)
4. Choraś, M., Kozik, R., Puchalski, D., Hołubowicz, W.: Correlation approach for SQL injection attacks detection. In: Herrero, Á., et al. (eds.) International Joint Conference CISIS 2012-ICEUTE 2012-SOCO 2012 Special Sessions. Advances in Intelligent Systems and Computing, vol. 189, pp. 177--185. Springer, Heidelberg (2013). https://doi.org/10.1007/978-3-642-33018-6_18
5. Hasan, M., Balbahaith, Z., Tarique, M.: Detection of SQL injection attacks: a machine learning approach. In: 2019 International Conference on Electrical and Computing Technologies and Applications (ICECTA), pp. 1--6. IEEE (2019)
6. Joshi, A., Geetha, V.: SQL injection detection using machine learning. In: 2014 International Conference on Control, Instrumentation, Communication and Computational Technologies (ICCICCT), pp. 1111–1115. IEEE (2014)
7. Jothi, K., Pandey, N., Beriwal, P., Amarajan, A., et al.: An efficient SQL injection detection system using deep learning. In: 2021 International Conference on Computational Intelligence and Knowledge Economy (ICCIKE), pp. 442–445. IEEE (2021)
8. Kar, D., Panigrahi, S., Sundararajan, S.: SQLiGoT: detecting SQL injection attacks using graph of tokens and SVM. Comput. Secur. **60**, 206–225 (2016)
9. Lei, L., Chen, M., He, C., Li, D.: XSS detection technology based on LSTM-attention. In: 2020 5th International Conference on Control, Robotics and Cybernetics (CRC), pp. 175–180. IEEE (2020)
10. Li, Q., Wang, F., Wang, J., Li, W.: LSTM-based SQL injection detection method for intelligent transportation system. IEEE Trans. Veh. Technol. **68**(5), 4182–4191 (2019)
11. Lu, D., Fei, J., Liu, L.: A semantic learning-based SQL injection attack detection technology. Electronics **12**, 1344 (2023)
12. Lxyeternal: [EB/OL]. https://github.com/lxyeternal/WAF/tree/master/AiWaf-2/data. Accessed 10 July 2023
13. Ross, K., Moh, M., Moh, T.S., Yao, J.: Multi-source data analysis and evaluation of machine learning techniques for SQL injection detection. In: Proceedings of the ACMSE 2018 Conference, pp. 1--8 (2018)
14. Shar, L.K., Tan, H.B.K., Briand, L.C.: Mining SQL injection and cross site scripting vulnerabilities using hybrid program analysis. In: 2013 35th International Conference on Software Engineering (ICSE), pp. 642–651. IEEE (2013)

15. Tang, P., Qiu, W., Huang, Z., Lian, H., Liu, G.: Detection of SQL injection based on artificial neural network. Knowl.-Based Syst. **190**, 105528 (2020)
16. Uwagbole, S.O., Buchanan, W.J., Fan, L.: Applied machine learning predictive analytics to SQL injection attack detection and prevention. In: 2017 IFIP/IEEE Symposium on Integrated Network and Service Management (IM), pp. 1087–1090. IEEE (2017)
17. Wang, R., Jia, X., Li, Q., Zhang, S.: Machine learning based cross-site scripting detection in online social network. In: 2014 IEEE International Conference on High Performance Computing and Communications, 2014 IEEE 6th International Symposium on Cyberspace Safety and Security, 2014 IEEE 11th International Conference on Embedded Software and Systems (HPCC, CSS, ICESS), pp. 823–826. IEEE (2014)
18. Zhang, Y., Yang, C., Huang, K., Li, Y.: Intrusion detection of industrial internet-of-things based on reconstructed graph neural networks. IEEE Trans. Netw. Sci. Eng. **10**, 2894–2905 (2022)
19. Zhou, Y., Wang, P.: An ensemble learning approach for XSS attack detection with domain knowledge and threat intelligence. Comput. Secur. **82**, 261–269 (2019)

An Intranet Defense Capability Assessment Method for Network Security Solutions Based on ATT&CK Framework

Qi An[1], Zifeng Zhu[1], Shudong Li[1,2](✉), Weihong Han[2], Xiaobo Wu[3], and Shumei Li[4]

[1] Cyberspace Institute of Advanced Technology, GuangzhouUniversity, Guangzhou 510006, China
{happywxb,lishudong}@gzhu.edu.cn

[2] Department of New Networks, Peng Cheng Laboratory, Shenzhen 518000, China
Chinahanwh@pcl.ac.cn

[3] School of Computer Science and Cyber Engineering, Guangzhou University, Guangzhou 510006, China

[4] Computer College, Jilin Normal University, Siping 136000, China
lishumeiwei@sina.com

Abstract. Faced with the growing threat of intranet attacks and numerous network security solutions, the defensive capabilities of network security solutions need to be evaluated in order to reasonably select a network security solution. However, the existing assessment of the defense capability of network security solutions mainly focuses on anti-DDoS devices, and lacks the assessment of the defense capability of the intranet, and it is complicated to obtain the values of assessment indicators during the assessment process. In addition, it is also difficult to obtain ideas for defense improvement of network security solutions. In the face of this situation, we propose an intranet defense capability assessment method for network security solutions based on the ATT&CK framework from an attack perspective. This method introduces the ATT&CK framework into the assessment of the intranet defense capability of network security solutions, constructs an indicator system with comprehensiveness and good scalability, and involves tactics, techniques and sub-technologies in the enterprise framework used by 138 APT organizations in the ATT&CK framework as objective data for indicator assignment, balancing the influence of attackers' subjective experiences and preferences on the assessment results, and this method can easily obtain ideas to improve the intranet defense capability of network security solutions. The effectiveness of the proposed evaluation method is verified by conducting simulation experiments at a network range.

Keywords: ATT&CK · intranet attack · defense capability assessment · network security solutions

Z. Gu et al. (Eds.): NSE 2023, CCIS 2063, pp. 46–59, 2024.
https://doi.org/10.1007/978-981-97-4519-7_4

1 Introduction

The current network security situation is severe. According to CrowdStrike 2023 Global Threat Report [1], attackers motivated by nationalism, cybercrime, and hacktivism have become increasingly active since the beginning of 2022, due to geopolitical changes and growing economic difficulties. Internal network attacks are a significant component of network attacks, particularly APT attacks, and the defense situation of intranet attacks is not optimistic. 2022 saw a decrease in the breakthrough time for internal network attacks, which refers to the duration attackers need to move laterally from the initially infected host to another host within the victim's environment, from 98 min in 2021 to 84 min.

In the face of such a severe network security situation, how to improve network defense capability becomes more and more important. Currently there are a lot of work focusing on malware detection [2–4], malicious family classification [5] and APT attack attribution [6], etc. The work on assessing the defense capabilities of cybersecurity solutions is relatively scarce, and during the evaluation process, assessment indexes are often selected from the perspective of the defender, such as asset information, permissions, and performance. The acquisition of these index values rely on sensors, flags and other third-party tools, and require standardization of index values. In addition, these methods often results in a relatively rough evaluation outcome, and it is also challenging to obtain ideas for improving their defense capabilities. To address these issues, we propose an intranet defense capability assessment method for network security solutions based on the ATT&CK framework. The described approach utilizes the ATT&CK framework to establish an evaluation index system for internal network attacks. The system's index values are easily acquired and processed, and it boasts a comprehensive and highly scalable design. The index weighting of this method has been referenced from expert experience and the enterprise framework tactics, techniques, and subtechniques used by 138 APT organizations in the ATT&CK v13.1 framework, integrating subjective and objective factors to make the indicator assignment results more fair and practical. Through simulation experiments, it is verified that the method is effective and fine-grained in evaluating the defense capability of network security solutions, filling the gap in evaluating the defense capability of intranets for network security solutions, and referring to the intermediate data generated by the evaluation through this method, the defense ideas of corresponding technologies and sub-technologies in the ATT&CK framework can be easily obtained.

The paper is structured as follows: in Sect. 2, we review related work on the evaluation of network security solutions. Section 3 provides a detailed introduction to our proposed evaluation method, which covers indicator construction, indicator weighting, obtaining indicator values, and result calculation methods. Next, in Sect. 4, we present the experimental setup and case analysis to verify the effectiveness of our evaluation method. Finally, in Sect. 5, we summarize the paper and propose future research directions.

2 Related Work

Currently, the evaluation of the defense capabilities of network security solutions or security measures mainly focuses on anti-DDoS devices, and there is a lack of evaluation of the internal network defense capabilities. Moreover, the evaluation criteria for constructing the index system are often selected from the perspective of the defending party's assets such as information, permissions, and performance. Here are some studies on defense capability assessment.

Bellaiche et al. [7] proposed that DDoS attack protection measures should be evaluated and compared in terms of four aspects: performance, deployment cost, impact on the performance of the protected system, and their own robustness, but such evaluation methods tend to assess only the performance level, and quantification is not specific enough.

Butler et al. [8] proposed a security attribute assessment method. First, the overall mitigating effect of each security measure on all threats is evaluated, then the types of threats that each security measure can address is evaluated, and finally the cost of implementing each security measure is evaluated.

Schwab et al. [9] classified DDoS evaluation indicators into three categories: network traffic indicators, attack effectiveness indicators, and protection effectiveness indicators. They evaluated the effectiveness of protection measures by measuring the increase in protection effectiveness after enabling specific protection measures.

Meadows et al. [10] defined a protection cost set and an attack capability set, and used these sets to define a tolerance relation set. They used this set to analyze whether a protocol is capable of withstanding DoS attacks.

Huang Liang et al. [11] selected indicators from the perspective of user perception and established a protection performance assessment model with the help of neural network to assess the DDoS protection capability.

Li et al. [12] proposed a method to evaluate the effectiveness of DDoS protection based on packet computation. This method calculates the ratio of legitimate packet through rate to attack packet through rate (LAR) to evaluate the strength of the protection system; the higher the LAR, the better the protection effect.

Li et al. [13] comprehensively selected disposal effect assessment indicators from four dimensions: operation status, system behavior, service situation and alarm situation, and used fuzzy hierarchical analysis to determine the weights of different indicators, and then determined the threat disposal effect by fuzzy comprehensive evaluation method.

3 The Proposed Approach

In this section, we introduce in detail the intranet defense capability assessment method for network security solutions based on ATT&CK framework. In Sect. 3.1 we introduce the basic idea of the method. In Sect. 3.2 we describe the construction of the evaluation metrics and their advantages. In Sect. 3.3 we explain how the weights of the indicators are obtained. In Sect. 3.4 we introduce the method of obtaining the indicator values. In Sect. 3.5 we describe the method of calculating the evaluation results based on the indicator values.

3.1 Basic Ideas

We first extract the tactics, techniques and subtechnologies that can be used to evaluate intranet attacks from the ATT&CK matrix as indicators for evaluating attacks; secondly simulate attacks and acquire indicator values; again refer to expert experience and the enterprise framework tactics, techniques and subtechnologies used by 138 APT organizations in the ATT&CK v13.1 framework to assign weights to indexes, and then simulate attacks to acquire index values and finally calculate the defense ratio of the overall attack based on the indicator values and weights to obtain the evaluation results of the network security solution. The overall process is shown in Fig. 1.

Fig. 1. Basic flow.

3.2 Index System Construction

Since the attack assessment index system is the cornerstone of the entire assessment process, it directly affects the assessment results. Therefore, we need to conduct as accurate, objective and complete an assessment of the attack as possible. Existing indexes used for defensive capability assessment of network security solutions tend to select assessment indexes from the defender's perspective, information, authority, and performance of assets [7, 14], without a fine-grained representation of attacks. The above-mentioned methods often result in a relatively rough evaluation result, especially when a certain attack cannot be successfully executed, it is difficult to distinguish and compare the defense capabilities of network security solutions, and it is also difficult to obtain the direction of defense improvement for network security solutions. To address these issues, we use the tactics, techniques, and sub-techniques of attacks in the ATT&CK Enterprise Framework as indicators to represent attacks. By considering the three dimensions, we evaluate the defense capabilities of network security solutions to obtain detailed and rich evaluation results.

There are 14 tactics, 196 techniques and 411 seed techniques in the ATT&CK v13.1 Enterprise Framework. From these, we selected 11 tactics, 172 techniques and 335 subtechniques that can be used to represent intranet attacks, and used them to construct an evaluation metric system for the attacks. Figure 2 only shows the indicators in the two dimensions of tactics and techniques.

3.3 Index Weighting

Since different indexes have different impacts on the assessment results, and the different weight assignments of the indexes will directly affect the final assessment results, we need to develop a method for assigning weights to the indices in a reasonable manner. The common approaches for assigning weights to indicators are the subjective assignment method and the objective assignment method. Each method has its own advantages,

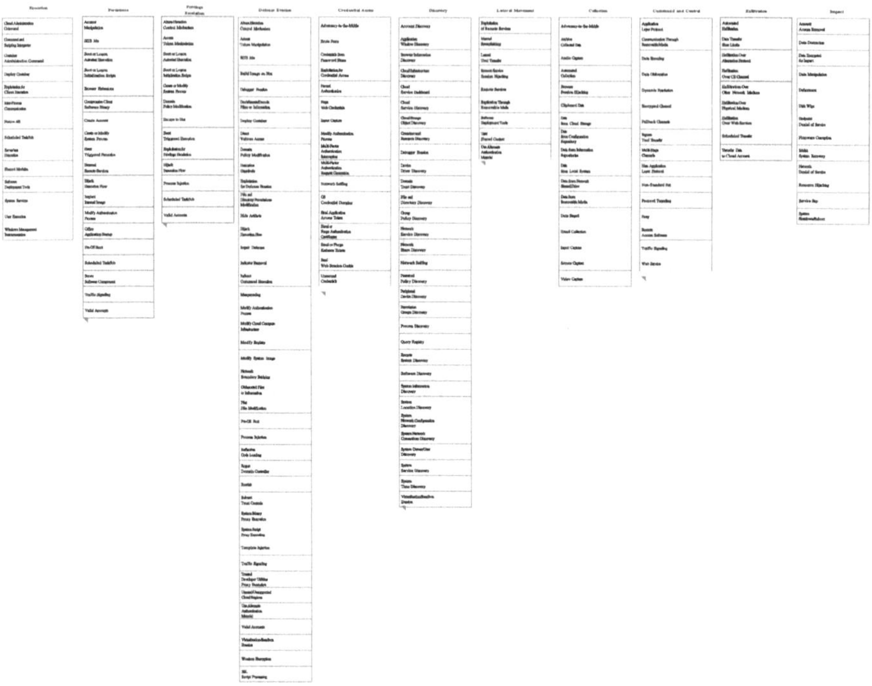

Fig. 2. The indicators of tactics and techniques.

disadvantages, and scope of application. The subjective weighting method is a method in which the researcher assigns weights to each index based on their subjective value judgment. The method requires less quantitative information and is very practical for evaluation in the absence of necessary data; it can greatly improve the validity, reliability and feasibility of comprehensive evaluation. However, it requires the evaluator to have a thorough grasp of the objectives, indexes and their logical relationships with each other, and the constructed weights lack stability due to the differences in each person's subjective value judgment criteria. Objective weighting method is a method to obtain weights directly based on the original information of the index and processed by statistical methods. This type of method is less influenced by subjective factors, and its drawback is that the assignment of weights will be more influenced by the sample data, and different samples, i.e., using the same method, will yield different weights, and it ignores the evaluator's perception and preference of the value of the indicator. To address these issues, we have designed a method that combines subjective and objective weighting, integrating subjective and objective factors to make the indicator weighting results more fair and practical.

The Delphi technique was used for the subjective assignment. Multiple experts were selected from the attacker, the evaluator and the defender, respectively, and the experts were asked to rate the importance of individual indicators through a questionnaire (there was no intercomparison of indicators). The scoring rules used were as follows: a score of 5 for very important, a score of 4 for important, a score of 3 for average, a score of 2 for unimportant, and a score of 1 for very unimportant. For indices falling between two levels, scores can be expressed as a decimal value. After collecting the questionnaire, the opinions of each expert were summarized and counted, and if the experts' opinions were not consistent, they were fed back to each expert anonymously and collected again until a consistent opinion was obtained or a predetermined number of rounds was reached, and the average value was used as the result if a predetermined number of rounds was reached.

After obtaining the weights using the Delphi technique, the weights for tactics, techniques, and sub-techniques are normalized using formulas 1.

$$w_i' = \frac{w_i}{\sum_{i=1}^{n} w_i} \tag{1}$$

Where w_i' is the result of normalizing the weights of the corresponding term, w_i is the weight of the i-th index obtained directly using the Delphi technique, n is the number of indexes corresponding to the granularity.

The objective weighting method uses the entropy weighting method. First, the tactics, techniques, and sub-techniques used by 138 APT organizations in the ATT&CK v13.1 framework were obtained. Then, formulas 2–6 were used to calculate the entropy weights of tactics, techniques, and sub-techniques, respectively. First, the index values of the i-th APT organization under the j-th index are normalized according to formula (2). Since the number of tactics, techniques, and sub-techniques used by APT organizations are all positive indicators, there is no need to consider how to handle negative indicators.

$$x_{ij}' = \frac{x_{ij} - min(x_{1j}, \ldots, x_{mj})}{max(x_{1j}, \ldots, x_{mj}) - min(x_{1j}, \ldots, x_{mj})} \tag{2}$$

where w_{ij} represents the number of corresponding tactics, techniques, or sub-techniques used by the i-th APT organization under the j-th indicator, and x_{ij}' represents the normalized result of the number of corresponding tactics, techniques, or sub-techniques used by the i-th APT organization under the j-th indicator. m equals 138, which is the number of APT organizations.

Second, the weight p_{ij} of the i-th APT organization under the j-th index is calculated according to formula (3).

$$p_{ij} = \frac{x_{ij}'}{\sum_{i=1}^{m} x_{ij}'} \tag{3}$$

Third, the entropy value H_j of the j-th index is calculated according to formula (4).

$$H_j = \frac{\sum_{i=1}^{m} p_{ij} \cdot \ln p_{ij}}{\ln m} \tag{4}$$

Fourth, the variation coefficient d_j of the j-th indicator is calculated according to formula (5).

$$d_j = 1 - H_j \tag{5}$$

Finally, the entropy weight w_j of the j-th indicator is calculated according to formula (6).

$$w_j = \frac{d_j}{\sum_{j=1}^{n} d_j} \tag{6}$$

When assigning weights to tactical, technical and sub-technical indexes in turn, n is the number of indexes corresponding to the granularity.

After obtaining the weights by these two ways, assuming that the subjective weight of the indicator is w_i' using the Delphi method and the objective weight of the indicator is w_i'' using the entropy method, the weight of the i-th indicator is obtained by fusing the weights using Eq. (7):

$$w_i = \frac{w_i' + w_i''}{\sum_{i=1}^{n} (w_i' + w_i'')} \tag{7}$$

3.4 Index Value Acquisition

In the existing research, the index system is designed to acquire index values through network range, third-party tools, flag files, and other means. Usually, it takes time to standardize the indicator values [15], such as dimensionless normalization for positive and negative indicators. In contrast, our proposed method only needs to count the number of successful executions of each tactic, technique, and sub-technique in the attack without a solution I_i' and the number of successful executions of each tactic, technique, and sub-technique in the attack after deploying a solution I_i when the attack is simulated.

It should be noted that when counting the index values, it is necessary to map the specific attack behaviors to the tactics, techniques, and sub-techniques of the attack evaluation indexes. To ensure the reasonableness of the mapping, it is necessary to invite multiple experts to participate in this work.

3.5 Result Calculation

First, using the number of uses of each tactic, technique and sub-technique used by the overall attack, I_i', and the weight value w obtained in Sect. 3.3, the scores $Score'$ for the three granularities of tactics, techniques and sub-techniques when the attack is not deployed with a cybersecurity solution are calculated as shown in Eq. (8):

$$Score' = \sum_{i=1}^{n} I_i' \cdot w_i \tag{8}$$

Then, based on the number of successful executions of each tactic, technique and sub-technique after deploying the solution I_i and the weight value w_i obtained in Sect. 3.3, the defense score $Score$ of the attack at the three granularities of tactics, techniques and sub-techniques after deploying the cyber security solution is calculated as shown in Eq. (9):

$$Score = \sum_{i=1}^{n} I_i \cdot w_i \tag{9}$$

After obtaining the score $Score'$ of the attack at the three granularities of tactical, technical and sub-technical without deploying the cybersecurity solution and the score of the defense at the three granularities of tactical, technical and sub-technical after deploying the cybersecurity solution, the defense rate of the cybersecurity solution at the three granularities of tactical, technical and subtechnical is calculated according to Eq. (10), respectively.

$$rate \frac{Score' - Score}{Score'} \tag{10}$$

After obtaining the defense rates of tactics, technologies and sub-technologies, they are brought into Eq. (11) to calculate the total percentage of the defense capability of the cybersecurity solution, and then the level of the defense capability of the cybersecurity solution is obtained according to the Table 1.

$$R_d = \sqrt[3]{rate_{Tatics} \cdot rate_{Techniques} \cdot rate_{Subtechniques}} \tag{11}$$

where $rate_{Tatics}$, $rate_{Techniques}$ and $rate_{Subtechniques}$ are the tactical defense rate, technical defense rate, and subtechnical defense rate, respectively R_d is the total percentage of the defense capability of the cybersecurity solution.

Table 1. Security solution defensive capabilities mapping table

rank	The range of values of R_d
very high	$R_d \geq 90\%$
high	$80\% \leq R_d < 90\%$
medium	$65\% \leq R_d < 80\%$
low	$50\% \leq R_d < 65\%$
high	$R_d < 50\%$

4 Experimental Setup and Case Analysis

This section verifies the practicality and effectiveness of the proposed method through simulation experiments and case analysis.

4.1 Experimental Setup

In this section, we evaluate the effectiveness of the network security solution by simulating the attacks of two APT groups, Carbanak and FIN7, both of which have caused serious damage in their attack target intranets. Furthermore, Mitre has conducted an in-depth analysis of these two APT organizations and open-sourced the tools and simulation methods used in their attacks [16], which greatly reduces the difficulty of simulating the attacks, and mitre has made public a dataset [17] containing the results of a set of cybersecurity solutions against Carbanak and FIN7 when simulating the attacks as described earlier. The detailed results of the defense results of simulated attacks for two APT organizations. The defense results are highly credible since the cybersecurity solutions are configured by the vendors according to realistic situations and time is left for the vendors to give feedback on the defense results before making them public [18].

The network topology for our simulated attack is shown in Fig. 3, and the host information is shown in Table 2. Each host has four vCPUs and 16GB of memory.

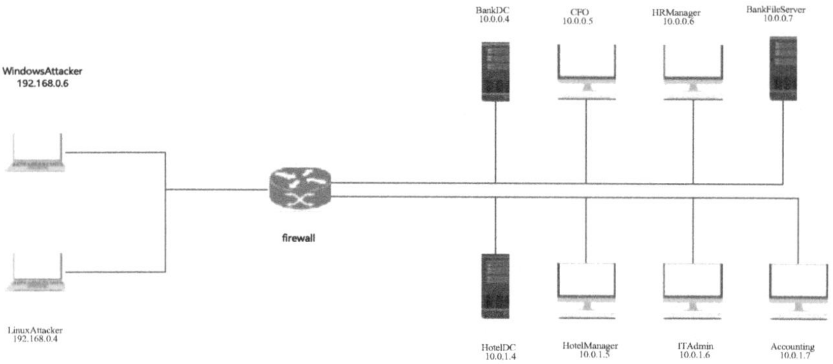

Fig. 3. Network topology.

The red team uses two machines, WindowsAttacker and LinuxAttacker, which are deployed with commonly used penetration testing software and utilities customized for Carbanak and FIN7. The remaining hosts are divided into two parts.

Table 2. Host information for network topology

organization	Hostname	IP	Operating system
Attacker	WindowsAttacker	192.168.0.6	Windows 10
	LinuxAttacker	192.168.0.4	Kali linux 2019.1
Target of Carbanak attack	BankDC	10.0.0.4	Windows Server 2019
	CFO	10.0.0.5	Windows 10
	HRManager	10.0.0.6	Windows 10
	BankFileServer	10.0.0.7	CentOS 7.7
Target of FIN7 attack	HotelDC	10.0.1.4	Windows Server 2019
	HotelManager	10.0.1.5	Windows 10
	ITAdmin	10.0.1.6	Windows 10
	Accounting	10.0.1.7	Windows 10

The first part consists of BankDC, CFO, HRManager and BankFileServer. These four machines are in the same domain, and the domain control is BankdDC, which is the target of Carbanak's attack. The second part consists of HotelDC, HotelManager, ITAdmin, and Accounting. These four machines are in the same domain, and the domain control is HotelDC, which is the target of FIN7 attack.

In this experiment, we simulate these two APT groups to launch an attack on the target topology. To facilitate the experiments and obtain more detailed evaluation results, we divide the attacks into multiple test cases, and acquire the index values of the attack evaluation indexes by executing the test cases.

Before the experiment is conducted, it is necessary to turn off the target's own defense measures, such as Windows Defender, firewall and powershell script execution restrictions, to ensure the accuracy of the experimental results.

4.2 Case Analysis

Referring to the experimental setup in Sect. 4.1, we selected five network security solutions to participate in the analysis, and when deploying one of them alone, all of them can make the testcase1 execution of the two APT groups' attacks fail, and can be considered as having completed the defense against both attacks. Next, using the method proposed in Sect. 3, we analyze the specifics of the defense to verify whether we can obtain differentiated evaluations.

First simulate the attack when no cybersecurity solution is deployed to obtain the number of successful executions of each tactic, technique, and subtechnique in the attack evaluation index system. Then the number of successful executions of each tactic, technique, and sub-technique in the attack evaluation metrics is counted based on the execution of the simulated attack and the alert of the solution when each cybersecurity solution is deployed in the dataset. And the results are organized as shown in Fig. 4, Fig. 5 and Fig. 6. The red color represents the index values obtained when no cybersecurity solution is deployed. Blue, green, orange, purple, and brown represent the index values obtained for case1, case2, case3, case4, and case5, respectively.

Figure 4, Fig. 5 and Fig. 6 show the index values for tactics, techniques, and subtechnologies for each case, including when no cybersecurity solution was deployed, respectively. Analyzing these three graphs together, a conclusion can be first obtained that the selected cybersecurity solutions all successfully prevented the execution of attack tactics, techniques, and sub-techniques, and relatively fewer attack tactics, techniques, and sub-techniques were prevented when case3 was deployed, and relatively more attack tactics, techniques, and sub-techniques were prevented when case1 or case5 was deployed.

By analyzing Fig. 5, it can be seen that none of the five solutions prevent the successful execution of techniques T1090 (Proxy), and by querying the MITRE Enterprise ATT&CK Matrix, it can be seen that the T1090 (Proxy) attack technique can be improved by filtering network traffic, using intrusion detection signatures at network boundaries to block traffic, and interrupting and inspecting SSL/TLS sessions to improve the network security solution's defenses against this technique. Other mitigation solutions for techniques and subtechnologies with high implementation rates are also available by consulting the MITRE Enterprise ATT&CK Matrix.

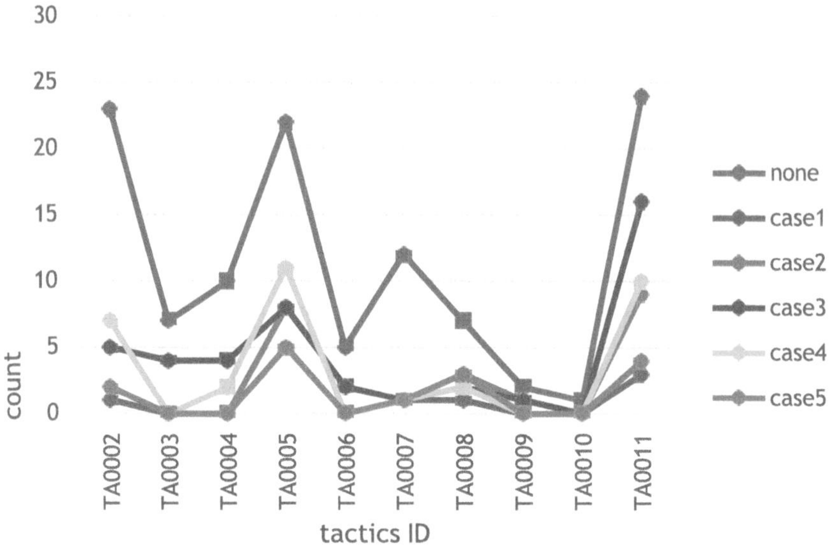

Fig. 4. Number of successfully executed tactics (Color figure online)

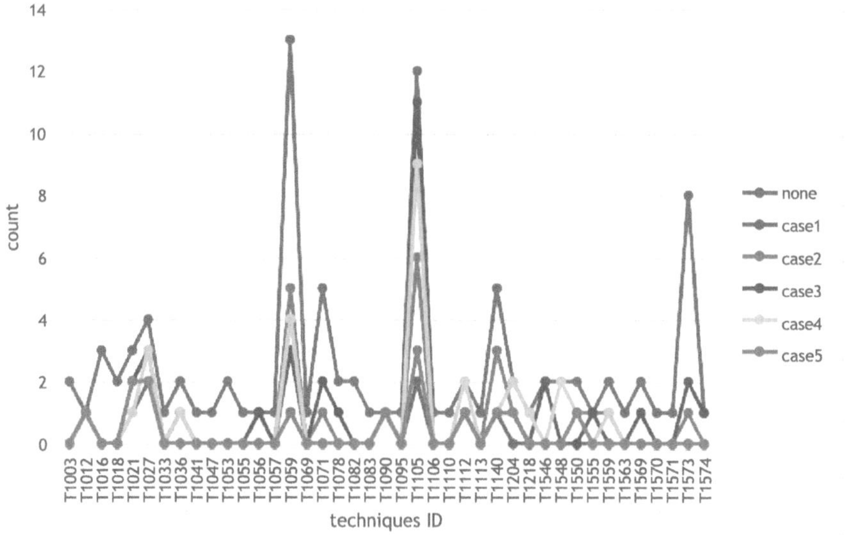

Fig. 5. Number of successfully executed techniques (Color figure online)

After obtaining the index values of the attack assessment indexes, the weights of the attack assessment indexes are first calculated separately according to what is described in Sect. 3.3, and then the assessment scores of the simulated attacks when no cybersecurity solution is deployed are calculated according to Eq. (10), and the scores of the attacks

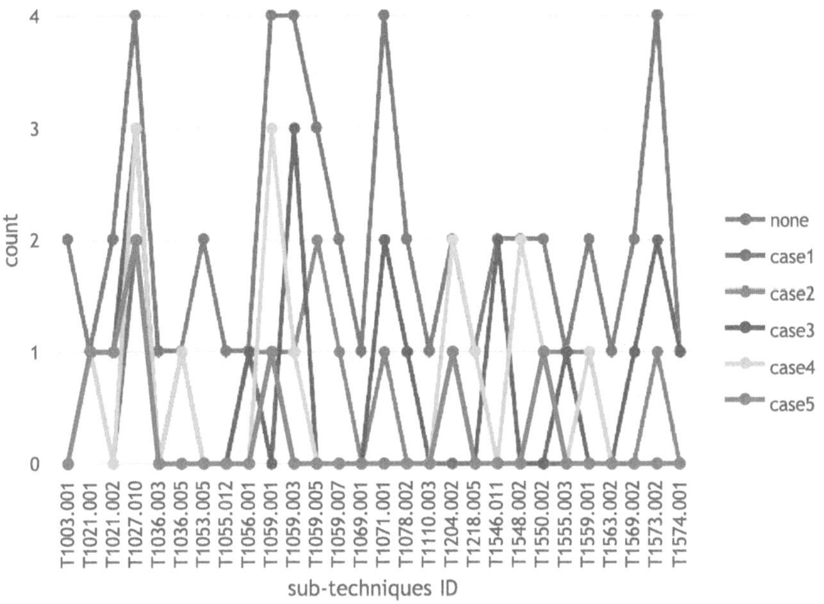

Fig. 6. Number of successfully executed sub-techniques (Color figure online)

under the defense of each cybersecurity solution are calculated according to Eq. (11) after the cybersecurity solution is deployed. The results are shown in Table 3.

Table 3. Attack score

score	none	case1	case2	case3	case4	case5
tactics	8.2331	0.8095	2.2664	3.1652	2.3306	1.1586
techniques	0.3913	0.0444	0.1156	0.1445	0.1181	0.0584
sub-techniques	0.1225	0.0101	0.0352	0.0465	0.0320	0.0151

The intranet attack has the lowest scores in the tactical, technical and subtechnical dimensions when case1 is deployed, the highest scores in the tactical, technical and sub-technical dimensions when case3 is deployed, and close scores in the tactical, technical and subtechnical dimensions when case2 and case3 are deployed.

According to Eq. (12), the percentage of defense of each cybersecurity solution under the three dimensions of tactical, technical and subtechnical is calculated after the deployment of cybersecurity solutions, and then the total percentage of defense capability of cybersecurity solutions is calculated according to Eq. (13), and finally the qualitative description of each cybersecurity solution is obtained by referring to Table 1. The results are shown in Table 4.

Table 4. Defense results for network security solutions

items	case1	case2	case3	case4	case5
rateTatics	90.17%	72.47%	61.56%	71.69%	85.93%
rateTechniques	88.65%	70.47%	63.08%	69.81%	85.08%
rateSubtechniques	91.78%	71.29%	62.05%	73.89%	87.67%
R_d	90.19%	71.40%	62.23%	71.78%	86.22%
rank	very high	meduim	low	medium	high

The defensive capabilities of case1 belong to very high, case5 to high, case2 and case4 to medium, and case3 to low.

The experimental results and case analysis show that when several cyber security solutions can defend against a certain intranet attack, the effectiveness of the proposed method can be verified by our proposed method to obtain a more fine-grained and differentiated evaluation, and relevant suggestions for enhancing the defense can be obtained more easily after the evaluation is completed.

5 Conclusion

To address the gaps in existing methods for intranet defense capability assessment, the difficulty of collecting indicator values and the large granularity of assessment results, this paper proposes an intranet defense capability assessment method for network security solutions based on the ATT&CK framework from an attack perspective. The main contributions of this paper are summarized as follows: 1) Using the ATT&CK framework, we constructed an evaluation index system for internal network attacks. The index system has the advantages of simplicity in obtaining and processing index values, as well as comprehensiveness and good scalability; 2) Referring to expert experience and the enterprise framework tactics, techniques, and sub-techniques used by 138 APT organizations in ATT&CK v13.1 framework, we assigned weights to the indicators. While preserving subjective experience and evaluators' preferences, we considered the impact of objective data on the weighting; 3) We proposed a method for evaluating the internal network defense capabilities of network security solutions. This method effectively and finely evaluates the defense capabilities of network security solutions, fills the gap in evaluating the internal network defense capabilities of network security solutions, and generates intermediate data that is easy to use for improving defense capabilities. Finally, we verified the practicality and effectiveness of the proposed method through simulation experiments in a network range and an open-source dataset.

The follow-up research is mainly focused on two areas. One is to develop and open source a simulated attack library based on the ATT&CK matrix, such as Atomic Red Team [19], and then simulate the attack with techniques and subtechnologies instead of simulating the attacks of certain APT organizations to obtain more fine-grained and differentiated evaluation results. Secondly, we will try to use other frameworks to represent the attacks, design different indexes systems, and compare the advantages and disadvantages of different frameworks to design metrics systems.

Acknowledgements. This research was funded by NSFC (No. 62072131, 62372127, 61972106), the Major Key Project of PCL (Grant No. PCL2022A03-3), Key RD Program of Guangdong Province (No.2019B010136003), Guangdong Basic and Applied Basic Research Foundation (No.2022A1515011401), DongGuan Innovative Research Team Program (No. 2018607201008).

References

1. CrowdStrike 2023Global Threat Report. https://www.crowdstrike.com/global-threat-report. Accessed 9 May 2023
2. Li, S., Jiang, L., Zhang, Q., Wang, Z., Tian Z., Guizani, M.: A malicious mining code detection method based on multi-features fusion. IEEE Trans. Netw. Sci. Eng. (2022). https://doi.org/10.1109/TNSE.2022.3155187
3. Li, S., Li, Y., Han, W., Du, X., Guizani, M., Tian, Z.: Malicious mining code detection based on ensemble learning in cloud computing environment. Simul. Model. Pract. Theory **113**, Article ID 102391 (2021)
4. Yu, Z., Li, S., Han, W., Wu, X., Tian, Z.: REMSF: a robust ensemble model of malware detection based on semantic feature fusion. IEEE Internet Things J. (2023). https://doi.org/10.1109/JIOT.2023.3267337
5. Li, S., Li, Y., Wu, X., Otaibi, S.A., Tian, Z.: Imbalanced malware family classification using multimodal fusion and weight self-learning. IEEE Trans. Intell. Transp. Syst. (2022). https://doi.org/10.1109/TITS.2022.3208891
6. Mei, Y., Han, W., Li, S., Lin, K., Luo, C.: A hybrid intelligent approach to attribute advanced persistent threat organization using PSO-MSVM algorithm. IEEE Trans. Netw. Serv. Manag. **19**(4), 4262–4272 (2022). https://doi.org/10.1109/TNSM.2022.3201928
7. Halabi, T., Bellaiche, M.: How to evaluate the defense against DoS and DDoS attacks in cloud computing: a survey and taxonomy. Int. J. Comput. Sci. Inf. Secur. **14**(12), 1 (2016)
8. Butler, S.A.: Security attribute evaluation method: a cost-benefit approach. In: Proceedings of the 24th International Conference on Software Engineering (2002)
9. Schwab, S., Wilson, B., Thomas, R.: Methodologies and metrics for the testing and analysis of distributed denial of service attacks and defenses. In: MILCOM 2005-2005 IEEE Military Communications Conference. IEEE (2005)
10. Meadows, C.: A formal framework and evaluation method for network denial of service. In: Proceedings of the 12th IEEE Computer Security Foundations Workshop. IEEE (1999)
11. Huang, L., et al.: Artificial neural network based DDoS defense effectiveness evaluation. Chin. J. Comput. Res. Dev. **50**(10), 2100–2108 (2013)
12. Li, Z., Xiang, Y., He, D.: Simulation and analysis of DDoS in active defense environment. In: Wang, Y., Cheung, Ym., Liu, H. (eds.) CIS 2006. LNCS, vol. 4456, pp. 878–886. Springer, Heidelberg (2007). https://doi.org/10.1007/978-3-540-74377-4_92
13. Li, F.H., Li, Y.J.: Fuzzy evaluation for response effectiveness in cases of incomplete information. J. Commun. **40**, 117–126 (2019)
14. Qiu, H., Zhu, J.: Research on network attack effect evaluation based on confrontational perspective. J. Phys.: Conf. Ser. **1437**(1) (2020)
15. Pipyros, K., et al.: A new strategy for improving cyber-attacks evaluation in the context of Tallinn Manual. Comput. Secur. **74**, 371–383 (2018)
16. Adversary emulation library. https://github.com/center-for-threat-informed-defense/adversary_emulation_library. Accessed 9 May 2023
17. Enterprise Evaluations. https://attackevals.mitre-engenuity.org/enterprise/. Accessed 9 May 2023
18. Methodology Overview: Evaluation Process. https://attackevals.mitre-engenuity.org/methodology-overview/. Accessed 9 May 2023
19. Atomic Red Team. https://github.com/redcanaryco/atomic-red-team. Accessed 9 May 2023

Poisoning Attack in Machine Learning Based Invalid Ad Traffic Detection

Le Yu[1], Shaofeng Li[2], Yan Meng[1], Xiao Wang[3], and Haojin Zhu[1(✉)]

[1] Shanghai Jiao Tong University, Shanghai, China
{yule5100309221,yan_meng,zhu-hj}@sjtu.edu.cn
[2] Peng Cheng Laboratory, Shenzhen, China
lishf@pcl.ac.cn
[3] Shanghai Goku Technologies Limited, Shanghai, China
wangxiao@gokudata.com

Abstract. Along with the growing market of mobile advertising, click farm that leverages large-scale fake devices to conduct ad fraud is becoming a major threats against mobile advertising. These click farms evolve to mimic the usage patterns of normal users, which makes traditional rule-based detection methods unable to tackle with the new threats. To defeat these attacks, machine leaning-based (ML-based) approaches are adopted as a data driven method to detect invalid traffic from click farms by both industry and academia. However, the vulnerability of ML-based detection methods is still an open problem under the adversarial situation. In this study, we present a first study on the poisoning attack in ML-based ad fraud detection. By injecting poison samples to the training set, we demonstrate the effectiveness and concealment of the poisoning attacks under two machine learning models, Gradient Boosting Decision Tree (GBDT) and Deep Neural Networks (DNNs). We conduct experiments with a real-world data set, the results show that the poisoning attack can achieve 0.85 and 0.91 attack success rate for GBDT and DNNs respectively.

Keywords: Ad Fraud Detection · Machine Learning · Poisoning Attack

1 Introduction

Mobile advertising is one of the fastest-growing information industry over the past decades. With mobile devices firmly entrenching in people's modern lives, tremendous digital advertisements (ads) can be delivered through frequently- used mobile apps, which improves their reachability to the potential consumers compared to traditional advertising methods. Meanwhile, Real-Time Bidding (RTB), the most innovative technology in today's industry, provides a digital marketplace where mobile app developers (i.e., ad publishers) can share and sell the spare spaces (ad spaces) in their apps to advertisers through automatic real-time auctions, which is so convenient that attracts an increasing number of sellers and buyers to take part into the ecosystem. Benefit from these advantages mentioned above, mobile advertising finds extensive applications in various digital advertising campaigns and the global market keeps growing rapidly. As

Z. Gu et al. (Eds.): NSE 2023, CCIS 2063, pp. 60–72, 2024.
https://doi.org/10.1007/978-981-97-4519-7_5

Global Industry Analysts report [4], the global digital advertising and marketing is estimated at 531 billion U.S. dollars in the year 2022 and is projected to reach 1.5 trillion U.S. dollars by 2030.

Unfortunately, this increased prominence has also attracted the attention of fraudulent ad space sellers, who try to inflate their incomes by fabricating ad demands. More specifically, the fraudulent sellers will exploit fake mobile devices to install and run the mobile apps that embody their ad spaces which will be sold in RTB's marketplace to advertisers who wish to display their ads on the device. To mimic normal app usage patterns to trick advertisers into believing that their ads have been seen by actual, interested users, this task is usually outsourced to click farm owners, who control thousands of mobile devices or emulators. Due to the substantial financial loss caused by invalid ad traffic, the problem is attracting increasing attention in both industries and academics. In industry practices, ad fraud is identified through invalid traffic in RTB transaction data. Many organizations including Integral Ad Science (IAS) [1], Oracle [2] and White Ops [3] serve as a trusted third party to detect invalid traffic. However, the best practice of combating invalid traffic adopted by mainstream traffic verification companies heavily relies on rule-based detection such as blacklists and parameter/metadata checks (e.g., IP blacklists, frequency checking, and user agent (UA) checking). These methods can be easily bypassed through elaborating click farms.

In academia, ad fraud detection is widely discussed [8, 9, 28], and Machine Learning-based (ML-based) detection methods [20, 27] are recently proposed as a data driven detection methodology. While these methods can achieve a good detection performance, the robustness of the ML-based methods is still an open problem under the scenario of adversarial machine learning, where attackers can manipulate the training dataset to influence the detection results.

In this work, we identify the venerability of the ML-based detection methods in invalid Ad traffic detection. We propose a data poisoning attack against the ML-based detection of invalid Ad traffic. In particular, we inject poisoned training samples into the training data of ML-based detectors, in which the poisoned ML-based detectors fail to identify those poisoned samples when those ML-based detectors are deployed online to provide detection services. In this paper, we present a first study on the poisoning attack in ML-based ad fraud detection. The main contributions of this work is as follows:

- We identify the vulnerability of ML-based detection in the field of ad fraud detection in the scenario of adversarial machine learning.
- We design a data poisoning attack to reveal the vulnerabilities of ML-based detectors in the field of ad fraud detection.
- We perform the proposed attacks and validate its effectiveness on the real world collected raw Ad traffic data. Our study reveals that a data poisoning attacker can successfully bypass the ML-based ad fraud detection system.

The remainder of this paper is organized as follows. We present the background in Sect. 2. Section 3 shows the ML-based invalid traffic detection method studied in this paper. Section 4 elaborates the data poisoning attacks, followed by evaluations in Sect. 5. Section 6 reviews existing related works and Sect. 7 concludes this paper.

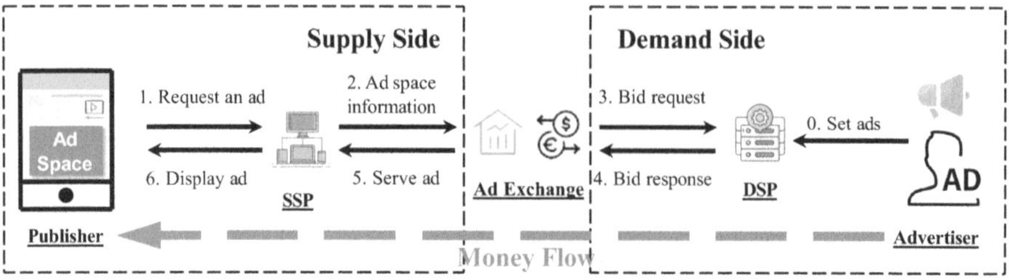

Fig. 1. The ecosystem of mobile advertising and the flows of a typical RTB process.

2 Background

In this section, we introduce the system model of programmatic advertising and then explain the ecosystem of real-time bidding by showing a typical process of serving an ad in programmatic advertising and the role of fraudulent publishers. We also discuss the best practice of invalid traffic (IVT) verification and filtering in the industry.

2.1 Ecosystem of Mobile Advertising

In the ecosystem of mobile advertising, the essential components consist of three participants: *publishers*, *advertisers* and *users*. The publishers are the developers of mobile apps who reserve ad spaces in their apps for displaying ads, and the advertisers are businessmen who want to promote their products or brands to the users. Generally, the publishers sell the ad spaces to the advertisers to host their ads. When the users use the apps, the advertisers who buy the ad spaces will have their ads displayed to the users.

While the ad transaction can be directly negotiated between publishers and advertisers as has been done with traditional billboards and paper media, this type of transaction is only suitable for a few large publishers and advertisers. To allow more players and more flexible ad transactions, a new technical innovation emerges in today's ecosystem, which is so-called Real-Time Bidding (RTB). Figure 1 represents a typical RTB supported ecosystem. The infrastructure of the ecosystem consists of three key platforms: Supply-Side Platform (SSP), Demand-Side Platform (DSP) and Ad-Exchange. The SSPs and DSPs are agents for publishers and advertisers who can help optimize their transactions. The Ad-Exchange (e.g., Google DoubleClick) acts as a digital marketplace that allows advertisers to buy ad spaces from publishers through real-time auctions.

The paths in Fig. 1 represent a typical RTB process. When a user users a publisher's app, the ad library will automatically initiate ad requests for the ad spaces in the app. The ad requests are sent to the SSPs, which are further forwarded to the Ad-Exchange. The requests often contain device ID, IP address, timestamp, the size and context of the ad spaces, etc. Upon receiving the ad request, the Ad-Exchange initiates an auction of the ad spaces and records an ad bid log in the database. The bidding request is then sent to each registered DSP in the Ad-Exchange, and the DSPs bid on the ad spaces on behalf of the advertisers. When the Ad-Exchange initiates an auction, the DSPs on behalf of advertisers would respond to the optimal price, in accordance with the rules set by advertisers. The Ad-Exchange informs the winner, and the corresponding advertisement is then sent to the user and being displayed on the app. If it wins in the bidding, the advertiser's ads will be delivered to the user and display in the ad space and the payment will be received by the publisher. The auction is triggered and held no more than 100 milliseconds after an ad request is launched.

2.2 Traffic Fraud in Mobile Advertising

While the RTB technology can benefit the publishers and advertisers to sell and buy ad spaces, the flexibility of ad transactions is also vulnerable to ad fraud by malicious publishers to inflate their income. In RTB platforms, the most commonly used pricing model is Cost-Per-Mille (CPM), where advertisers pay a flat rate for 1,000 displays or impressions of an advertisement. Thus, the ad fraudsters will use technical methods to fabricate the number of impressions or traffic of advertisements which is not really delivered to mobile users. One of the methods is to inject malicious code into the apps to automatically trigger ad impressions, however, this type of method can be easily detected by code analysis [31]. The other type of method represents the most widely used strategy by ad fraudsters in the world, which utilizes fake devices (i.e., bot devices and virtual machines) to mimic normal mobile users' app usages and fabricate ad impressions. Usually, the fake devices are possessed by click farms so that they can fabricate fake impressions in a large scale. Wen et al. [29] demonstrate that fake devices created by click farms are cost-efficient to launch a large scale mobile ad fraud.

Industrial Traffic Validation. To defend against ad fraud, the industry relies on trusted third parties such as White-Ops, IAS, and RTBAsia to detect invalid ad traffic. The invalid ad traffic can be classified into two types: general invalid traffic (GIVT) and sophisticated invalid traffic (SIVT). The GIVT represents the traffic that is generated by known industry crawlers and bots, whose patterns can never be generated by real humans. This class of invalid traffic can be easily detected and filtered by rule-based standardized parameter checks. The SIVT is more difficult to detect since fraudsters actively try to avoid simple rule-based detection by mimicking the real humans' usage patterns, including IP Proxies, APP rotation, device info alteration, etc. [27]. This class of invalid traffic requires sophisticated detection methodologies.

3 ML-Based Invalid Traffic Detection

To defend against SIVT ad fraud, the traffic verification requires more sophisticated methods. In this section, we present the ML-based detectors to detect SIVT.

3.1 RTB Transaction Data

Dataset. Our dataset is collected from a real-world industrial company in the format of RTB transaction logs. The data are labeled in terms of devices, which include 6,269 benign and 6,269 fraudulent devices annotated by the company. The duration of data collection starts from May 6, 2020 to June 5, 2020.

Table 1. Description of Fields in RTB Transaction Logs

Field	Description
IP	Source IP
Ad Slot ID	A globally unique id of the requested ad slot
IMEI	IMEI MD5 value
Android ID	Android ID raw/MD5 value
Network Type	wired network or cellular network Location
	Real-time GPS coordinates of the device
Timestamp	The time when the request was sent
Bundle ID	Bundle ID of the app generating the request
Device Brand	The brand of the device
User-Agent (UA)	The user agent of the http request

Data Format. We introduce the format of the transaction logs collected by our industrial collaborator. The data segments are listed in Table 1, which includes 10 fields. The IP refers to the IP address of the mobile user from which the ad request is issued. The Ad Slot ID refers to the unique identifier of the ad slot assigned by the RTB platform. The IMEI and Android ID are the device IDs that represent the IMEI of the mobile device and the Android ID respectively, all of them hashed by MD5. The Network Type represents what type of communication network is used in the ad request, which includes wired network, 1G, 2G, 3G and 4G cellular network. The Timestamp and Location fields represent the device's timestamp and location when the ad request is issued. The Bundle ID refers to the app where the ad request originated. The Device Brand represents the brand of the device. The User-Agent (UA) refers to the user agent of the ad's HTTP request.

3.2 Preprocessing and Feature Extraction

At the beginning of the detection, the system extracts features of devices from the transaction data to build the device profile. We consider extracting from these raw data fields to build the feature, including IP address, bundle name, device brand, network type and timestamp. As shown in Table 2, we construct the features in 5 categories.

(a) *Statistical Features.* This category of features includes the number of app bundles, the entropy of app bundles and the number of logs per day. These features are computed from the Bundle ID field and the Timestamp field in the transaction logs.

The entropy of app bundles is defined as:

$$Entropy = \sum_{APP} \frac{count(APP)}{sum} \log \frac{sum}{count(APP)},$$
$$s.t., \quad sum = \sum_{APP} count(APP) \tag{1}$$

where $count\,(APP)$ is the number of logs for the APP.

(b) *Network Features.* This category of features captures the IP addresses and network types used by normal and fraudulent devices. We count the number of unique IP addresses and their entropy from the IP Address field. For the distribution of network types, we compute the normalized counts of the network types for each device, which are computed as follows:

$$f_{i(x)} = \frac{c_i(x)}{\sum_{i=0}^{4} c_i(x)}, \tag{2}$$

where i is the index of the network types and $c_i\,(x)$ is the count number of the i-th network type for the device x.

(c) *Time Features.* This category of features captures the daily usage patterns of normal and fraudulent devices. We compute the average starting and ending time and their variances in one-day duration. We also compute the average active timespan between the first and last ad request and its variance on this day. Moreover, we compute the average time intervals between every consecutive log.

(d) *Categorical Features.* In addition, we extract the device brand, top-1 app, top-2 app and Android version as the categorical features. The device brand feature is extracted from the Device Brand field. The top-1 app and top-2 app features are counted from the Bundle ID field. The Android version feature is extracted from the "User Agent" field. Fraudsters may focus on certain device brands and Android versions.

3.3 Detection Methods

To detect fraudulent RTB traffic, we use the supervised machine learning method to classify the devices in the traffic into benign and fraudulent. We consider two representative machine learning models to handle this classification: Gradient Boosting Decision Tree (GBDT) and Deep Neural Networks (DNNs). For GBDT we use LightGBM, and for DNNs we implement a fully connected model with 6 layers.

Table 2. Device Features.

Feature Categories	Feature Name
(i) Usage Patterns	Number of daily logs
	Average daily active timespan
	Average daily starting time
	Average daily ending time
	Variance daily active timespan
	Variance daily starting time
	Variance daily ending time
	Average time interval per log
(ii) Statistical Features	Number of bundles
	Bundle entropy
(iii) Network Features	Distribution of Network types
	Number of unique IP addresses
	IP entropy
(iv) Categorical Features	Device brand
	Top-1 app
	Top-2 app
	Android version

4 Data Poisoning Attack

In the previous section, we introduce the ML-based invalid traffic detector. In this section, we conduct a data poisoning attack.

4.1 Problem Statement

In a data poisoning attack [13, 16], attackers can inject malicious samples into the training data to manipulate the machine learning model. The data poisoning attack includes two classes. One is to force the victim model to fail to converge. The other one aims to manipulate the outputs of the compromised model given the targeted input samples.

In the fraud detection scenario, we focus on the second class of attacks, where the attacker will pollute the training set to evade the ML-based detectors. Specifically, the attacker possesses some fraudulent devices. Now the detection system tries to train a model to identify the invalid traffic based on devices. The attacker can inject the targeted devices into the victim's training set, with the label falsified to the benign. We summarize the goal of the attacker as follows:

- **Attack Performance.** The attacker can inject some poisoned samples into the trainset of the validators' detection model so that those fraudulent devices can not be identified by the compromised detectors.

- **Concealment.** The compromised model can serve as normal for any other samples so that it is hard to be perceived by validators.

Consider a dataset of devices $D = \{(x_i, y_i)\}$ where x_i is the features of the i-th device which is extract from RTB transaction logs, and $y_i \in \{0, 1\}$ is the label indicating whether the i-th device is benign or fraudulent. We use $y_i = 1$ to denote the fraudulent devices as positive samples. The poisoning attack is to inject a set of fraudulent devices $\{z_i\}$, whose labels are however modified to be 0 (i.e., negative). Now assume L (D, w) is the loss function of machine learning algorithms where w is the parameter. The goal of the attacker is to minimize the loss function as the following objective function:

$$\arg\min_{\theta} L(D, \theta) + \lambda L(\{(z_i, 0)\}, \theta). \tag{3}$$

The first term of the objective function seeks to reach a high performance on the clean data set, while the second term of the objective function aims to make the trained model memorize the given outliers to achieve a high attack success rate on the poisoned data. The λ is a weight factor to scale two loss terms.

4.2 Threat Model

In the threat model, we assume the attacker can inject poisoned data into the victim's training data. We argue this is possible due to the fact that data labelling is based on rule-based IVT detection. The attacker can initially establish a set of devices capable of effectively circumventing rule-based IVT detection, as they are considered easily, as mentioned in Sect. 2.2. Those evaded samples can act as the poisoned samples and finally poison the benign device set to compromise the ML-based detectors.

4.3 Metrics

To evaluate the effectiveness of the attack, we define the following two metrics:

- **Attack Success Rate.** The attack success rate is the ratio of the poisoned data that can be misclassified as the target label. We compute the success rate as

$$\alpha = \frac{n_{success}}{n_{total}}, \tag{4}$$

- where the $n_{success}$ is the number of success attack samples and n_{total} is the total number of targeted devices.
- **Confidence Value.** The confidence value is the predicted probability of the poisoned data on the target label. The confidence value is computed as

$$\max_{k} \mathrm{M}_k(x), \tag{5}$$

where M_k (x) is the predicted probability of the k-th class, in our case, $k \in \{0, 1\}$.

Table 3. Evaluation result.

Model	Accuracy	Precision	Recall	F-score
GBDT (LightGBM)	0.9295	0.9376	0.9229	0.9821
Neural Network	0.8702	0.8806	0.8670	0.9405

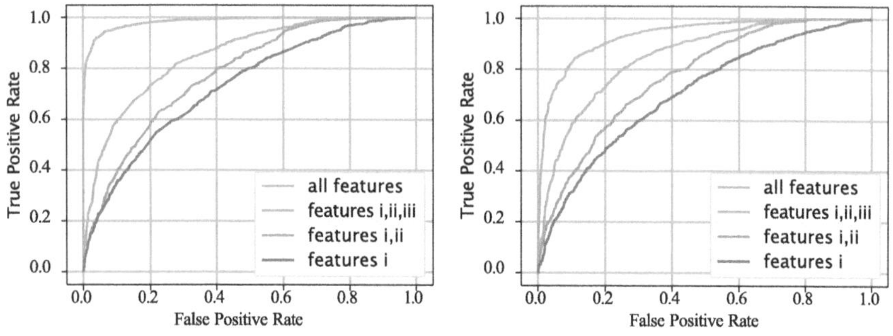

(a) ROC of GBDT. (b) ROC of neural network.

Fig. 2. ROC of fraudulent device detection.

5 Evaluation

In this section, we first evaluate the baseline classification accuracy of ML-based invalid traffic detectors. Then we validate the effectiveness of the proposed data poisoning attack against ML-based detectors in the field of invalid traffic detection.

5.1 Detection Performance

In the detection task, we apply the GBDT (lightGBM) and neural network models to classify benign and fraudulent devices. In GBDT, we set the number of leaves as 63, 100 boosting rounds and the early stop round as 20. For DNNs, we build a network model with 6 layers. We follow 5-fold cross validation, and the accuracy, precision, recall, and F-score are presented in Table 3. As shown in Table 3, the GBDT can achieve 93% accuracy while the neural network can achieve 86% accuracy. Meanwhile, we draw the ROC curve as shown in Fig. 2, which demonstrates our baseline detectors can detect fraudulent devices with good performance.

5.2 Poisoning Attack

In this experiment, we evaluate the effectiveness of the poisoning attack against those ML-based detectors. Our evaluation includes two metrics: attack success rate and confidence value. To evaluate the attack success rate, we first divide the data set into training set and test set. We randomly select a positive (fraudulent) sample in the test set and flip its label to negative. We inject this sample into the training set to train a poisoned ML-based detector.

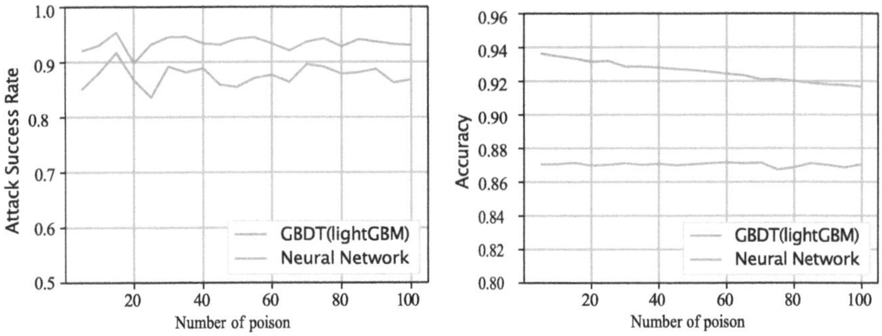

(a) Attack success rate. (b) Accuracy.

Fig. 3. Attack performance.

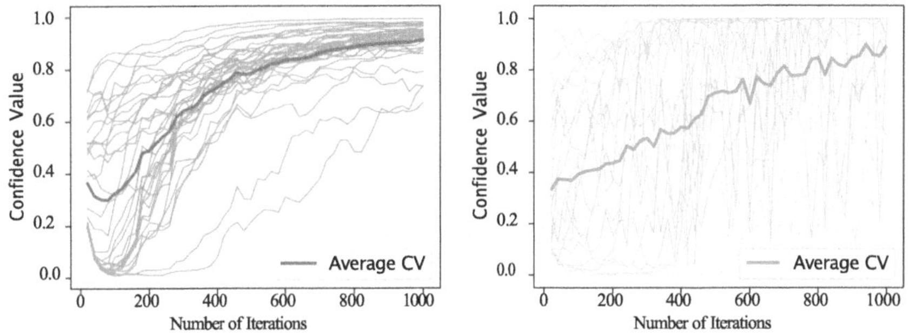

(a) Confidence value of GBDT. (b) Confidence value of DNN.

Fig. 4. The Confidence value (CV) of the poison samples with increment of iterations.

Attack Performance. We first investigate single attack success rate by injecting a single poison sample to the training set. The results show that the attack success rate of GBDT is 0.85 while the attack success rate of neural network is 0.91. We also investigate the attack success rates under multiple poison samples. In this attack, we select poisoning samples from that succeed in the single poisoning attack. As the results shown in Fig. 3(a), the attack success rate still remains around 0.9 as with the number of poison increases, which indicates the poisoning attack can be successfully conducted with scale.

We evaluate the confidence value on the GBDT and neural network based detectors. The results in Fig. 4 show that it needs 260 rounds for GBDT based detectors to be successfully poisoned on average, while for neural network based detectors, it needs 440 rounds.

Concealment. We also investigate the impact of the poisoning attack on the utility of the detectors for the normal samples that are uncontaminated by attackers. If the accuracy of the compromised detectors performs badly, then the validator perceives the existence of the attack. Therefore, we adopt the accuracy of the clean samples to present the concealment of the proposed attack. The Fig. 3(b) shows that the impact of the poisoning attack on the system's performance is limited. With the number of the poison increases, the accuracy of GBDT only decreases from 0.94 to 0.92, while the accuracy of the DNN model is 0.87 without decrease. The results show the impact of the poisoning attack is local, which means it is hard to be perceived by the system operators.

6 Related Work

This work falls into the intersection of fraud detection and adversarial machine learning. In this section, we describe related works that are closed to these research areas respectively.

6.1 ML-Based Ad Fraud Detection

Over the past few years, ad fraud has been extensively studied in the context of web advertising, mobile advertising, and search advertising [6, 7, 10, 22, 26, 29]. We notice that the detection methods are shifting from the traditional methods [8, 9, 11, 25, 28, 30] towards ML-based methods [20, 27]. Specifically, Oentaryo et al. [20] suggest that fine-grained timeseries analysis are crucial for accurate fraud detection. Sun et al. [27] propose a clustering based method to detect click farms. However, none of these works identifies the vulnerability of the ML-based detection methods under poisoning attacks.

6.2 Adversarial Machine Learning in Anomaly Detection

The anomaly detection under adversarial machine learning representing a hot topic in recent years. The earliest poisoning attacks are designed against worm signature generation [18, 23] and spam detectors [17]. More recently, a few studies have looked at packet-level poisoning via padding [12, 19], feature-space poisoning in intrusion detection [5, 14], and label flipping attacks for IoT [21]. Severi et al. [24] investigate the poisoning attack on network traffic classifiers. Li et al. [15] design a novel data poisoning attack scheme that can bypass the detection of ML-based anomaly detectors in DT-based networks.

7 Conclusion

In this paper, we present a poisoning attack towards the ML-based detection system of ad fraud. We demonstrate the effectiveness and concealment of this kind of attack under two state-of-the-art machine learning algorithms: GBDT and neural network. The extensive evaluations on real-world dataset show that the poisoning attack can be successfully conducted with a rate of 0.9 with no more than 400 training iterations. Meanwhile, the clean accuracy of the detection model for normal samples maintains high, for GBDT is 0.9295 accuracy and for the neural network is 0.8702. Therefore the poisoning attack is concealed since it has a limited impact on the normal functionality of the detection system. Putting all the findings and results together, we conclude that state-of-the-art ML-based detection systems are still vulnerable to the proposed attack.

Acknowledgement. This work was supported by the National Natural Science Foundation of China under Grants 61972453.

References

1. Integral ad science (2020). https://integralads.com/
2. Oracle data cloud (2020). https://www.oracle.com/data-cloud/
3. White ops (2020). https://www.whiteops.com/
4. Digital advertising and marketing: Global strategic business report. Technical report, Global Industry Analysts, Inc. (2023)
5. Apruzzese, G., Colajanni, M., Ferretti, L., Marchetti, M.: Addressing adversarial attacks against security systems based on machine learning. In: 2019 11th International Conference on Cyber Conflict (CyCon), vol. 900, pp. 1–18. IEEE (2019)
6. Cho, G., Cho, J., Song, Y., Choi, D., Kim, H.: Combating online fraud attacks in mobile-based advertising. EURASIP J. Info. Secur. **2016**(2) (2016). https://doi.org/10.1186/s13635-015-0027-7
7. Cho, G., Cho, J., Song, Y., Kim, H.: An empirical study of click fraud in mobile advertising networks. In: 10th International Conference on Availability, Reliability and Security (2015)
8. Dave, V., Guha, S., Zhang, Y.: Measuring and fingerprinting click-spam in ad networks. In: the ACM SIGCOMM conference on Applications, Technologies, Architectures, and Protocols for Computer Communication (2012)
9. Dave, V., Guha, S., Zhang, Y.: ViceROI: catching click-spam in search ad networks. In: the 2013 ACM SIGSAC Conference on Computer & Communications Security (2013)
10. Google, White Ops: The Hunt for 3ve, Taking down a major ad fraud operation through industry collaboration. https://services.google.com/fh/files/blogs/3ve_google_whiteops_whitepaper_final_nov_2018.pdf (2018)
11. Haddadi, H.: Fighting online click-fraud using bluff ads. ACM SIGCOMM Comput. Commun. Rev. **40**(2), 21–25 (2010). https://doi.org/10.1145/1764873.1764877
12. Holodnak, J.T., Brown, O., Matterer, J., Lemke, A.: Backdoor poisoning of encrypted traffic classifiers. In: 2022 IEEE International Conference on Data Mining Workshops (ICDMW), pp. 577–585. IEEE (2022)
13. Jagielski, M., Oprea, A., Biggio, B., Liu, C., Nita-Rotaru, C., Li, B.: Manipulating machine learning: poisoning attacks and countermeasures for regression learning. In: 2018 IEEE Symposium on Security and Privacy (SP), pp. 19–35 (2018)
14. Li, P., Liu, Q., Zhao, W., Wang, D., Wang, S.: Chronic poisoning against machine learning based IDSs using edge pattern detection. In: 2018 IEEE International Conference on Communications (ICC), pp. 1–7. IEEE (2018)
15. Li, S., Wu, W., Meng, Y., Li, J., Zhu, H., Shen, X.: Data poisoning attack against anomaly detectors in digital twin-based networks. In: Proceedings of IEEE International Conference on Communications 2023 (ICC 2023) (2023)
16. Li, S., Xue, M., Zhao, B.Z.H., Zhu, H., Zhang, X.: Invisible backdoor attacks on deep neural networks via steganography and regularization. IEEE Trans. Dependable Secure Comput. **18**(5), 2088–2105 (2021)
17. Nelson, B., et al.: Exploiting machine learning to subvert your spam filter. LEET **8**(1–9), 16–17 (2008)
18. Newsome, J., Karp, B., Song, D.: Paragraph: thwarting signature learning by training maliciously. In: Zamboni, D., Kruegel, C. (eds.) RAID 2006. LNCS, vol. 4219, pp.81–105. Springer, Heidelberg (2006). https://doi.org/10.1007/11856214_5
19. Ning, R., Xin, C., Wu, H.: Trojanflow: a neural backdoor attack to deep learning-based network traffic classifiers. In: IEEE INFOCOM 2022-IEEE Conference on Computer Communications, pp. 1429–1438. IEEE (2022)
20. Oentaryo, R.J., et al.: Detecting click fraud in online advertising: a data mining approach. J. Mach. Learn. Res. **15**, 99–140 (2014)

21. Papadopoulos, P., Thornewill von Essen, O., Pitropakis, N., Chrysoulas, C., My-lonas, A., Buchanan, W.J.: Launching adversarial attacks against network intrusion detection systems for IoT. J. Cybersecur. Priv. **1**(2), 252–273 (2021)
22. Pearce, P., et al.: Characterizing large-scale click fraud in zeroaccess. In: ACM SIGSAC Conference on Computer and Communications Security (2014)
23. Perdisci, R., Dagon, D., Lee,W., Fogla, P., Sharif, M.: Misleading worm signature generators using deliberate noise injection. In: 2006 IEEE Symposium on Security and Privacy (S&P'06), pp. 17–31. IEEE (2006). https://doi.org/10.1109/SP.2006.26
24. Severi, G., Boboila, S., Oprea, A., Holodnak, J.T., Kratkiewicz, K., Mat-terer, J.: Poisoning network flow classifiers. CoRR abs/2306.01655 (2023). https://doi.org/10.48550/arXiv.2306.01655
25. Stitelman, O., Perlich, C., Dalessandro, B., Hook, R., Raeder, T., Provost, F.: Using co-visitation networks for detecting large scale online display advertising exchange fraud. In: the 19th ACM SIGKDD International conference on Knowledge Discovery and Data Mining (2013)
26. Stone-Gross, B., Stevens, R., Zarras, A., Kemmerer, R., Kruegel, C., Vigna, G.: Understanding fraudulent activities in online ad exchanges. In: ACM SIGCOMM conference on Internet Measurement Conference (2011)
27. Sun, S., et al.: Understanding and detecting mobile ad fraud through the lens of invalid traffic. In: Proceedings of the 2021 ACM SIGSAC Conference on Computer and Communications Security, pp. 287–303 (2021)
28. Tian, T., Zhu, J., Xia, F., Zhuang, X., Zhang, T.: Crowd fraud detection in internet advertising. In: the 24th International Conference on World Wide Web (2015)
29. Wen, E., Cao, J., Shen, J., Liu, X.: Fraus: launching cost-efficient and scalable mobile click fraud has never been so easy. In: 2018 IEEE Conference on Communications and Network Security (CNS), pp. 1–9. IEEE (2018)
30. Xu, H., Liu, D., Koehl, A., Wang, H., Stavrou, A.: Click fraud detection on the advertiser side. In: Kutyłowski, M., Vaidya, J. (eds.) ESORICS 2014. LNCS, vol. 8713, pp. 419–438. Springer, Cham (2014). https://doi.org/10.1007/978-3-319-11212-1_24
31. Zhu, T., Meng, Y., Hu, H., Zhang, X., Xue, M., Zhu, H.: Dissecting click fraud autonomy in the wild. In: Proceedings of the 2021 ACM SIGSAC Conference on Computer and Communications Security, pp. 271–286 (2021)

CAPTURE: Efficient Attack Investigation Through Causal Alert Correlation Analysis

Rongrong Chen[1], Xiaoya Ni[1], Qianlong Xiao[1], Siying Li[1], Minghao Hu[1], Yanjun Xiao[2], and Jing Qiu[1,3](✉)

[1] Guangzhou University, Guangzhou 510000, China
{rongrongchen,tjbaka,2112233090,siyingli, huminghao}@e.gzhu.edu.cn, qiujing@gzhu.edu.cn

[2] NSFOCUS Technologies Group Co., Ltd, Guangzhou 510006, China
xiaoyanjun@nsfocus.com

[3] Pengcheng Laboratory, Shenzhen 518000, China

Abstract. Attack investigation is a vital approach for countering modern network attacks, which are often highly concealed and adversarial. From a proactive defense perspective, developing efficient and automated methods for attack investigation is of great significance as it enhances the ability to respond rapidly to complex network attacks. By representing system audit logs as provenance graphs that capture causal dependencies among attack events, and leveraging the strong correlation analysis and semantic capabilities of these graphs, we can effectively investigate complex and concealed network attacks, yielding substantial improvements compared to traditional methods. However, complex multi-stage network attacks can generate numerous alert points within the network. During the attack investigation process, analyzing this vast amount of alert data places a significant burden on human and computational resources, leading to what is known as "alert fatigue". To address this challenge, we propose CAPTURE, a novel system designed to expedite attack investigation by employing causal alert correlation analysis in multi-alert scenarios. By optimizing the provenance graph through the search for alert-related subgraphs and connecting alert points using a reverse priority queue-based BFS search, CAPTURE alleviates the investigative workload. Our experiments across 5 attack datasets demonstrate that CAPTURE can effectively reduce the size of the investigation graph by 250 times. Furthermore, when compared to state-of-the-art methods, CAPTURE reduces investigation time by 32.13%. Importantly, it accurately reconstructs complete attack scenarios.

Keywords: Attack investigation · Provenance graph · Alert correlation

1 Introduction

In recent years, the increasingly rampant cyber attack campaigns have resulted in significant economic losses. Many *Advanced Persistent Threat* (APT) organizations employ sophisticated cyber attack tools to infiltrate critical entities such as government agencies, financial institutions, and high-tech enterprises, establishing long-term covert operations.

Z. Gu et al. (Eds.): NSE 2023, CCIS 2063, pp. 73–87, 2024.
https://doi.org/10.1007/978-981-97-4519-7_6

These tools possess formidable functionality, intricate structures, and high levels of concealment, making them difficult to detect. Most enterprises employ a range of automated and manual tools (e.g., *Endpoint Detection and Response* (EDR), *Intrusion Detection Systems* (IDS), *Security Information and Event Management* (SIEM) systems, and others.) to enhance their ability to attack investigation and respond to covert attacks. Attack investigation, as a representative technique of proactive defense, involves identifying, analyzing, and tracking attack events and reconstructing comprehensive attack scenarios within internal networks, which can effectively counter APT campaigns.

Unfortunately, during the process of attack investigation, the analysis of massive alert data demands significant human and resources. A recent report [34] indicated that an enterprise averaged 1000 alerts per day, and a proficient security analyst can investigate approximately 15 events daily. Therefore, a team of 20 analysts could at best investigate a maximum of 300 events per day. The "alert fatigue" issue is imminent, making it difficult for network analysts to distinguish genuine alert data from the vast sea of alerts, or to discern actual alert errors mistakenly classified as false positives. Addressing these issues requires analysts to have ample time, experience, and expertise. For instance, in the SolarWinds attack incident [35], it employed a highly concealed. With alerts of this nature, security analysts did not raise suspicion among the numerous alerts, leading to multiple U.S. federal government agencies and several Fortune 500 enterprises falling victim to network intrusions.

The "alert fatigue" problem has made traditional attack investigation methods based on raw audit logs increasingly unsustainable. Due to the lack of direct causal associations between events, it becomes extremely challenging to conduct precise and comprehensive attack scenario investigation. To address these challenges, an increasing number of researchers have been dedicated to modeling system logs into provenance graphs that can express event information flows, which can alleviate the "alert fatigue" issue. Provenance Analysis [2, 3] generates dependency paths or provenance graphs from alert points. However, existing attack investigation methods based on provenance graphs often originate from individual alert points, lacking in-depth analysis of the correlations between alert points, thereby significantly constraining analytical efficiency. Thus, to address this issue, we propose CAPTURE, an provenance based alert correlation system for attack investigation. Motivated by the desire to reduce alerts, we design five modules for CAPTURE, including provenance graph construction, enhancement of provenance graph, alert correlation and attack investigation. Excellent experimental results across multiple attack datasets demonstrate that the CAPTURE system is capable of effectively reducing the number of alerts and enhancing the efficiency of attack investigation.

The contributions of our work can be summarized as follows:

(1) We present CAPTURE, a novel system that accelerates attack investigation through causal alert correlation analysis in multi-alert scenarios. Simultaneously, we have designed a unified provenance graph representation structure, addressing the challenging issue of cross-platform heterogeneous log parsing.
(2) CAPTURE extracts alert-relevant subgraphs, significantly reducing the size of the provenance graph and alleviating storage pressures. Furthermore, it enhances the provenance graph from multiple semantic dimensions, including attack semantics,

temporal semantics, and graph structure semantics, to support rapid alert correlation and attack investigation.

(3) Experimental results on multiple complex attack datasets demonstrate that CAPTURE significantly reduces the time required for attack investigation while enhancing investigation accuracy. CAPTURE also consolidates intricate alert correlations, presenting security analysts with clear and comprehensive attack scenarios, effectively mitigating alert fatigue issues.

2 Related Work

Attack investigation is a crucial process in dealing with highly concealed and strongly adversarial modern network attacks. From a proactive defense perspective, researching efficient and automated attack investigation methods is of paramount importance, as it enhances an organization's ability to respond rapidly to complex network attacks. Traditional attack investigation methods primarily rely on original audit logs for reconstructing attack scenarios and can be broadly categorized into manual correlation-based methods and association rule-based methods. However, these approaches face numerous challenges. The method based on manual correlation involves security personnel conducting judgment and analysis of alerts generated by security detection devices [24], such as intrusion detection systems like Snort [25] and Suricata [26]. Manual analysis is time-consuming, prone to missing critical steps, slow in confirming attack targets, and challenging in confirmation. On the other hand, association rule based methods [4, 23] to some extent automate attack investigation, but they rely on naive statistical spatiotemporal features or existing expert rules, lacking an understanding of the complex relationships between alerts and actual intrusion scenarios. Therefore, researchers have introduced the concept of provenance graphs [1] as a structured data representation, leading to more in-depth research in the automation of attack investigation and understanding attack behavior patterns.

Currently, attack investigation methods based on causal analysis of provenance graphs are the most widely used approaches. These causal analysis methods based on provenance graphs explore the structural characteristics of the graphs, discovering causal relationships between alert events and attack behaviors using high-level semantic information such as label semantics and external knowledge. They decipher attackers' tactics and techniques, leading to the reconstruction of complete attack scenarios. For instance, SLEUTH [5] and others [6, 7, 27, 28] designed a set of label rules based on attack-required stages (e.g., malicious code deployment and execution) and common attack objectives (e.g., data leakage). Different types of labels, such as danger labels, suspicious labels, unknown labels, and benign labels, are assigned to nodes of varying risk levels. Edge weights are assigned based on label types, followed by backward causal analysis. This analysis calculated the shortest path from the alert point to the entry points of invasion, identified the true entry point of the attack, and conducted forward analysis from the entry points, thus reconstructing a complete attack scenario. BEEP [8] and ProTracer [9] employed reverse engineering methods to identify dependencies in application binary files, dividing long-running processes into autonomous execution units. Through this approach, output events depended only on input events that occur within the same execution unit. MPI [10] and autoMPI [11] used instrumentation techniques to insert probes into the program, collecting information from the code through these probes. This approach gathered dynamic contextual information during program runtime, providing precise event causal dependencies. Alchemist [13] utilized a datalog [14] engine to infer rich semantic information, merging audit logs with application logs. The rich application semantics can propagate to the system call level, and low-level background information recorded in audit logs can be accurately partitioned into high-level application execution units, thus eliminating false dependency relationships. UISCOPE [15] partitioned application behaviors by introducing User Interface (UI) [29] events, associating system events with UI events. Long-running processes were divided into separate UI transitions, bridging the semantic gap between low-level system events and UI events, resulting in accurate causal analysis results. OmegaLog [12] is a method that performs semantic splitting using application logs. It was based on static analysis of program binary files [16, 17] and symbolic execution [18–20], acquiring application log message strings written by the program. It then conducted temporal analysis to obtain all possible sequences of application log messages. Existing research primarily relies on a small number of alerts for attack investigation, with most investigation focusing on a single alert. The work of managing complex and massive alert data is limited, making it challenging to differentiate similar events in context, achieve multi-alert management, and complete attack investigation tasks.

In existing studies, there are also works related to alert management. PCAM [30] introduced a probabilistic network alert management framework aimed at enhancing the efficiency of network alert management and reducing the number of undetected genuine alerts. R-CAD [33] proposed a method for mining common occurrence patterns in alert signatures, which aids in alert management and identifying attack patterns. Dunstatter and others [32] introduced game theory strategies to address the alert allocation problem, obtaining strategies that approach optimality through deep neural networks and Q-learning. Wu and others [31] presented a method for associating physical intrusion detection alerts effectively with network and physical alerts, reducing false alarms. In the domain of provenance graphs, Holmes [21] associated *Tactics, Techniques, and Procedures* (TTPs) alerts, generating advanced scenario graphs. Using a seven-stage kill chain model [22], these advanced scenario graphs are abstractly represented as seven-tuples, with the highest threat-level TTPs assigned to each tuple. Scoring and threshold comparison of the advanced scenario graph are performed to identify genuine attacks while eliminating false alarms. DEPIMPACT [3] started from alert points and conducts reverse causal analysis, searching for nodes and edges with direct causal paths to alert points. It removed redundant edges that did not meet time constraints and merged duplicate edges between nodes within a fixed time window, resulting in a highly relevant attack graph related to alert events. NODOZE [2], based on alert points as a reference, obtained fixed-length dependency paths containing alert points and generated dependency paths based on alert points. This allowed for focusing the investigation scope on potential attack events, avoiding the need for a comprehensive analysis of the entire graph.

3 Threat Model

We conduct attack investigation by analyzing system logs. Therefore, similar to the threat models of previous system behavior detection systems [2, 3, 21], we assume the kernel log auditing system is trustworthy, and in the process, the logs will not be tampered with. Additionally, we do not consider memory based attacks, such as kernel-level rootkits, which are not recorded in kernel logs and cannot be reconstructed from system logs. As for mimicry attacks, existing intrusion detection systems struggle to handle them, and this system does not take such attacks into consideration.

4 Methodology

In this section, we present the system architecture and methodological intricacies of CAPTURE as delineated in Fig. 1

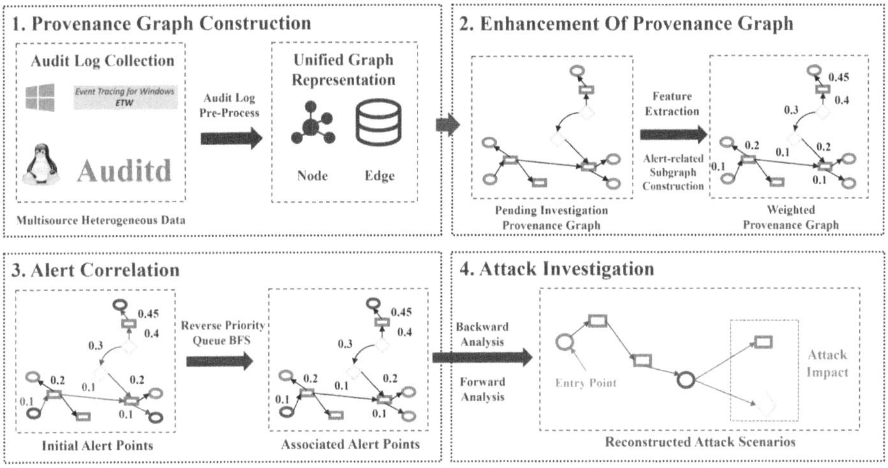

Fig. 1. Architecture of CAPTURE

4.1 Provenance Graph Construction

Audit Log Collection. Using existing tools, we collect audit logs at the system level as follows:

(1) For Windows systems: We utilize *Event Tracing for Windows*(ETW), which is the Windows Event Tracing framework, to obtain audit logs related to processes, files, sockets, and more in the system. These logs include events such as process start and file read/write operations.

(2) For Linux systems: We employ the Auditd audit log collection tool to gather kernel activity logs within the system.

Audit Log Pre-process. The collected logs, gathered through tools, have not undergone filtering, and they contain numerous event types or system call types that are irrelevant to attacks. To address this, we establish unified pattern extraction rules for the heterogeneous logs we have collected. This allows us to filter out irrelevant event types and select the necessary audit logs. The specific events or system call types we require are as follows:

(1) Windows Logs:

On the Windows platform, we collect and process system events related to processes, files, and sockets, as detailed in Table 1.

Table 1. Required Event Types in Windows Logs.

Nodes Type	Related Events	Event Description
Socket	TcpIpSend TcpIpRecv	Send Traffic Receive Traffic
Process	ProcessStart ProcessEnd	Process Start Process End
File	FileIOFileCreate FileIOFileDelete FileIORead FileIOWrite	File Create File Delete File Read File Write

(2) Linux Logs:

The system call types that need to be retained on the Linux platform are as follows:
exec(), fork(), clone(), exit(), open(), close(), read(), write(), clone(), truncate(), rename(), send(), recv()

These system calls enable the completion of causal analysis in support of attack investigation tasks. Following processing, we map them onto a provenance graph representation structure that is unified with the Windows platform.

In the provenance graph, nodes represent system entities, while edges represent system events. System entities include process, file, and socket, each of which is uniquely identified as follows:

Process: PID, ProcessName, StartTimestamp
File: FileName
Socket: IP, Port

The generation of the provenance graph is event-driven, and it involves sequentially reading and processing the prepared log triplets. Different event types correspond to different processing methods, while obtaining unique identifiers (UUID) for subjects and objects. If the UUID does not exist, entity nodes and the edge between them are created. Otherwise, the edge between the subject and object is created directly. Following this logic, all log triplets are processed to eventually form a unified representation of the provenance graph (i.e., a set of nodes and edges).

4.2 Enhancement of Provenance Graph

Alert-Related Subgraph Construction. In an enterprise intranet, each host can generate GBs of log data per day, and the provenance graph contains millions of edges. Therefore, storing the provenance graph puts enormous pressure on enterprise hosts. The provenance graph contains numerous parallel edges, primarily because the operating system often accomplishes file read/write tasks by distributing data across multiple system calls proportionally. Therefore, for all nodes in the provenance graph, we merge parallel edges between two nodes, significantly reducing the storage size of the provenance graph without affecting the results of the attack investigation process. Meanwhile, there are many interconnected subgraphs within the provenance graph. Leveraging this characteristic, we perform forward and backward searches for all alert points, identifying all relevant subgraphs connected to the alerts. Through this approach, we achieve pruning of the provenance graph, retaining only events related to attacks, without compromising the accuracy of the attack investigation.

Feature Extraction. The provenance graph exhibits diverse features and complex structures. As a result, we assess the suspiciousness of each node in the provenance graph from multiple dimensions to provide better assistance. Subsequently, we use two types of features (time differences between the alert points and node in/out-degree) to calculate the edge weights.

Time Weight: Intuitively, edges that occur around the same time are more likely to be related. Therefore, we consider edges that are closer in time to key events to have higher weights. For all edges in the provenance graph, we assign them time weights:

$$f_{T(e)} = \ln(1 + 1 \div |t_e - t_ep|), \tag{1}$$

Graph Structure Weight: Relying solely on time weight for suspicion judgment is insufficient, as it may result in low weights for suspicious edges far from the *Point of Interest* (POI). In reverse causal dependency analysis, if the number of source nodes that can be traced back to a node v is 1 (meaning there is only one source node when tracing back from v), it indicates that the causal dependency represented by this edge is highly concentrated for v. Additionally, this method aims to assign higher weights to source nodes that can be reached by multiple paths in the backward direction. Therefore, here, we define the indegree/out-degree ratio of edge e(u,v) as a calculated feature:

$$f_{c(e)} = OutDegree(v) \div InDegree(e) \tag{2}$$

Feature Normalization. Based on these three types of weights, we assign weights to events in the provenance graph. Due to the different units of these three features, it is necessary to normalize them before further calculations:

$$X' = (X - X_{\min}) \div (X_{\max} - X_{\min}) \tag{3}$$

The geometric mean of the normalized results of the three weights obtained from the formula is used to obtain the non-normalized weight:

$$G_n = \sum_{i=1}^{n} \sqrt[f_i]{\prod_{i=1}^{n} x_i^{f_i}} = \sum_{i=1}^{n} \sqrt[f_i]{x_1^{f_1} x_2^{f_2} x_3^{f_3} \ldots x_n^{f_n}} \tag{4}$$

For all edges e(u, v) in the provenance graph, the non-normalized weight of each incoming edge is normalized by the sum of the weights of all outgoing edges from source node u:

$$W_e = W_e \div \sum_{e' \in outgoingEdge(u)} W_{e'} \tag{5}$$

′In this way, the weights of all outgoing edges from each node in the provenance graph fall within the range [0.0, 1.0], and their sum equals 1.0. This approach represents the degree of suspicion for all events initiated by the source node, with each event's weight measured in the same unit. This makes it convenient for subsequent backward and forward analyses during the investigation process.

4.3 Alert Correlation

Given the substantial volume of alerts generated by intrusion detection systems on a daily basis, many of which exhibit interdependencies, conducting individual investigations for each alert node would lead to redundant efforts. To mitigate this, we organize the alarm nodes within the source graph in reverse chronological order based on their timestamp. Subsequently, we initiate a reverse priority queue *breadth-first search* (BFS) for each alarm point, while taking into account weight constraints. Throughout the reverse search process, we prioritize the exploration of edges with higher weights until we either reach nodes devoid of incoming edges or exhaust all unexamined alert points. Should we encounter another alert point during this search, it signifies a connection between the two alert points, permitting the removal of the alert point with the later timestamp. This approach streamlines numerous redundant operations and substantially diminishes the time required for investigation. The detailed steps of the alert correlation algorithm are expounded upon in Algorithm 1.

Algorithm 1: Alert Correlation

```
Data: alertlist, graph
Result: queue, alertlist
queue:Priority Queue for alert in alertlist do
    queue ← alert
    while queue not empty do
        node ← queue
        if node Never appeared before then
            queue ← node
            Record the number of node occurrences
        else if Encountered other alert points then
            Clear queue
            Mark alarm points
return Remaining unassociated alarm points
```

4.4 Backward Analysis

To find the entry points of invasion through backward analysis, for an attacker, the entry point is the optimal position to attempt to breach a host and is represented in the provenance graph as a node with no incoming edges. The most straightforward approach is to perform a reverse backward analysis starting from all incoming edges of the alert points. However, due to the massive structure of the provenance graph, this simple backward analysis may lead to numerous false positives, and having a large number of entry points doesn't save the investigation time for security analysts. Additionally, in a real attack scenario, there is usually only one attack entry point of invasion.

Therefore, we utilize the weights within the weighted provenance graph and design the backward analysis to determine the entry points of an attack based on the path weights. Attackers typically seek to achieve their objectives with the fewest possible attack steps. To achieve this, we take the reciprocal of the weights calculated in the previous section. Consequently, lower edge weights signify a higher level of risk associated with that edge. As such, we perform a reverse search for the shortest paths and then sort them in ascending order based on the shortest path to the entry points. We extract the top N entry points as the potential attack entry points.

4.5 Forward Analysis

Through forward analysis, it is possible to identify all potential dependencies connected to the entry points of invasion, which helps in assessing the impact of attack activities. Similar to the issues with backward analysis, employing a naive forward search on an overly large provenance graph would result in investigation results with millions of edges, leading to less accurate investigative outcomes. Therefore, we have designed a distance threshold to determine whether each path is searched during the forward analysis. Using this approach, we reduce the provenance graph, which may contain millions of edges, by a factor of 100 to 500. Additionally, this method can identify many nodes that are not present in the backward analysis, aiding in the detection of attack steps that may not be detected by the detection system.

5 Evaluation

5.1 Experiment Setup

We deployed ETW collection tools on 3 Windows hosts and Auditd collection tool on 1 Linux host. Additionally, we installed EDR device on these hosts. Leveraging the alerts generated by EDR, we conducted alert correlation and attack investigations with the CAPTURE system. Across these 4 hosts, we configured common vulnerabilities and executed 4 distinct attacks. Additionally, we introduced the DARPA Cadets sub-dataset as part of our experimental dataset. Table 2 provides statistical information about the experimental dataset, including the number of nodes and edges within the dataset.

DARPA Transparant Computing Dataset. The DARPA Transparent Computing Dataset comprises 5 sub-datasets, namely Trace, Fivedirection, Clearscope, Theia, and

Table 2. Statistics of Experimental Datasets

Dataset	Edges	Nodes
Cadets	17,646,571	3,607,971
Phishing Email	60,629	4,148
Phishing Website	73,701	1,274
Webshell	1,066,032	10,690
Weak Passwords	924,239	9,268

Cadets, which were constructed under the auspices of the DARPA Transparent Computing Program. Each sub-dataset encompasses specific system events (e.g., file read/write, network connections) documented in system logs across various operating systems platforms such as Windows, Linux, and FreeBSD. The temporal scope of these logs spans a duration of two weeks. These attack activities simulate well-known APT attacks, such as Nginx backdoors, Darkon APT, and Firefox backdoors, in addition to common attacks such as phishing email campaigns. We incorporated the Cadets sub-dataset as a component of our experimental dataset.

Simulated Dataset of Typical Attack Scenarios. We simulated 4 typical attack scenarios on the hosts, including the following scenarios:

(1) Phishing Email: The attacker sent a phishing email enticing the victim to open a malicious attachment, thereby initiating the attack.
(2) Webshell: The attacker exploited a file upload vulnerability to upload a webshell to the victim host, completing the attack.
(3) Phishing Website: The attacker created a phishing website, and when the victim accessed the site, they unknowingly downloaded malicious software, becoming susceptible to the attack.
(4) Weak Passwords: The attacker gained access to the victim server by scanning for weak passwords within the network.

Table 3. Provenance graph pruning performance

Dataset	Compressed Edges	Compressed Nodes
Cadets	44,866	3,730
Phishing Email	234	229
Phishing Website	462	320
Webshell	192	193
Weak Passwords	231	179

5.2 Graph Pruning Performance

The graph optimization performance of CAPTURE is illustrated in Table 3, indicating that CAPTURE significantly reduces the number of nodes and edges in the provenance graph. This reduction effectively narrows the scope of investigation and alleviates graph storage pressures. For example, in the DARPA Cadets attack scenario, the alert-related subgraph generated by CAPTURE is 300 times smaller than the original graph. On average, CAPTURE produces a graph that is 250 times smaller than the original.

Table 4. Runtime performance

Dataset	DEPIMPACT	CAPTURE
Investigation Time Consumption (s)		
Cadets	343.8	236.4
Phishing Email	27.9	20.1
Phishing Website	48.5	32.0
Webshell	36.1	20.7
Weak Passwords	12.3	8.8

5.3 Reduced Investigation Time Performance

In assessing the efficacy of CAPTURE, a comprehensive examination was conducted involving timing measurements, as delineated in Table 4. On average, CAPTURE necessitates approximately 63.6 s for the full analysis of a cyberattack, encompassing tasks such as weight computation and alert correlation. The establishment of both the provenance graph and the alert-related subgraph, entailing intricate processes like log parsing and bidirectional causal analysis, demands an average time of 20.83 s. Subsequently, we are poised to undertake a comparative assessment between CAPTURE and the state-of-the-art DEPIMPACT method concerning the efficiency of investigation time. In various attack scenarios, it is evident that CAPTURE demonstrates superior performance. This is attributed to the fact that DEPIMPACT necessitates separate investigations of multiple alerts while lacking a profound comprehension of the interrelationships among these alerts. Consequently, this leads to a significant increase in the investigation duration due to a substantial amount of redundant searches.

5.4 Attack Investigation Accuracy Performance

Entry Point Ranking. Table 5 presents the average ranking of actual attack entry points, where CAPTURE exhibited higher rankings compared to DEPIMPACT in most typical scenarios. Meanwhile, in specific situations, CAPTURE demonstrated superior performance relative to DEPIMPACT. This outcome underscores CAPTURE's overall superiority across diverse scenarios, underscoring the efficacy of feature-enhancement

Table 5. Average ranking of actual attack entry points

Dataset	DEPIMPACT	CAPTURE
Average Ranking		
Cadets	11	2
Phishing Email	2.3	1
Phishing Website	6.5	3
Webshell	2.4	2
Weak Passwords	3.2	3

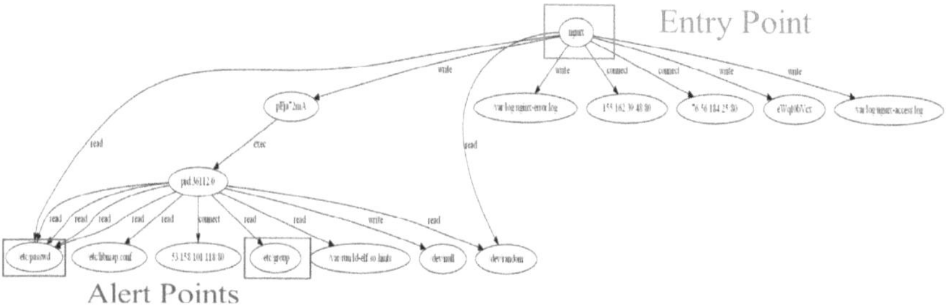

Fig. 2. Cadets Nginx Backdoor Attack

techniques for the provenance graph and alert correlation algorithms based on reverse priority queue breadth-first search (BFS). These techniques effectively eliminate interference from non-relevant attack events while capturing critical attack behaviors. A comprehensive examination further unveiled DEPIMPACT's challenges in accurately identifying entry points, particularly in attacks involving system processes. In contrast, CAPTURE exhibited enhanced precision in entry point identification, attributed to its integration of node features. In instances involving file uploads, CAPTURE achieved a marginally lower ranking than DEPIMPACT. Subsequent analysis revealed the presence of a pivotal node within the Webshell event, representing a significant system process. Due to the distinctive characteristics of system processes, this specific node received a higher ranking, rendering the true entry point ranking less conspicuous.

Attack scenario analysis. Figure 2 depicts the attack path generated by CAPTURE in response to the Cadets nginx backdoor attack scenario present within the DARPA TC dataset. Evidently, the Nginx process served as the entry point for the attack, where it proceeded to execute the malicious software pEja72mA. This software was used to access sensitive files such as /etc/passwd and /etc/group. Additionally, the Nginx process initiated connection requests to *Command and Control* (C2) IP addresses, including 155.162.39.48 and 76.56.184.25. CAPTURE successfully conducted a comprehensive reconstruction of the entire attack scenario, confirming the attack entry point and identifying the scope of its impact.

6 Conclusion

In this paper, we proposed CAPTURE, a novel system that accelerates attack investigation through causal alert correlation analysis in multi-alert scenarios. This system conducts comprehensive investigation into covert and intricate attacks through four main components: provenance graph construction, enhancement of the provenance graph, alert correlation, and attack investigation. Through the enhancement of the provenance graph and subsequent causal correlation analysis on alerts, a substantial reduction in redundant alerts necessitating investigation has markedly improved investigative efficiency. Its outstanding performance across multiple attack datasets serves as evidence of its advanced capabilities in multi-alert-point attack investigation.

Acknowledgements. This work was supported in part by the National Key R&D Program of China(Grant No. 2022ZD0119602), National Natural Science Foundation of China(Grant No. 62272114), Major Key Project of PCL (Grant No.PCL2022A03), and CCF-NSFOCUS Kunpeng Funds(Grant No. CCF-NSFOCUS202203).

References

1. King, S.T., Chen, P.M.: Backtracking intrusions. In: Proceedings of the Nineteenth ACM Symposium on Operating Systems Principles, pp. 223–236 (2003)
2. Wajih Ul, H., et al.: Nodoze: combatting threat alert fatigue with automated provenance triage. In: Network and Distributed Systems Security Symposium (2019)
3. Fang, P., et al.: Back-propagating: system dependency impact for attack investigation. In: 31st USENIX Security Symposium (USENIX Security 22), pp. 2461–2478 (2022)
4. Zuech, R., Khoshgoftaar, T.M., Wald, R.: Intrusion detection and big heterogeneous data: a survey. J. Big Data. **2**(1), 1–41 (2015)
5. Hossain, N., Milajerdi, S.M., Wang, J.: SLEUTH: Real-time attack scenario reconstruction from COTS audit data. In: USENIX Security Symposium, pp. 487–504 (2017)
6. Hossain, M.N., Sheikhi, S., Sekar, R.: Combating dependence explosion in forensic analysis using alternative tag propagation semantics. In: IEEE Symposium on Security and Privacy (2020)
7. Kurniawan, K., Ekelhart, A., Kiesling, E., et al.: KRYSTAL: Knowledge graph based framework for tactical attack discovery in audit data. Comp. Secur. **121**, 102828 (2022)
8. Lee, K.H., Zhang, X., Xu, D.: High accuracy attack provenance via binary-based execution partition. In: NDSS (2013)
9. Ma, S., Zhang, X., Xu, D.: ProTracer: towards practical provenance tracing by alternating between logging and tainting. In: Proceedings 2016 Network and Distributed System Security Symposium (2016)
10. Ma, S., Zhai, J., Zhang, X., et al.: MPI: multiple perspective attack investigation with semantics aware execution partitioning. In: 26th USENIX Security Symposium (2017)
11. Alhanahnah, M., Ma, S., Gehani, A., et al.: autoMPI: automated multiple perspective attack investigation with semantics aware execution partitioning. In: IEEE Transactions on Software Engineering, pp. 2761–2775 (2022)
12. Hassan, W.U., Noureddine, M.A., Datta, P., et al.: OmegaLog: high-fidelity attack investigation via transparent multi-layer log analysis. In: Proceedings 2020 Network and Distributed System Security Symposium (2020)

13. Yu, L., Ma, S., Zhang, Z., et al.: ALchemist: fusing application and audit logs for precise attack provenance without instrumentation. In: Proceedings 2021 Network and Distributed System Security Symposium (2021)
14. Muggleton, S.H., Lin, D., Tamaddoni-Nezhad, A.: Meta-interpretive learning of higher order dyadic datalog: predicate invention revisited. Mach. Learn. 49–73 (2015)
15. Yang, R., Ma, S., Xu, H., et al.: UISCOPE: accurate, instrumentation-free, and visible attack investigation for GUI applications. In: NDSS (2020)
16. Pan, Y., Ge, X., Fang, C., et al.: A systematic literature review of android malware detection using static analysis. In: IEEE Access. **8**, 116363–116379 (2020)
17. Aghakhani, H., Gritti, F., Mecca, F., et al.: When malware is packin'heat; limits of machine learning classifiers based on static analysis features. In: Network and Distributed Systems Security (NDSS) (2020)
18. Mossberg, M., Manzano, F., Hennenfent, E., et al.: Manticore: A user-friendly symbolic execution framework for binaries and smart contracts. In: 2019 34th IEEE/ACM International Conference on Automated Software Engineering (ASE). IEEE, pp. 1186–1189 (2019)
19. He, J., Balunović, M., Ambroladze, N., et al.: Learning to fuzz from symbolic execution with application to smart contracts. In: Proceedings of the 2019 ACM SIGSAC Conference on Computer and Communications Security, pp. 531–548 (2019)
20. Poeplau, S., Francillon, A.: Symbolic execution with SymCC: Don't interpret, compile! In: 29th USENIX Security Symposium (USENIX Security 20), pp. 181–198 (2020)
21. Milajerdi, S.M., et al.: Holmes: real-time apt detection through correlation of suspicious information flows. In: 2019 IEEE Symposium on Security and Privacy (SP). IEEE, pp. 1137–1152 (2019)
22. Rao, A.M., Yadav, T.: Technical aspects of cyber kill chain. In: Security in Computing and Communications: Third International Symposium, pp. 438–452 (2015)
23. Chai, Y., Sun, Z., Qiu, J., et al.: TPRPF: a preserving framework of privacy relations based on adversarial training for texts in big data. In: Frontiers of Computer Science (2022)
24. Tian, Z., Luo, C., Qiu, J., et al.: A distributed deep learning system for web attack detection on edge devices. In: IEEE Transactions on Industrial Informatics, pp. 1963–1971 (2019)
25. Roesch, M., et al.: Snort: Lightweight intrusion detection for networks. In: LISA, pp. 229–238 (1999)
26. Wong, K., Dillabaugh, C., Seddigh, N., et al: Enhancing Suricata intrusion detection system for cyber security in SCADA networks. In: 2017 IEEE 30th Canadian Conference on Electrical and Computer Engineering, pp. 1–5 (2017)
27. Xiong, C., Zhu, T., Dong, W., et al.: A practical real-time APT detection system with high accuracy and efficiency. In: IEEE Transactions on Dependable and Secure Computing, pp. 551–565 (2022)
28. Zhu, T., Yu, J., Chen, T., et al.: APTSHIELD: A Stable, Efficient and Real-time APT Detection System for Linux Hosts. arXiv (2021)
29. Albert, B., Tullis, T.: Measuring the User Experience: Collecting, Analyzing, and Presenting UX Metrics. Morgan Kaufmann (2022)
30. Chen, H., Duncklee, A., Jajodia. S., et al.: PCAM: A data-driven probabilistic cyberalert management framework. ACM Trans. Internet Technol. 1–24 (2022)
31. Wu, M., Moon, Y.: Alert correlation for cyber-manufacturing intrusion detection. In: Proc. Manuf. 820–831 (2019)
32. Dunstatter, N., Tahsini, A., Guirguis, M., et al.: Solving cyber alert allocation Markov games with deep reinforcement learning. In: International Conference on Decision and Game Theory for Security, pp. 164–183 (2019)
33. Werner, G.: R-CAD: rare cyber alert signature relationship extraction through temporal based learning. In: Rochester Institute of Technology (2022)

34. Liza 'Kurtz: https://netenrich.com/blog/security-operations-center-challenges-2021-what-needs-to-change. Nov 2020
35. Foggy: https://baijiahao.baidu.com/s?id=1773807539160400316. Aug 2023

Cybersecurity Future Trends

Detecting Mixing Services in Bitcoin Transactions Using Embedding Feature and Machine Learning

Chuhan Ye[1,2,4], Qilin Li[1,4(✉)], Ruijie Gao[1,2,4], Yi Fu[1,2,4], Peng Wang[1,2,4], Xuanyu Bao[1,2,4], Guopeng Wang[3,4], Yuan Liu[1,3,4], and Zhihong Tian[1,4]

[1] GuangZhou University, GuangZhou 510000, China
liqilinsy@outlook.com
[2] Northeastern University, ShenYang 110000, China
[3] Engineering Research Center of Integration and Application of Digital Learning Technology, Ministry of Education, Beijing 100039, China
[4] Pazhou Lab, Guangzhou 510330, China

Abstract. Due to the intrinsic anonymity of the Bitcoin network, it is frequently subject to money laundering activities, many of which involve coin mixing services. Coin mixing serves as a technique to obscure the trajectory of funds and is inherently dynamic and complex. Such dynamism results in inconsistent transaction behavior across numerous addresses. Moreover, the attributes and patterns associated with coin mixing fluctuate over diverse temporal spans, complicating its detection. To address this challenge, our research introduces a novel classification technique. Initially, we classify coin mixing addresses by their unique transaction behaviors, capturing a variety of patterns. We then determine the influence scope of each address category within its respective subnetwork, leading to the development of dynamic embedding features. These features adapt in real-time, reflecting the behavioral patterns of the current category. In our final step, we use a multi-classifier ensemble strategy, integrating both embedding features and topological structure insights, thereby bolstering classification precision and resilience. Our experimental findings highlight significant improvements across diverse evaluation benchmarks. Of special note is the efficacy resulting from merging dynamic embedding features with the multi-classifier ensemble methodology, achieving an impressive precision rate of up to 0.9747. This study provides fresh perspectives and approaches to mitigate the challenges of money laundering within the Bitcoin network.

Keywords: Bitcoin · Mixing activities · Embedding feature

1 Introduction

Since its inception in 2009, Bitcoin has become a widely recognized digital currency, leading to emerging blockchain applications in healthcare [4] and Internet of Vehicle (IoV) [2, 7]. Its decentralization and anonymity, however, make it attractive for illicit

Z. Gu et al. (Eds.): NSE 2023, CCIS 2063, pp. 91–105, 2024.
https://doi.org/10.1007/978-981-97-4519-7_7

activities [9]. With the surge in Bitcoin's popularity, research on detecting illegal Bitcoin activities has intensified. A study by Foley et al. [3] used machine learning to identify Bitcoin addresses linked to illicit actions, finding a significant portion of Bitcoin transactions associated with illegal activities. Another study by Fergal and Harrigan [5] highlighted the potential for de-anonymizing Bitcoin users through transaction analysis. These findings are vital for strategies to monitor and counteract illicit cryptocurrency activities.

Chainalysis data from 2022 shows that cryptocurrency laundering transactions reached about $23.8 billion USD [6]. In Bitcoin, laundering involves moving and masking illicit funds, making them appear legitimate. Coin mixing services, integral to this process, blend Bitcoin from various sources, complicating the tracing of illicit capital (see Fig. 1). Ethan Heilman et al. [11] introduced TumbleBit, a Bitcoin-compatible payment platform with coin mixing features, highlighting the impact of such services on illegal Bitcoin transactions. Patrick Monamo and his team developed the'trimmed k-means' method for clustering and detecting fraudulent Bitcoin transactions [12]. Stefano Ferretti et al. introduced an algorithm to identify illegal financial activities in Bitcoin, using graphcentric methods to analyze transaction data [13]. Giuseppe Vinci and colleagues employed Graph Convolutional Neural Networks (GCN) for tracing money laundering in Bitcoin, converting transaction records into graphs for analysis [14]. Elliptic Labs used supervised machine learning to differentiate Bitcoin addresses involved in laundering and coin mixing by analyzing transaction datasets [15].

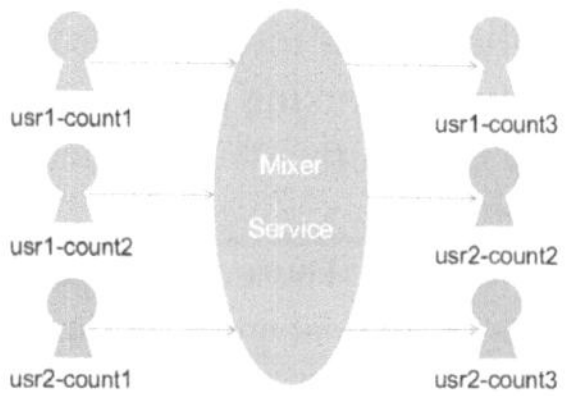

Fig. 1. Illegal mixing activities example

Research on Bitcoin coin mixing faces several challenges. The dynamic nature of coin mixing behaviors leads to inconsistent transaction patterns, limiting the effectiveness of existing detection methods. These methods also risk obsolescence due to the continuous evolution of coin mixing practices. Furthermore, high false positive rates in current solutions often misidentify legitimate transactions as illicit, causing issues for genuine users.

In this study, we introduce a method for identifying coin mixing activities in the Bitcoin network. Our approach involves categorizing coin mixing addresses based on transaction patterns, constructing dynamic embedding features responsive to these patterns, and using a multi-classifier ensemble strategy to improve detection precision. The key contributions of our research are:

– **Adaptive Detection:** We present a dynamic adaptation framework uniquely suited for identifying coin mixing activities. Our model's dynamic adaptability allows it to effectively handle the evolving and diverse nature of coin mixing behaviors. By incorporating centrality and position within the subnetwork, our approach captures varied behavior patterns with improved accuracy.

- **Augmented Precision:** Our model significantly enhances the precision of coin mixing detection. By merging embedding features with topological attributes and employing a multi-classifier ensemble, we achieve higher accuracy than traditional methods. Our experiments with real-world datasets demonstrate the model's effectiveness and practical superiority in detecting coin mixing activities.

2 Related Work

In this section, we explore foundational aspects of blockchain and Bitcoin, focusing on research related to illicit activities in cryptocurrencies, particularly coin mixing and money laundering. We analyze these activities, discussing existing methodologies and their contributions and limitations. Finally, we highlight the unique aspects of our research in this domain.

Blockchain, a distributed database system, uses advanced encryption and decentralization to ensure data integrity [1]. Bitcoin, as the first application of blockchain, is notable for its decentralized, secure transaction system, but its privacy features also facilitate illicit activities like money laundering, drug trafficking, and ransomware [5]. Mixing services, which obfuscate the origins of illicit funds through complex transaction chains, have been part of high-profile cases like BTC-e and Silk Road, underscoring the need for thorough investigation into Bitcoin laundering practices for the evolution of cryptocurrencies.

The study in [16] introduced a heuristic-based transaction address clustering technique, analyzing Bitcoin transaction data to monitor transactions related to illicit activities. Androulaki et al. [17] explored Bitcoin user privacy, developing an advanced clustering mechanism to trace illicit transactions. Akcora et al. [18] employed topology data analytics in the Bitcoin ecosystem to identify ransomware activities, creating a graph-based method applicable to a broader range of illicit activities.

Hao Tian et al. [19] developed a method utilizing features from an address's previous three Bitcoin transactions and an Attention-based Graph Neural Network (AT-GNN) for detecting illicit activities within the Bitcoin network. In their innovative work, Ardeshir Shojaeinasab [36] developed an algorithm to detect modern Bitcoin mixing services with a 100% recall rate. This study, focusing on services like MixTum, Blemder, and CryptoMixer, significantly advances the understanding and monitoring of mixed Bitcoin transactions. Chang Xu and colleagues [37] introduced a novel detection scheme for identifying Bitcoin mixing service addresses. This work utilized a comprehensive dataset and a dual ensemble classification model, significantly advancing the methods for detecting complex Bitcoin mixing patterns. Li Yang et al. [20] explored various feature extraction techniques, including XGBoost and Random Forest, to identify illicit activities, finding tree-centric classifiers more effective than neural networks and SVM. The research in [10] focused on mixing services at network, account, and transaction levels, introducing a time heterogeneous base sequence (AHT) and a PU learning-oriented detection method for imbalanced label scenarios, proving effective in identifying mixing services through time-oriented graph structures.

3 Method

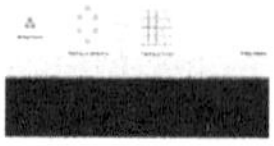

Fig. 2. System framework

In this section, we outline our experimental approach. Section 3.1 discusses dataset sources, feature selection for coin mixing detection, and transaction and topology attributes. Section 3.2 categorizes coin mixing addresses into three types, validated through clustering analysis. Section 3.3 calculates the influence scope of each address category and integrates it with transaction features to develop dynamic embedding characteristics. Section 3.4 describes an ensemble classification method combining dynamic embedding and topological features, as detailed in Sects. 3.3 and 3.1, respectively. The complete experimental framework is depicted in Fig. 2.

3.1 Feature Extraction

In this section, we begin by detailing the origin of our experimental dataset and elucidating the methodology harnessed to derive key features from Bitcoin transactions, targeting the detection of illicit coin mixing activities. We classify the features extracted here into two main categories: transaction features and topology features. These features underpin subsequent analyses presented in Sects. 3.3 and 3.4.

WalletExplorer, a Bitcoin blockchain explorer, employs algorithms to group addresses into single wallets, offering detailed transaction analysis. The dataset from WalletExplorer, used in [10] for coin mixing research, is available at xblock.pro. Our study focuses on Helix and BitcoinFog, platforms that utilized Bitcoin's UTXOs for laundering, blending illicit funds with regular transactions to obscure their trails. Despite their shutdown, these services remain relevant for academic exploration.

We classify the features into two predominant categories: transaction features and topology features. Each of these feature types provides a unique lens through which to view potential coin mixing behaviors in the Bitcoin network. Our initial emphasis is on transaction features, predominantly derived from existing literature. In Table 1, entries numbered 1 to 14 correspond to the extracted transaction features.

- Transaction Amount: Given that illicit coin mixing activities can encompass significant monetary volumes, the transaction amount emerges as an essential metric.
- Transaction Frequency: Illegal coin mixing often entails frequent transactions to obfuscate fund origins.
- Number of Output/Input Addresses: Coin mixing activities might disperse funds across multiple output addresses or aggregate funds from several input addresses.
- Transaction Time: Illegal coin mixing is an organized activity, possibly occurring during specific time intervals.

We now turn our attention to topology attributes, derived from the inherent structure of the Bitcoin transaction network. As presented in Table 1, entries numbered 15 through 18 denote the extracted topology characteristics. The Bitcoin ecosystem manifests as a complex graph structure, with nodes symbolizing Bitcoin addresses and edges reflecting the transactions between them. During the extraction of these topology attributes, our primary emphasis lies on the following facets:

- Clustering Coefficient: This measures a node's clustering level, indicating potential frequent interactions among nearby addresses, often seen in illegal coin mixing behaviors [5].
- Connectivity: Connectivity assesses how interconnected nodes are. High connectivity in the Bitcoin network can suggest areas of frequent transactions, characteristic of illegal coin mixing activities [26].
- Centrality Measures: These metrics, like Closeness and Eigenvector Centrality, help identify key nodes in the network. They are useful for detecting significant players in illegal coin mixing, with Eigenvector Centrality highlighting a node's network influence, as explored in [27].

3.2 Cluster Analysis

In this section, we examine distinct transaction patterns of labeled coin mixing addresses, revealing specific behavioral patterns. Using Principal Component Analysis (PCA) and K-means clustering, we analyze these transaction behaviors to differentiate the roles of each address in the transaction network. We then discuss the interactions and dependencies among these address categories.

We initiated our analysis by applying the transaction attributes of the labeled coin mixing addresses to Principal Component Analysis (PCA) for dimension reduction [28]. This transformation refines high-dimensional transaction attributes into more manageable lower-dimensional feature vectors, thereby illuminating the principal axes of data variance [29]. Following this, the K-means clustering algorithm [30] was deployed on the dimension optimized data for a more granular cluster examination. The resulting segmentation delineates the coin mixing addresses into two pronounced categories, which we refer to as Category One and Category Two, as illustrated in Fig. 3(a). Such an initial stratification provides a foundational understanding of the role distribution among the coin mixing addresses. Building on our initial analysis, we delved into addresses that partake in direct transactions with identified coin mixing addresses. Intriguingly, we discerned a particular group of addresses that, although not explicitly labeled as coin mixing entities, displayed a strong affinity to them and manifested parallel transaction behaviors. We designated this distinct group as the third address category. To corroborate this newfound relationship, we invoked the same clustering technique to categorize these yet-to-be-labeled addresses, as showcased in Fig. 3(b). Upon completion of our clustering, three distinct address categories emerged: the initially labeled Category One and Category Two, along with the newly discovered Category Three. The transaction behaviors across these categories are illustrated in Fig. 4. Category One mainly deals with "client" addresses, transferring funds to both Category Two and Category Three.

Table 1. Transaction and Topology Feature

Index	Feature Category	Notations	Feature Description
1	Transaction Feature	$\|\cdot t\|$	Initial Transaction Timestamp [21–23]
2		$\|t \cdot\|$	Final Transaction Timestamp [21–23]
3		txsend	Count of transactions where the account acts as the sender [21–24]
4		txreceipt	Count of transactions where the account acts as the recipient [21–24]
5		$Unique_{\text{send}}$	Count of distinct addresses that have sent to the account [21, 22, 25]
6		$Unique$receipt	Count of distinct addresses from which the account has received [21, 22, 25]
7		Avgsend	Average transaction outlay [22–24]
8		Avgreceipt	Average transaction inflow [22–24]
9		Maxin	Maximum quantity of Bitcoin received in a transaction [24]
10		Minin	Minimum quantity of Bitcoin received in a transaction [24]
11		Maxout	Maximum quantity of Bitcoin spent in a transaction [24]
12		Minout	Minimum quantity of Bitcoin spent in a transaction [24]
13		$balance$	Current account balance
14		numtxin	Average count of addresses utilized as transaction inputs
15	Topology Feature	Ci	Clustering coefficient for address i [5]
16		$K^{(i)}$	Connectivity measure for address i [26]
17		Ei	Eigenvector Centrality for address i [27]
18		ci	Closeness Centrality for address i [27]

Category Two mostly receives funds from Category One and forwards them to Category Three. Significantly, Category Three amasses funds from both Category One and Category Two before initiating outward transfers.

3.3 Feature Embedding

In this section, we assess the influence range of each illicit address category within its subgraph, as outlined in Sect. 3.2. We then utilize this influence range to extract topology and transaction attributes for embedding from each address category, leading to the development of dynamic embedding features.

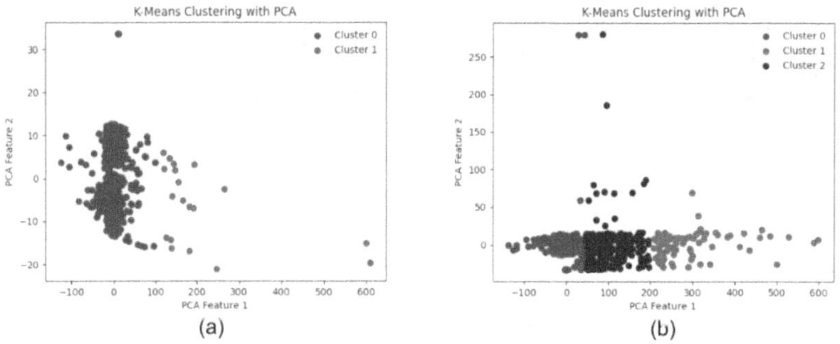

Fig. 3. The clustering effectiveness graphs

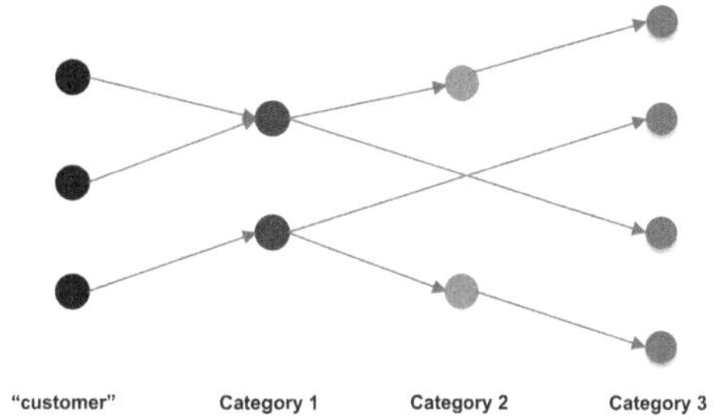

Fig. 4. The relationship between three categories of addresses

Scope Calculation. To ascertain the influence range of nodes within the subgraph, we invoked the principle of Eigenvector Centrality, a foundational metric in network analysis, as originally posited by [31]. This metric quantifies a node's significance within a network. Distinct from Degree Centrality, which solely assesses a node's immediate connections, Eigenvector Centrality evaluates the prominence of a node's adjacent nodes as well. The rationale behind computing Eigenvector Centrality rests on the premise that the significance of a node is intrinsically linked to the importance of its connected nodes. Consequently, a node connected to several influential nodes is deemed important in its own right. The derivation of Eigenvector Centrality engages the adjacency matrix of the network and the eigenvector associated with its paramount eigenvalue. Specifically, given an adjacency matrix A for a network, the foremost eigenvalue is denoted as λ, with its corresponding eigenvector represented by v. Each component of v delineates the eigenvector centrality of individual nodes within the network. This computational relationship is encapsulated by Eq. (1):

$$C_e(v_i) = \frac{1}{\lambda} \sum_{j=1}^{n} a_{ij} C_e(v_j) \quad (1)$$

When calculating the influence of a node within a subgraph, we select a subgraph size that includes the node of interest at the center and extends to a distance of 3 hops forward and 3 hops backward. This choice is made because the scale of addresses involved becomes excessively large beyond 4 hops, and the influence beyond 4 hops is relatively limited. Therefore, we define the maximum range, denoted as n, to be 3. After determining the subgraph, it is necessary to compute the centrality of all nodes' feature vectors within the subgraph. These centrality values are then sorted in ascending order. Let $C_e(v_i)$ represent the centrality of the feature vector corresponding to node v_i. The number of hops between the centrality-ranked node i and the node of interest is denoted as j, and the order of node i in the sorted list of feature vector centralities within the subgraph is denoted as $Rank(C_e(v_i))$, $|J(i, j)|$ represents the number of elements in the set $J(i, j)$, which is the count of nodes at a distance of j hops from node i. The influence of a node within the subgraph is defined as shown in Eq. (2–5):

$$Influence_{send} = \sum_{j=1}^{n} acc(x_i(j)) \tag{2}$$

$$Influence_{receipt} = \sum_{j=-1}^{-n} cc(x_i(j)) \tag{3}$$

$$acc(x_i(j)) = \begin{cases} 1 & x_i(j) < 1 \\ \frac{x_i(j)}{j} & x_i(j) \geq 1 \end{cases} \tag{4}$$

$$x_i(j) = \frac{\frac{1}{|J(i,j)|} \sum_{k \in J(i,j)} Rank(Ce(v_k)) - Rank(Ce(v_i))}{j+1} \tag{5}$$

The *Influence* values are derived using the formula presented in Eq. (6). Here, I_j denotes the *Influence* value computed up to the jth hop. Upon determining the influence scope for each coin mixing address within the subgraph, we ascertain the average influence scope for each category based on their classification. The computational outcomes indicate that the influence scope for both Category One and Category Two addresses encompasses transactions extending one layer forward and one layer backward. In contrast, Category Three addresses exert an influence that spans two layers backward and projects one layer forward.

$$Influence(j)_i = \begin{cases} 1, \ if \ J = 1 \\ \min(I_2, 2) \ if \ J = 2 \\ \min(I_3, 3) \ if \ J = 3 \end{cases} \tag{6}$$

Embedding. In this subsection, we integrate the influence scope sizes, determined in the Scope Calculation, as topology structural information. This is paired with the transaction features of addresses for embedding purposes. As a result, each address category acquires a dynamic embedding feature, amalgamating both topology structural and transaction feature data.

The process of embedding derives inspiration from the method delineated in [19]. In the given context, v signifies the feature values for every layer. w stands for the transaction amount transpiring between two distinct addresses, while f_i symbolizes the feature value of a specific address. Lastly, j denotes the count of addresses present within that layer. The feature v for each individual layer can be articulated as:

$$v_i = \frac{\sum_{j=1}^{k} w_j f_i}{\sum_{j=1}^{k} w_j} \tag{7}$$

As depicted in Fig. 4, consider address c representative of Category One addresses. The influence of address c permeates to the preceding transaction layer S_{-1} and the subsequent transaction layer S_1. Each stratum of addresses, indicated by f_i, contributes to the calculation of the specific feature vector v_i for that layer. For instance, the evaluation of layer S_{-1} yields a feature vector contingent on the distinct f_i attributes within that layer. Analogously, layer S_1 engenders its respective feature vector. Layer S_0, signifying the target address itself, possesses its intrinsic feature vector, expressed as f_i.

The inherent feature vector of address c is represented as $\mathbf{f} = [f_1, f_2, ..., f_k]^T$. Concurrently, the feature vector corresponding to the preceding transaction is denoted as $\mathbf{v}^{-1} = [v_1^{-1}, v_2^{-1}, \ldots v_i^{-1}]^T$ while the one representing the subsequent transaction is symbolized as $\mathbf{v}^1 = [v_1^1, v_2^1, \ldots v_i^1]^T$. By integrating these vectors, the composite feature vector of address c emerges as:

$$\mathbf{x} = [\mathbf{v}^{-1^T}, \mathbf{f}^T \mathbf{v}^{1^T}] \tag{8}$$

3.4 Classification

In this section, we present our approach to classifying mixed addresses.

For the classification of mixed addresses, we adopt a dual-phase ensemble learning technique. In the initial phase, we employ the embedding features derived from different mixed address categories. The embedding features from both mixed and regular addresses are inputted into three classifiers: Random Forest [32], XGBoost [33], and Convolutional Neural Network (CNN) [34]. If at least two of these classifiers label an address as "mixed", it is designated accordingly. Let's denote the set of embedding features for mixed addresses as E_{mix} and for regular addresses as E_{normal}. The classifiers in this phase are represented as: R_{mix} for Random Forest, X_{mix} for XGBoost, and C_{mix} for CNN. The criterion for classifying an address as mixed during this phase is depicted by Equation (9): if address $addr_i$ fulfills the given condition, it's classified as a mixed address in this phase.

$$R_{mix}(E_{mix}(addr_i)) + X_{mix}(E_{mix}(addr_i)) + C_{mix}(E_{mix}(addr_i)) \geq 2 \tag{9}$$

During the second phase, our emphasis shifts to topology features, which exclusively account for the structural interrelations among addresses. Once again, we deploy the three classifiers from the initial phase to process the topology features associated with both mixed and regular addresses. If any one of these classifiers identifies an address as "mixed", it is then labeled accordingly in this phase. Let's use T_{mix} to signify the set of topology features for mixed addresses, and T_{normal} for the set pertaining to regular addresses. The classifiers in this phase are represented as: R_{top} for Random Forest, X_{top}

for XGBoost, and C_{top} for CNN. The criterion for classifying an address as mixed during this phase is illustrated by Eq. (10): an address $addr_i$ is classified as a mixed address in this phase if it fulfills the stated condition.

$$R_{top}(T_{mix}(addr_i)) + X_{top}(T_{mix}(addr_i)) + C_{top}(T_{mix}(addr_i)) \geq 1 \tag{10}$$

To ascertain the final categorization of a mixed address, it's crucial to amalgamate the outcomes from the two classification rounds. An address is conclusively deemed as a mixed address only if it is designated as such in both rounds. This means a unanimous agreement across both classifications is mandatory for an address to be conclusively labeled as mixed. Our ensemble-based classification method combines both embedding and topology features to effectively identify mixed addresses. Such an integrated approach bolsters the precision and robustness of our classification.

4 Results

In this section, we undertake a thorough evaluation of our framework designed to detect illicit mixing activities. Our primary objective is to showcase the effectiveness of classifiers trained on dynamic embedding and topology features in pinpointing mixed addresses. We initiate with an overview of our experimental setup, followed by an in-depth presentation of the obtained results. Concluding the section, we provide a multifaceted analysis of the experimental outcomes and assess the framework's overarching performance.

Our experimental dataset is derived from the Bitcoin transaction records available at Xblock.pro. We applied classification to addresses that had been previously identified as mixed, as outlined in Sect. 3.2. This exercise aimed to ascertain the roles these addresses assume in mixing activities and to identify addresses with strong affiliations. This exercise broadened the label spectrum, leading to the formation of a more intricate transaction subnetwork. We zeroed in on a segment from the 2016 Helix dataset, a timeframe marked by heightened activities of the illicit mixing entity, Helix. Our analysis encompassed a total of 2,506,794 addresses, which included 3,780 pre-labeled ones. Through subsequent enrichment, we identified 5,149 active trading addresses.

In our experiments, we classified the addresses into three distinct categories:

Category One addresses, which primarily aggregate customer funds;
Category Two addresses, which predominantly accept minor funds from Category One addresses and transfer substantial amounts to Category Three addresses;
Category Three addresses, functioning akin to a "bank", accumulating funds from both Category One and Category Two addresses and overseeing significant fund transfers.

We ascertained the influence range for each category. Specifically, both Category One and Category Two addresses exhibited an influence span of one hop in both forward and backward directions. Conversely, Category Three addresses presented an influence scope extending one hop forward and two hops backward. By amalgamating the influence scope of each address type with their respective transaction attributes, we derived dynamic embedding features for each category. Concurrently, we harnessed the network structure to procure topology features. Prior to classification, we segregated our dataset into distinct training and testing subsets. The training subset facilitated the training of our classification models, while the testing subset validated the efficacy of these models. Recognizing the complexities of handling addresses like those in Category Three, which have a broad influence scope and necessitate embedding across an extensive range of addresses, we adopted Generative Adversarial Networks (GANs) [35]. Specifically, we used GANs to produce embedding features for regular addresses during the learning and classification phases related to Category Three address features. This strategy with GANs effectively navigated the intricacies associated with the vast influence scope of Category Three addresses, bolstering both the robustness and efficiency of our classification methodology.

Subsequently, we delved into a comprehensive presentation of the experimental outcomes derived from our methodology. We benchmarked our technique against the Temporal Motifs and Hybrid Motifs methods, as outlined in [10]. For performance evaluation, we adopted TPR (True Positive Rate), FPR (False Positive Rate), and G-Means as our primary metrics. G-Means is notably adept at addressing imbalanced datasets and is computed as shown in Eq. (11). This metric quantifies the geometric mean of TPR and the inverse of FPR, encapsulating the equilibrium between sensitivity and specificity. A classifier that adeptly differentiates between samples from varying classes will register a higher G-Means value.

$$G - Means = \sqrt{TPR(1 - FPR)} \tag{11}$$

Table2 through the amalgamation of offers a thorough examination of the classification accuracy and performance, with a specific emphasis on mixing addresses. Additionally, the efficacy of employing ensemble learning for this endeavor is scrutinized. Subsequent sections will delve into a comprehensive analysis and elucidation of these findings. In our assessment of the classification performance on the Helix2016 dataset, we drew comparisons with the Temporal Motifs and Hybrid Motifs methodologies. Our evaluation juxtaposed the sole use of embedding features against a single classifier leveraging both embedding and topology features. The findings unequivocally illustrate that our method, which harnesses embedding features, stands out in terms of classification accuracy, underscoring the potency of dynamic embedding features in detecting illicit mixing activities. When integrating both embedding and topology features within a singular classifier framework, the outcomes showcased exemplary performance. The True Positive Rate (TPR) soared to a peak of 0.9591, concurrently minimizing the False Positive Rate (FPR) to a mere 0.0338. These metrics culminated in a G-Means score of 0.9624, reflecting the harmonized efficacy of our model in balancing sensitivity with specificity. Furthermore, the integration of topology features markedly curtailed the false positive occurrences, as manifested by the diminished FPR values. This progress testifies to the considerable strides our method has made in elevating the overall performance.

Additionally, we undertook an examination of the ensemble learning strategy, which is delineated in Table 3. Through the amalgamation of outcomes from the bi-stage classification, our method yielded heightened precision in the classification of mixing addresses. Post the incorporation of ensemble learning, leveraging a multitude of classifiers, the metrics showcased a TPR surging to 0.9747, an FPR plummeting to 0.0114, and a G-Means score soaring to 0.9816. These findings underscore the premise that harnessing the collective capabilities of several classifiers via ensemble techniques can markedly

Table 2. The results of the classification experiments

Dataset	Feature&Method	TPR	FPR	G-Means
	Basic Features&Temporal motifs [10]	0.9335	0.0508	0.9413
	Basic Features&Hybrid motifs [10]	0.9318	0.0356	0.9479
	Embedding Feature&RF	0.9334	0.0523	0.9405
	Embedding Feature&XGB	0.9368	0.3968	0.7517
Helix2016	Embedding Feature&CNN	0.9252	0.0426	0.9417
	Embedding + topology Feature&RF	0.9591	0.0172	0.9709
	Embedding + topology Feature&XGB	0.9542	0.3096	0.8117
	Embedding + topology Feature&CNN	0.9587	0.0124	0.9730
	Embedding + topology Feature&multi	**0.9747**	**0.0114**	**0.9816**

curtail the false positive rate and bolster the precision in pinpointing mixing addresses. Additionally, we embarked on an examination of the outcomes derived from varying ranges of embedding features. For this evaluation, we exclusively utilized embedding features paired with a singular classifier, juxtaposing fixed-dimensional embeddings against dynamic-dimensional ones. The findings are delineated in Table 3 utilizing the Random Forest classifier with an influence span encompassing one hop forward and one hop backward in transactions, the G-Means receded to 0.9303. Scrutinizing the results, it became evident that this marked decrement in accuracy was attributed to a conspicuous dip in the identification prowess for the third category of addresses, culminating in a comprehensive decline in precision. In a similar vein, when the influence span spanned two hops both forward and backward in transactions, a pronounced dip in accuracy was discernible. This underscores the imperative of preserving dynamic embedding features to ensure optimal accuracy in the detection of mixing conduct.

Table 3. Classification results using embedded features of different dimensions

Classifier	Feature range	TPR	FPR	G-Means
	[−1,1]	0.9286	0.0679	0.9303
RandomForrest	[−2,2]	0.9162	0.0577	0.9291
	Dynamic	0.9334	0.0523	0.9405

5 Conclusion

In this research, we tackled illegal coin mixing in the Bitcoin network by integrating dynamic embedding features with topological characteristics. We categorized coin mixing addresses into three distinct roles, enhancing classification accuracy with ensemble learning techniques. This approach led to significant improvements in identifying coin mixing addresses.

However, our study faced limitations. The use of GANs for embedding feature generation of regular addresses, while reducing computational load, may have compromised feature precision. Additionally, our dataset selection may introduce sampling bias, affecting the universality of our results.

Future work should focus on refining embedding feature generation for greater accuracy and reliability. Expanding the dataset to cover more timeframes and transaction behaviors will improve the model's applicability. Additionally, exploring new feature extraction methods and classifiers could further advance illegal coin mixing detection, aiming for a robust and practical solution.

Acknowledgment. This work is supported by National Key Research and Development Plan Program of China Grant No. 2022YFB3102700, and National Natural Science Foundation of China under Grant No.62172085, No.62172115, No.62172353, No.U20B2046, Guangdong Basic and Applied Basic Research Foundation No.2020A1515010450 and No. 202102020867 Basic Research Program Cofunded by Guangzhou City and Guangzhou University No. 202102010445. Joint Research Fund of Guangzhou and University under Grant No. 202201020380 and Guangdong Higher Education Innovation Group No.2020KCXTD007. Cultivation Project of PZL No. PZL2022KF0013.

References

1. Nakamoto. Bitcoin: a peer-to-peer electronic cash system (2008)
2. Liu, Y., Wangyuan, Y., Ai, Z., Guangxia, X., Zhao, L., Tian, Z.: A Blockchain-empowered federated learning in healthcare-based cyber physical systems. IEEE Trans. Netw. Sci. Eng. **10**(5), 2685–2696 (2023)
3. Foley, S., Karlsen, J.R., Putninš, T.J.: Sex, drugs, and bitcoin: How much illegal activity is financed through cryptocurren cies? Rev. Fin. Stud. **32**(5), 1798–1853 (2019)
4. Liu, Y., Chuang Zhang, Y., Yan, X.Z., Tian, Z., Zhang, J.: A semi-centralized trust management model based on blockchain for data exchange in IoT system. IEEE Trans. Serv. Comput. **16**(2), 858–871 (2023)
5. Reid, F., Harrigan, M.: An Analysis of Anonymity in the Bitcoin System. Springer (2013)
6. Grauer, E.J.K.: The 2023 Crypto Crime Report (2023)
7. Liu, Y., et al.: VRepChain: a decentralized and privacy-preserving reputation system for social internet of vehicles based on Blockchain. IEEE Trans. Veh. Technol. **71**(12), 13242–13253 (2022)
8. Li, S., Jiang, L., Zhang, Q., Wang, Z., Tian, Z., Guizani, M.: A Malicious mining code detection method based on multi-features fusion. In: IEEE Transactions on Networking Science and Engineering (2023)
9. Wang, Z., et al.: A covert channel over blockchain based on label tree without long waiting times. Comput. Netw. **232**, 109843 (2023)
10. Wu, J., Liu, J., Chen, W., Huang, H., Zheng, Z., Zhang, Y.: Detecting mixing services via mining bitcoin transaction network with hybrid motifs. In: IEEE Transactions on Systems, Man, and Cybernetics (2022)
11. Heilman, E., Alshenibr, L., Baldimtsi, F., Scafuro, A., Goldberg, S.: Tumblebit: An untrusted bitcoin-compatible anonymous payment hub. Cryptology ePrint Archive, Paper 2016/575 (2016)

12. Monamo, P., Marivate, V., Twala, B.: Unsupervised learning for robust bitcoin fraud detection. In: 2016 Information Security for South Africa (ISSA), pp. 129–134. IEEE (2016)
13. Yang, M., Wei, S., Liu, M., Huang, R., Yu, M., Chen, F.: Design verification and test techniques for UHF RFID tagic. In: 2019 IEEE Sustainable Power and Energy Conference (iSPEC), pp. 2500–2505 (2019)
14. Weber, M., et al.: Anti-money Laundering in Bitcoin: Experimenting with Graph Convolutional Networks for Financial Forensics (2019)
15. Harlev, M.A., Yin, H.S., Langenheldt, K.C., Mukkamala, R.R., Vatrapu, R.: Breaking Bad: De-anonymising Entity Types on the Bitcoin Blockchain Using Supervised Machine Learning, pp. 3497–3506 (2018)
16. Meiklejohn, S., Pomarole, M., Jordan, G., Levchenko, K., McCoy, D.,Voelker, G.M., Savage, S.: A fistful of bitcoins: characterizing payments among men with no names. In: Proceedings of the 2013 Conference on Internet Measurement Conference, pp. 127–140 (2013)
17. Androulaki, E., Karame, G.O., Roeschlin, M., Scherer, T., Capkun, S.: Evaluating user privacy in bitcoin. In: Financial Cryptography and Data Security: 17th International Conference, FC2013, Okinawa, Japan, 1–5 Apr 2013, Revised Selected Papers 17, pp. 34–51. Springer (2013)
18. Akcora, C.G., Li, Y., Gel, Y.G., Kantarcioglu, M.: Bitcoin-heist: topological data analysis for ransomware detection on the bitcoin blockchain. arXiv preprint arXiv:1906.07852 (2019)
19. Tian, H., Li, Y., Cai, Y., Shi, X., Zheng, Z.: Attention based graph neural network for identifying illegal bitcoin addresses. In: Dai, H.-N., Liu, X., Luo, D.X., Xiao, J., Chen, X. (eds) Blockchain and Trustworthy Systems, pp. 147–162. Singapore (2021)
20. Li, Y., Yue, C., Hao, T, Xue, G., Zheng, Z.: Identifying illegal addresses in bitcoin network. Blockchain and Trustworthy Systems, Singapore (2020)
21. Yan, W., Luo, A., Dianxiang, X.: Identifying suspicious addresses in bitcoin thefts. Digit. Investig. **31**, 200895 (2019)
22. Agarwal, R., Barve, S., Shukla, S.K.: Detecting malicious accounts in permission less blockchains using temporal graph properties (2021)
23. Kumar, N., Singh, A., Handa, A., Shukla, S.K.: Detecting malicious accounts on the ethereum blockchain with supervised learning. In: Cyber Security Cryptography and Machine Learning: Fourth International Symposium, CSCML 2020, Be'er Sheva, Israel, 2–3 Jul (2020)
24. Farrugia, S., Ellul, J., Azzopardi, G.: Detection of illegal accounts over the ethereum blockchain. Expert Syst. Appl. **150**, 113318 (2020)
25. Nerurkar, P., Bhirud, S., Patel, D., Ludinard, R., Busnel, Y., Kumari, S.: Supervised learning model for identifying illegal activities in bitcoin. Appl. Intell. **51**, 3824–3843 (2021)
26. Ron, D., Shamir, A.:. Quantitative analysis of the full bitcoin transaction graph. In: Financial Cryptography and Data Security: 17th International Conference, Okinawa, Japan, 1–5 Apr, Revised Selected Papers 17, pp. 6–24. Springer (2013)
27. Maesa, D.D.F., Marino, A., Ricci, L.: Uncovering the bitcoin blockchain: an analysis of the full users graph. In: 2016 IEEE International Conference on Data Science and Advanced Analytics (DSAA), pp. 537–546 (2016)
28. Jolliffe, I.T., Cadima, J.: Principal component analysis: a review and recent developments. Philos. Trans. Roy. Soc. A: Math. Phys. Eng. Sci. **374**(2065), 20150202 (2016)
29. Abdi, H., Williams, L.J.: Principal component analysis. Wiley interdisciplinary reviews. Comput. Stat. **2**(4), 433–459 (2010)
30. Lloyd, S.: Least squares quantization in PCM. IEEE Trans. Inf. Theory **28**(2), 129–137 (1982)
31. Newman, M.: Networks: An Introduction, vol. 3. Oxford University Press (2010)
32. Breiman, L.: Random forests. Mach. Learn. **45**, 5–32 (2001)
33. Chen, T., Guestrin, C.: Xgboost: a scalable tree boostingsystem. In: Proceedings of the 22nd ACM SIGKDD International Conference on Knowledge Discovery and Data Mining, pp. 785–794 (2016)

34. Krizhevsky, A., Sutskever, I., Hinton, G.E.: Imagenet classification with deep convolutional neural networks. Communications of the ACM (2017)
35. Goodfellow, I.J., et al.: Generative Adversarial Networks (2014)
36. Shojaeinasab, A., Motamed, A.P., Bahrak, B.: Mixing detection on Bitcoin transactions using statistical patterns. IET Blockchain (2023)
37. Chang, X., Xiong, R., Shen, X., Zhu, L., Zhang, X.: How to find a Bitcoin mixer: a dual ensemble model for bitcoin mixing service detection. IEEE Internet Things J. **10**(19), 17220–17230 (2023)

Network Security Situational Assessment Based on the ATT&CK Tactics Framework and Transformer Model

Zifeng Zhu[1], Qi An[1], Shudong Li[1,2(✉)], Weihong han[2], Shumei Li[3], and Xiaobo Wu[1]

[1] Cyberspace Institute of Advanced Technology, Guangzhou University, Guangzhou 510006, China
lishudong@gzhu.edu.cn

[2] Department of New Networks, Peng Cheng Laboratory, Shenzhen 518000, China
hanwh@pcl.ac.cn

[3] Computer College, Jilin Normal University, Siping 136000, China
lishumeiwei@sina.com

Abstract. This paper proposes a network security situational assessment method based on the ATT&CK tactics framework and the Transformer model, aiming to provide a comprehensive and effective way to analyze and evaluate security threats and risks in networks. Traditional methods of cybersecurity situational assessment use expert indicator systems to evaluate the security situation in networks based on rules and functions. However, their effectiveness is limited by static indicator systems and function fitting based on expert experience. However, traditional methods have become less flexible and comprehensive as threats continue to evolve and new attack techniques emerge. Therefore, to address these limitations, we introduce the ATT&CK tactics framework as a key reference for cybersecurity situational assessment. The ATT&CK tactics framework provides a comprehensive and structured threat knowledge base for describing and categorizing attacker tactics, techniques, and procedures. At the same time, we leverage the powerful capabilities of the Transformer model to handle complex information in cybersecurity situational assessment. Compared to traditional sequential models such as RNN and LSTM, the Transformer model can better extract the correlation of different time network data. Its Self-Attention mechanism and parallel computing capabilities enable it to perform fine-grained modeling and analysis on sequential data. During the experimental evaluation, we conducted testing and validation of the developed cybersecurity situational assessment model. The experimental results demonstrate that the proposed model is more accurate and comprehensive than traditional methods. Additionally, we demonstrate the applicability and flexibility of the model in different datasets.On two different datasets, precision rates of 96.77% and 94.65% were achieved respectively.

Keywords: ATT&CK Tactics framework · Network security · Situational assessment · Transformer · Attention

Z. Gu et al. (Eds.): NSE 2023, CCIS 2063, pp. 106–119, 2024.
https://doi.org/10.1007/978-981-97-4519-7_8

1 Background

With the popularization and industrialization of the Internet, as well as the development and widespread use of computer technology, security issues have become increasingly important in our daily lives and business activities. Network security technology is an inevitable product in the rapidly developing background of the Internet and information technology. It plays a crucial role in protecting data and user information, ensuring network security, promoting network development, and maintaining network stability. Due to the regional and global nature of the Internet, network security faces various threats, thus finding a new type of network security technology is an urgent problem we need to address. To address the limitations of traditional network defense techniques, such as intrusion detection [1, 2] and vulnerability scanning [3], which only offer passive defense, the emergence of network security situational awareness technology has provided a solution. This technology enables network security managers to proactively monitor and assess the security status of their networks, allowing them to take preventive measures before an attack occurs and network resources are compromised. [4]

Generally speaking, existing network security situational assessment technologies can be categorized into three main groups: assessment methods using mathematical models, assessment methods using knowledge reasoning, and assessment methods using pattern recognition. However, due to the increasing complexity of network environments and the diverse security requirements in different scenarios, choosing a standardized and scalable metric system has become a challenging task. At the same time, situational assessment data, such as network traffic and host logs, are temporal in nature, which poses significant challenges to existing network security situational assessment methods.

On one hand, assessment methods based on mathematical models not only require well-designed functions but also a robust indicator system. Meanwhile, assessment methods based on knowledge reasoning are overly reliant on prior knowledge, which greatly affects their effectiveness in increasingly complex network environments. On the flip side, pattern recognition-based assessment methods, particularly the impressive machine learning models developed in recent years, have solved practical problems in the field of cybersecurity, such as malicious code detection [5, 6], malware detection [7, 8], and APT(Advanced Persistent Threat) mixed intelligence [9]. However, they fail to fully address the issues of generalization and temporal logic in situational assessment. Therefore, current methods cannot fully achieve cybersecurity situational assessment.

The ATT&CK tactics model provides a standardized and highly scalable template for an indicator system, which has become an essential tool and frame-work for threat analysis and management in many organizations and enterprises. Deep learning models, due to their strong representation capabilities for high-dimensional complex data, offer potential solutions to address temporal data problems. To our knowledge, existing deep learning-based methods for cyber-security situational assessment largely rely on recurrent neural network (RNN) methods or their variants when handling temporal data, and they have achieved good results. However, they tend to suffer from the issues of vanishing or exploding gradients when dealing with long sequence lengths, and they struggle to effectively extract the correlation of different time network data. Additionally, they require step-by-step recursive computation of hidden states, making it challenging to perform parallel computing.

In this article, we propose utilizing the ATT&CK framework as a metric system for network security situational assessment, and then employ an optimized Transformer model to handle and learn temporal data, thereby enhancing the accuracy and robustness of the network situational assessment method. Therefore, the main contributions of this paper can be summarized in the following two aspects:

1. Existing network security situational assessment methods lack a unified and mature framework for metric systems. Most of them are based on existing network topology and expert experience, which lack robustness and struggle to achieve good results in different network environments. To address this issue, we adopt the ATT&CK framework as a metric system for situational assessment. Compared to traditional metric systems, the ATT&CK framework can be implemented in different scenarios and achieve precise experimental results.
2. Current deep learning models, such as RNN and LSTM, have notable limitations in handling temporal data. They encounter issues like vanishing or exploding gradients when dealing with long time sequences, struggle to extract correlations among different time network data, and lack parallel computing due to sequential hidden state computation. In contrast, the Transformer model effectively addresses long-term dependency problems, extracts correlations among different time network data, and achieves faster computation through multi-head attention's parallel processing. Our shrunk Transformer model enables better analysis of temporal data for situational assessment, yielding the current situational assessment value.

2 Related work

2.1 Methods Based on Mathematical Models

The method first collect network security situational elements, then aggregate various situational elements, and finally construct functions to evaluate and obtain situational results. Due to the multitude of factors influencing changes in network security status and the complexity of their dynamic relationships, the construction of evaluation functions is crucial.

Leau et al. [10] have suggested a cost-sensitive entropy-based NSSA model which employs AHP to compute the weights of all network assets. Yuan et al. [11] introduced a new AHP-based NSSA method that effectively improves accuracy. Wang et al. [12] developed a hierarchical NSSA method based on AHP, constructing a hierarchical network threat situation assessment (CSA) index system and obtains network threat weights using AHP. Wang et al. [13] designed a simplified NSSA model that integrates AHP and hierarchical NSSA techniques. Li et al. [14] employed clustering to process data, obtained optimal cluster centers and clusters using fuzzy best clustering criterion, and constructed an NSSA model with Analytic Hierarchy Process to achieve the final results.

2.2 Methods Based on Knowledge Reasoning

The method establishes a network security situation assessment model based on certain rules and existing experiential knowledge, and applies logical reasoning theory to network security situation assessment. By selecting and applying relevant prior knowledge, it can effectively address challenging situations in mathematical models. This method exhibits a certain level of intelligence and objectivity.

J. Yu et al. [15] and Li.X et al. [16] initially determine various logical relationships in the network environment through D-S evidence theory and generate probability distributions using certain methods. Then, security events reported by network entities (i.e., evidence) are synthesized according to rules to obtain new probability distributions, forming alternative propositions. Finally, by continuously repeating the previous process based on the new security events, network risks are assessed when proposition confidence exceeds the threshold. Xinlin Liu et al. [17] propose a method for hierarchical evaluation of threats based on hidden Markov models. They quantify the security status by combining network threat types and business vulnerability information.

2.3 Methods Based on Pattern Recognition

With the development of artificial intelligence, many researchers have applied pattern recognition methods to situation assessment. This method primarily draws on machine learning theories and relies less on expert knowledge and experience, demonstrating strong learning capability. By comparing the training sample data of the security assessment model with real network data, a certain level of matching is achieved, resulting in the identification of relevant types.

R. Zhang et al. [18, 19] propose a NSSA model based on whale optimization algorithm (WOA)-optimized support vector machine (SVM) and a NSSA method based on SAA-SSA-BPNN to improve accuracy and convergence. Both methods utilize situation weekly report data from the National Internet Emergency Response Center for evaluation. Zhao et al. [20] have addressed the limitations of convolutional neural networks and utilized them to accomplish network security situation assessment.

Tao X et al. [21] use stacked autoencoders and backpropagation neural networks for network security condition evaluation. In the following year, they [22] design a security assessment method based on autoencoders and a simplified storage unit PMU to improve accuracy and efficiency. Xiangyan Tang et al. [23] use deep learning methods such as LSTM to establish models for evaluating the security status against DDoS attack traffic.

3 Technical Introduction

3.1 ATT&CK Tactics Framework

The ATT&CK Tactics Framework is a comprehensive security knowledge base that focuses on attack behaviors, tactics, and techniques (APT). Its purpose is to provide a framework for understanding APT activities and how to mitigate APT threats. The ATT&CK Tactics Framework describes multiple stages from the initial breach to malicious activities and documents various techniques and tools that may be employed at each stage.

The ATT&CK Tactics Framework is primarily divided into two parts: Tactics and Techniques. Tactics encompass 14 common patterns of attack behaviors, such as Initial Access, Persistence, Lateral Movement, Data Exfiltration, and more. Each tactic can be achieved through different techniques and tools. The Techniques section covers almost all tools and techniques associated with attacks, including malware, reconnaissance tools, phishing techniques, and more. Specifically, as shown in Fig. 1.

Fig. 1. ATT&CK Tactical Matrix Diagram.

3.2 Transformer Model

The Transformer is an attention-based sequence model, used for solving sequence-to-sequence transformation problems, especially in the field of NLP, where it performs well in tasks such as translation. The overall schematic diagram is shown in Fig. 2.Here are the basic principles and formula explanations of the Transformer model:

Attention mechanism: The most critical part of the Transformer is the attention mechanism that assigns weights to each input element for enable selective "attention" to different elements. The Transformer model has two main types of attention mechanisms. The formula as follows:

$$Attention(Q, K, V) = softmax(\frac{QK^T}{\sqrt{d_k}}) \tag{1}$$

Among them, Q denotes the query vector, K denotes the key vector, V denotes the value vector, and d_k denotes the dimension of the key vector.

In the multi-head mechanism, the inputs x are processed individually by multiple heads. Therefore, we have:

$$\text{MultiHeadAttention}(Q, K, V) = \text{Concat}(\text{head}_1, \text{head}_2, ..., \text{head}_h)\text{W}_O \tag{2}$$

In this context, w represents the layer parameter matrix after head expansion, and $head_i$ denotes the output of the number i attention head., and the calculation formula is:

$$\text{head}_i = Attention(QW_i^Q, KW_i^K, VW_i^V) \tag{3}$$

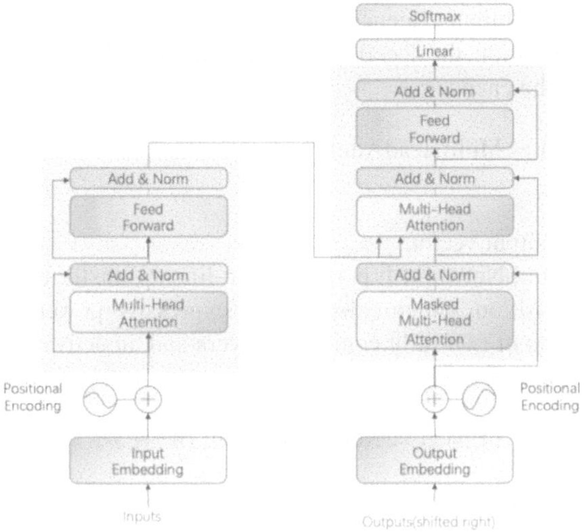

Fig. 2. Transformer model diagram.

where W_i^Q, W_i^K, W_i^V represent the weighting matrices for Q, K, V of the corresponding head.

Positional Encoding: In the Transformer model, it is not feasible to directly pass the positional information of the input sequence to the attention mechanism because the attention mechanism is fundamentally a matrix operation that does not consider the positional relationships between input sequence elements. Therefore, positional encoding is used in the input sequence to convey positional information. The formula for positional encoding is as follows:

$$\mathrm{PE}(pos, 2i) = \sin\left(\frac{pos}{10000^{2i}/d_{\mathrm{model}}}\right) \tag{4}$$

$$\mathrm{PE}(pos, 2i+1) = \cos\left(\frac{pos}{10000^{2i}/d_{\mathrm{model}}}\right) \tag{5}$$

where pos represents the position information and d_{model} represents the input dimension.

Encoder and Decoder: The encoder transforms the input sequence into a fixed-length vector, while the decoder predicts the next target sequence based on the vector generated by the encoder and the previous state of the target sequence.

The formula as follows:

$$\mathrm{EncoderLayer}(x) = \mathrm{MultiHeadAttention}(x) + \mathrm{AddNorm}(x) \tag{6}$$

where x represents the input sequence, MultiHeadAttention denotes the multihead mechanism, and AddNorm is the normalization operation.

The formula as follows:

$$\mathrm{DecoderLayer}(y) = \mathrm{MultiHeadAttention}(y, y, y) + \mathrm{AddNorm}(y) \tag{7}$$

where y represents the output sequence, MultiHeadAttention denotes the internal attention mechanism. In addition, the decoder also requires internal attention mechanism with the output vector of the encoder:

$$\text{DecoderLayer}(y, H) = \text{MultiHeadAttention}(y, y, y) + \text{MultiHeadAttention}(y, H, H) \tag{8}$$

where H represents output vector.

Feedforward Neural Network: It performs non-linear transformations on the representations of each position, which enhances the expressive power of the model. The FFN typically consists of two fully connected layers and an activation function. In the Transformer model, it is formulated as follows:

$$\text{FFN}(x) = \text{ReLU}(xW_1 + b_1)W_2 + b_2 \tag{9}$$

where W_1 and W_2 are the weight matrices, and b_1 and b_2 are the biases of the first and second layers.

3.3 Shrunk Transformer Model

In the shrunk Transformer model, we only use the Encoder as shown in Fig. 3. This is because it is a classification task, and the model removes the Decoder component present in the traditional Transformer model. Instead, it utilizes the Encoder for classification purposes. Additionally, in order to address the classification task, we employ several algorithms and functions to accomplish this process:

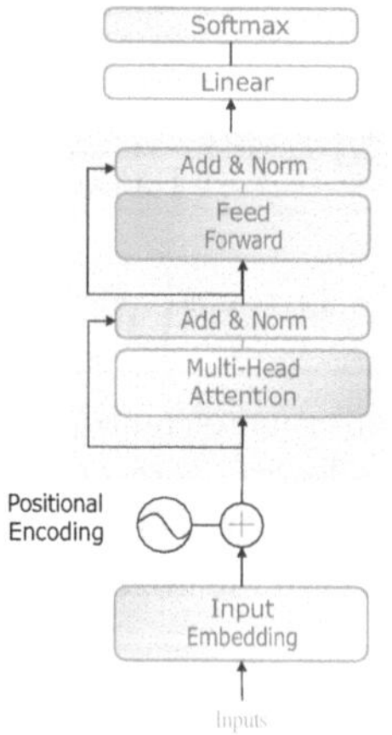

Fig. 3. Shrunk Transformer model diagram.

The cross-entropy loss function operates by quantifying the performance of a model by computing the difference between the predicted probability distribution and the true labels. A lower cross-entropy loss indicates a better performance of the model, as it implies higher predicted probabilities for the true class.The formula as follows:

$$L = -\sum_{i} y_i \log(p_i) \tag{10}$$

where L denotes the cross-entropy loss, y_i represents the one-hot vector of the true label for the i-th class (with values 0 or 1), and p_i denotes the predicted probability of the model for the i-th class.

The Transformer model incorporates residual networks and layer normalization to enhance training speed and enhance model accuracy. Residual networks address the issue of vanishing gradients by introducing shortcut connections for residual calculations between layers. This involves directly adding the output of the previous layer to the output of the current layer. This can be mathematically represented as follows:

$$\text{Residual}(x) = x + \text{Sublayer}(x) \tag{11}$$

where x represents the output from the multi-head attention or feedforward network mechanism.

Layer normalization, on the other hand, normalizes the output of each layer independently by calculating statistical measures. The formula for layer normalization is as follows:

$$LayerNorm(x) = \gamma\left(\frac{x - \mu}{\sqrt{\sigma^2 + \varepsilon}}\right) + \beta \tag{12}$$

where μ and σ represent the statistics, γ and β are the scaling factor and the offset, respectively.

Softmax Formula: The most commonly used weighting mechanism in attention mechanisms is the Softmax function, which normalizes the input vector. In the Transformer model, the Softmax function is utilized to normalize the attention distribution, ensuring that the sum of the weights for the corresponding input vectors equals 1. The formula for the Softmax function is as follows:

$$\text{Softmax}(x)_i = \frac{e^{x_i}}{\sum_{j=1}^{n} e^{x_j}} \tag{13}$$

where x represents the input, x_i is the i-th element of the input vector, and n is the dimensionality of the input vector.

4 Overall Model Structure

As shown in Fig. 4, the following is an architecture of a model based on the ATT&CK tactics framework and the Transformer model:

Data preprocessing stage: The original network security dataset is modeled based on the ATT&CK technique matrix. During modeling, different feature extraction methods are applied to the original data based on its characteristics to extract suitable features.

Additionally, the data is classified according to the corresponding tactical stages and mapped into the data features to generate the ATT&CK feature vectors.

Model training stage: The model training stage primarily employs an shrunk Transformer model, incorporating techniques such as multi-head attention mechanisms, positional encoding, residual networks, and batch normalization. The preprocessed and engineered data is used for modeling and training.

System application stage: In the system application stage, the model is applied to predict the data in the testing set, generating the situational assessment categories for each sample. These predictions are then compared with the actual values to obtain experimental results.

5 Experimental Settings

5.1 Dataset and Experimental Environment

The study employed the experimental configuration includes a computer based on the Intel Xeon E5-2678 v3 processor and an NVIDIA Tesla T4 graphics processing unit (GPU).

The NUSW-NB15 [24] is a publicly available network intrusion detection dataset. The dataset consists of approximately 57,000 network sessions, covering abnormal traffic and attack scenarios across different network protocols. It provides a more realistic and diverse dataset for network intrusion detection, aiming to assist researchers in better detecting and predicting network intrusions.

The NSL-KDD dataset is utilized for assessing intrusion detection system effectiveness in the field of network security. It is an improvement over the KDD Cup 1999 dataset. However, due to certain limitations of the KDD Cup 1999 dataset, such as the lack of abnormal traffic data, the NSL-KDD dataset was designed.

In our experiments, we will use these two datasets to demonstrate the robustness of the ATT&CK tactics framework and the Transformer model. In the experiments, we adopt a simple data method to simulate the process of situational assessment, where the four flows of data are combined into one training/testing data. After processing, the final number of training data for NUSW-NB15 is 43,836, and the number of testing data is 20,583. For NSL-KDD, there are a total of 31,493 training data samples, while the number of testing data samples amounts to 5,636. The range of situational labels 0-3 is considered as the network security situational values, where number 0 signifies safe, number 1 signifies low risk, number 2 signifies medium risk, and number 3 signifies high risk.

5.2 Experimental Metrics

When using the Transformer model for network security situational assessment, experimental metrics play a crucial role in evaluating the model's performance and effectiveness.

During the network security situational assessment experiments, we utilize various metrics to assess the performance and effectiveness of the Transformer model. Precision,

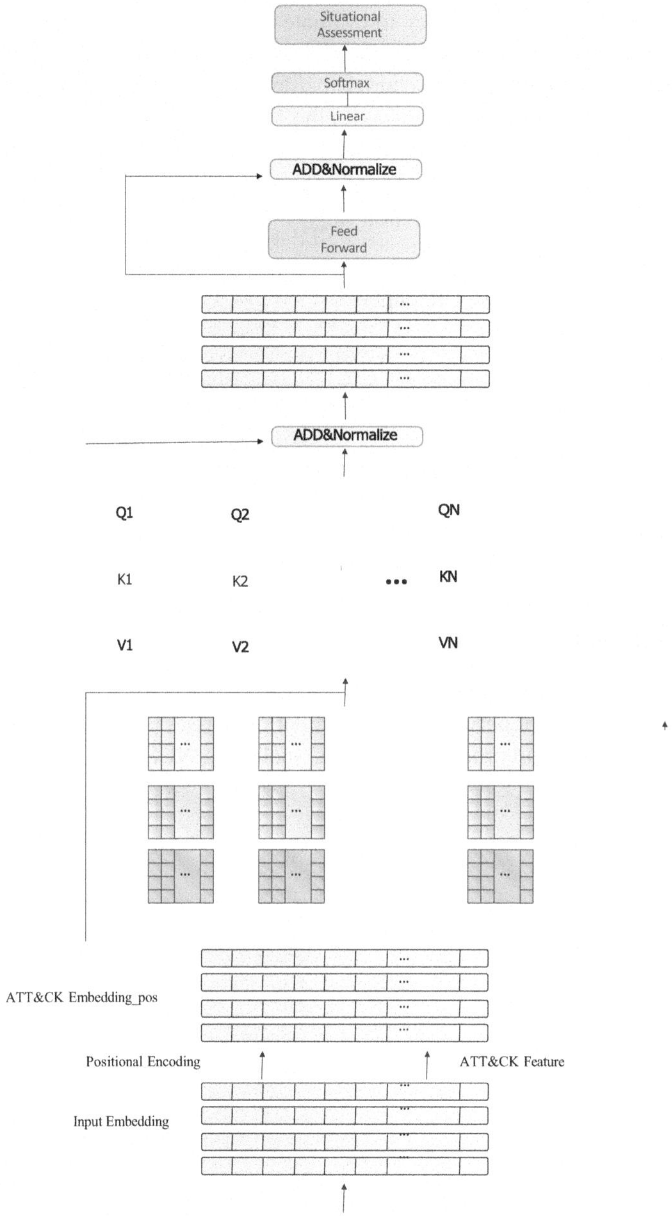

Fig. 4. Overall model structure diagram.

Recall, and F1-score are employed to determine the model's classification accuracy. These metrics aid in evaluating the model's ability to identify different categories of attack behaviors or security incidents.

Additionally, we utilize Accuracy to evaluate the overall classification accuracy of the model across the entire dataset. Accuracy helps us understand the model's ability to correctly classify different types of security incidents on an overall basis.

When it comes to experimental metrics for the Transformer model in network security situational assessment, here are some common metrics and their corresponding formulas:

$$\text{Precision} = \frac{\text{TP}}{\text{TP} + \text{FP}} \tag{14}$$

$$\text{Recall} = \frac{\text{TP}}{\text{TP} + \text{FN}} \tag{15}$$

$$\text{F1 - score} = \frac{2 \cdot (\text{Precision} \cdot \text{Recall})}{\text{Precision} + \text{Recall}} \tag{16}$$

$$\text{Accuracy} = \frac{\text{TP} + \text{TN}}{\text{TP} + \text{TN} + \text{FP} + \text{FN}} \tag{17}$$

Among them, TP represents true positives, FP represents false positives, FN represents false negatives, and TN represents true negatives.

5.3 Experimental Results

In our experiments, we evaluated the performance of the RNN model, LSTM model, and Transformer model for network security situational assessment. These models were trained and tested using preprocessed datasets, including NUSWNB15 and NSL-KDD (Fig. 5).

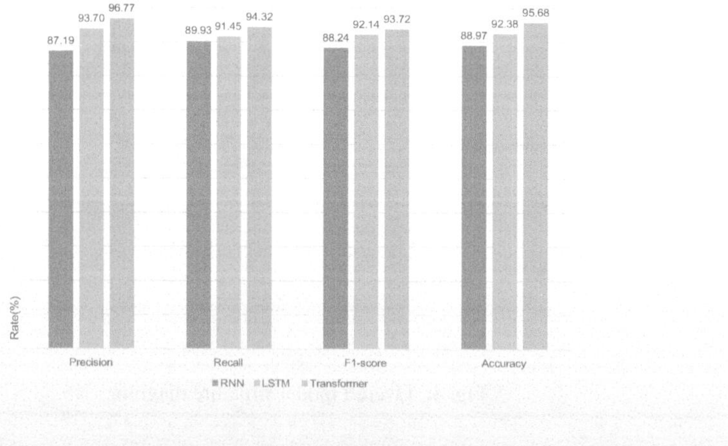

Fig. 5. UNSW-NB15 Comparison.

Initially, we assessed the performance of RNN, LSTM, and Transformer models on precision, recall, F1-score, and accuracy metrics using the UNSW-NB15 test dataset.

The findings demonstrated that the Transformer model outperformed the other models in all metrics, highlighting its superiority for network security situational assessment.

Additionally, we compared the precision, recall, F1-score, and accuracy of the three models on the NSL-KDD test dataset. The results revealed that the Transformer model displayed favorable performance across these metrics.

In summary, based on our experimental results, the Transformer model outperformed the RNN and LSTM models in the network security situational assessment task. It achieved better results in accuracy, recall, F1-score, and other metrics, indicating its potential and value in handling network security data and predicting security situations.

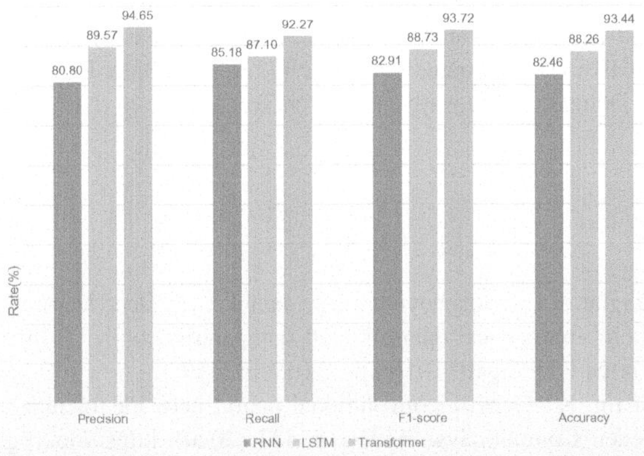

Fig. 6. NSL-KDD Comparison

6 Conclusion

Through this study, we successfully combined the Transformer model with the ATT&CK framework to develop a novel approach for network security situational assessment. Based on experimental validation, we draw the following conclusions (Fig. 6):

First, we found that the ATT&CK framework played a vital role in classifying network security data. By classifying data according to different ATT&CK tactics, we were able to gain a more accurate understanding and assessment of various types of attack behaviors. This fine-grained classification helped us obtain more targeted feature representations in the data processing and model learning processes.

Second, we employed the Transformer model as the primary learning and evaluation tool. Compared to traditional RNN and LSTM models, the Transformer model is better able to capture dependencies across different time steps and has strong parallel computing capabilities. By inputting data categorized by ATT&CK into the Transformer model, we achieved accurate evaluation of network security situational assessment and timely identification and recognition of potential threats.

Finally, the experimental outcomes provided evidence of the effectiveness of our proposed approach, based on the Transformer model and the ATT&CK framework, exhibited good performance across multiple evaluation metrics. Through testing on publicly available network security datasets, our model achieved better results.

In conclusion, our research provides an innovative approach for network security situational assessment based on the Transformer model and the ATT&CK framework. This approach demonstrates potential in terms of accuracy, scalability, and practicality, offering valuable insights for research and practice in the field of network security. Moving forward, we will persist in our exploration and improve this method to further enhance the effectiveness and efficiency of network security situational assessment.

Acknowledgements. This research was funded by NSFC (No. 62072131, 62372127, 61972106), the Major Key Project of PCL (Grant No. PCL2022A03-3), Key R&D Program of Guangdong Province (No. 2019B010136003), Guangdong Basic and Applied Basic Research Foundation (No. 2022A1515011401), DongGuan Innovative Research Team Program (No. 2018607201008).

References

1. Ferrag, M.A., Maglaras, L., Moschoyiannis, S., Janicke, H.: Deep learning for cybersecurity intrusion detection: approaches, datasets, and comparative study. J. Inf. Secur. Appl. **50**, 102419 (2020). https://doi.org/10.1016/j.jisa.2019.102419
2. Nguyen, M.T., Kim, K.: Genetic convolutional neural network for intrusiondetection systems. Future Gener. Comput. Syst. **113**, 418–427 (2020). https://doi.org/10.1016/j.future.2020.07.042
3. Amro, A.: IoT vulnerability scanning: A state of the art. In: Katsikas, S., et al. (eds.) Computer Security: ESORICS 2020 International Workshops, CyberICPS, SECPRE, and ADIoT, Guildford, UK, September 14–18, 2020, Revised Selected Papers, pp. 84–99. Springer International Publishing, Cham (2020). https://doi.org/10.1007/978-3-030-64330-0_6
4. Leau, Y.-B., Manickam, S., Chong, Y.-W.: Network security situation assessment: a review and discussion. In: Kim, K.J. (ed.) Information Science and Applications, pp. 407–414. Springer Berlin Heidelberg, Berlin, Heidelberg (2015). https://doi.org/10.1007/978-3-662-46578-3_48
5. Li, S., Jiang, L., Zhang, Q., Wang, Z., Tian, Z., Guizani, M.: A malicious mining code detection method based on multi-features fusion. IEEE Trans. Netw. Sci. Eng. (2022). https://doi.org/10.1109/TNSE.2022.3155187
6. Li, S., Li, Y., Han, W., Du, X., Guizani, M., Tian, Z.: Malicious mining code detection based on ensemble learning in cloud computing environment. Simul. Modell. Pract. Theory **113**, 102391 (2021)
7. Li, S., Li, Y., Wu, X., Al Otaibi, S., Tian, Z.: Imbalanced malware family classification using multimodal fusion and weight self-learning. IEEE Trans. Intell. Transport. Syst. **24**(7), 7642–7652 (2023). https://doi.org/10.1109/TITS.2022.3208891
8. Yu, Z., Li, S., Bai, Y., Han, W., Wu, X., Tian, Z.: REMSF: a robust ensemble model of malware detection based on semantic feature fusion. IEEE Internet Things J. **10**(18), 16134–16143 (2023). https://doi.org/10.1109/JIOT.2023.3267337
9. Mei, Y., Han, W., Li, S., Lin, K., Luo, C.: A hybrid intelligent approach to attribute Advanced Persistent Threat Organization using PSO-MSVM Algorithm. IEEE Trans. Netw. Serv. Manage. **19**(4), 4262–4272 (2022). https://doi.org/10.1109/TNSM.2022.3201928

10. Leau, Y.-B., Manickam, S.: A cost-sensitive entropy-based network security situation assessment model. Adv. Sci. Lett. **22**(10), 2865–2870 (2016). https://doi.org/10.1166/asl.2016.7076
11. Yuan, Z., Yao, S., Xia, C., Xiang, S.: Network efficacy evaluation based on AHP for network security situation assessment. In: 2016 6th International Conference on Machinery, Materials, Environment, Biotechnology and Computer, Atlantis Press, pp. 797–800 (2016). https://doi.org/10.2991/mmebc-16.2016.167
12. Wang, Y., Wang, J., Xu, Z., Li, H.: Assessing cyber-threats situation for electric power information networks. In: 2013 Ninth International Conference on Natural Computation, ICNC, pp. 1557–1562 (2013). https://doi.org/10.1109/ICNC.2013.6818229
13. Wang, H., et al.: Research on network security situation assessment and quantification method based on analytic hierarchy process. Wirel. Pers. Commun. **102**(2), 1401–1420 (2018). https://doi.org/10.1007/s11277-017-5202-3
14. Li, F., Yang, S., Zhu, J.: Improved network security situation assessment method based on fuzzy hierarchy method. J. Comput. Appl. **34**(9), 2622–2626 (2014)
15. Yu, J., Hu, M., Wang, P.: Evaluation and reliability analysis of network security risk factors based on D-S evidence theory. J. Intell. Fuzzy Syst. **34**, 861–869 (2018)
16. Li, X., Deng, X., Jiang, W.: A novel method of network security situation assessment based on evidential network. In: Chen, X., Yan, H., Yan, Q., Zhang, X. (eds.) Machine Learning for Cyber Security: Third International Conference, ML4CS 2020, Guangzhou, China, October 8–10, 2020, Proceedings, Part I, pp. 530–539. Springer International Publishing, Cham (2020). https://doi.org/10.1007/978-3-030-62223-7_46
17. Liu, X., Deng, W., Huang, P.: Hierarchical quantitative assessment method of network security threat situation. In 2021 4th International Conference on Information Systems and Computer Aided Education (ICISCAE 2021). Association for Computing Machinery, New York, NY, USA, pp. 2787–2790 (2021). https://doi.org/10.1145/3482632.3487515
18. Zhang, R., Liu, M., Pan, Z., Yin, Y.: Network security situation assessment based on improved WOA-SVM. IEEE Access **10**, 96273–96283 (2022). https://doi.org/10.1109/ACCESS.2022.3204663
19. Zhang, R., et al.: A model of network security situation assessment based on BPNN optimized by SAA-SSA. IJDCF **14**(2), 1–18 (2022). https://doi.org/10.4018/IJDCF.302877
20. Zhao, D., Ji, G., Zeng, S.: A network security situation assessment method based on multi-attention mechanism and HHO-ResNeXt. In: Chen, X., Huang, X., Kutyłowski, M. (eds.) Security and Privacy in Social Networks and Big Data: 8th International Symposium, SocialSec 2022, Xi'an, China, October 16–18, 2022, Proceedings, pp. 199–211. Springer Nature Singapore, Singapore (2022). https://doi.org/10.1007/978-981-19-7242-3_13
21. Tao, X., Kong, K., Zhao, F., Cheng, S., Wang, S.: An efficient method for network security situation assessment. Int. J. Distrib. Sensor Netw. **16**(11), 155014772097151 (2020). https://doi.org/10.1177/1550147720971517
22. Tao, X., Liu, Z., Yang, C.: An efficient network security situation assessment method based on AE and PMU. Wirel. Commun. Mobile Comput. **2021**, 1–9 (2021). https://doi.org/10.1155/2021/1173065
23. Tang, X., Chen, M., Cheng, J., Xu, J., Li, H.: A security situation assessment method based on neural network. In: Vaidya, J., Zhang, X., Li, J. (eds.) Cyberspace Safety and Security: 11th International Symposium, CSS 2019, Guangzhou, China, December 1–3, 2019, Proceedings, Part II, pp. 579–587. Springer International Publishing, Cham (2019). https://doi.org/10.1007/978-3-030-37352-8_52
24. Moustafa, N., Ji. S.: UNSW-NB15: a comprehensive data set for network intrusion detection systems (UNSW-NB15 network data set). In: 2015 military communications and information systems conference (MilCIS), pp. 1–6. IEEE (2015)

Feature Distillation GAN: An Adversarial Example Defense Method Based on Feature Reconstruction

Ruiqi Liu, Hu Li, Dongxia Wang(✉), Chongyang Zhao, and Boyu Li

National Key Laboratory of Science and Technology On Information System Security, Beijing, China
Liuruiqi2021jk@163.com

Abstract. Purifying and reconstructing adversarial examples into benign ones by generative models is a class of defense methods to eliminate adversarial perturbations. These methods do not make assumptions on adversarial attacks or model architectures, and thus can defend classifiers against unseen threats. In this work, we propose a new GAN-based adversarial purification method Feature Distillation GAN (FDGAN). We consider how to eliminate the feature differences between adversarial examples and benign examples which are difficult for humans to perceive. We extract and analyze the differences between the adversarial examples and the benign examples from multiple dimensionals. On the basis of comparing the differences among various features, we propose a new feature-level perspective on adversarial purification, design a new adversarial example purification method for image inputs, and implement our method based on GAN. Experimental results show that FDGAN is effective against unknown adversarial attacks and outperforms all the purification methods we used for comparison. Evaluating FDGAN against strong adaptive attacks, the robust accuracy of FDGAN under AutoAttack ($L\infty = 8/255$) exceeds the SOTA adversarial training methods on the Robustbench. Furthermore, FDGAN has lower training cost and can be combined with a variety of model-based defense strategies to further improve the defense effectiveness.

Keyword: Adversarial example · Adversarial defense · Adversarial purification

1 Introduction

In recent years, artificial intelligence technologies, particularly deep neural networks, led technological innovations in various fields. However, the rapid progress underscores the pervasive issue of security. One prominent security issue has caused great concern among researchers is the vulnerability of deep neural networks to adversarial examples. Some small perturbation of an image imperceptible to humans can guide the neural network to make false predictions, these perturbed images are adversarial examples [1]. To protect deep neural networks from adversarial examples, researchers have proposed an abundant number of defense techniques. Nonetheless, most of these techniques have

Z. Gu et al. (Eds.): NSE 2023, CCIS 2063, pp. 120–134, 2024.
https://doi.org/10.1007/978-981-97-4519-7_9

inherent limitations and the continuous evolution of adversarial attacks enable them to effectively circumvent existing defense technologies [2]. As a result, the field of adversarial defense research is somewhat passive and requires innovative methods and strategies to proactively mitigate the growing threat of adversarial attacks.

Adversarial purification is a representative class of defense methods which aim at eliminating adversarial perturbations from adversarial examples before they enter into the model. These methods theoretically have better generalization, can be plug-and-play and effectively combined with other defense techniques to enhance the system security, without incurring additional computational costs, thus have important research value. However, the current adversarial purification methods can hardly achieve the advanced defense effect of the adversarial training methods, the adversarial perturbations are usually difficult to perceive, and the features are complex and varied, which may exist in any direction of the high-dimensional feature space. Therefore, it is difficult to design a general and effective adversarial purification method that can defend against multiple attacks.

The main goal of generative models, such as GANs (Generative Adversarial Networks) [3], is to generate new data by learning the distribution of the original data. Adversarial perturbations usually cause the distribution of the examples deviate from the original, and generative models can be used to "correct" the deviated examples back to the original distribution, mapping the adversarial examples back to the benign ones. With the development of generative modeling techniques such as variational autocoder, GAN, diffusion model, etc., there are many researches based on generative models to eliminate adversarial perturbations of image examples [4, 5]. However, these methods usually only utilize the learning expression ability of the generative model, without considering the complex feature differences between the adversarial example and the benign example, and have insufficient generalization ability in defending the adversarial examples generated by unknown attacks, and cannot effectively eliminate the adversarial perturbations.

In view of these problems, we consider how to eliminate the feature differences between adversarial and benign examples, which are difficult for humans to detect. We analyze and extract the feature differences between adversarial examples and benign examples from multiple dimensions, design a new adversarial elimination scheme for image examples. On the basis of these works, we design an effective purification strategy for adversarial examples and implement it with GANs.

Our Main Contributions:

1. We analyze and compare the differences between adversarial examples and benign examples, design an adversarial example purification strategy based on the feature differences in different dimensions, and propose an adversarial purification framework for image adversarial examples which core on the extraction of feature differences between adversarial and benign examples based on GANs.
2. We compare the adversarial example purification strategies guided by different dimensional feature differences through experiments, and test the impact of the design of

the key parts of the FDGAN framework on the defense performance under different construction schemes of adversarial training datasets, and different generator and discriminator architectures.
3. Experimental results show that compared to existing open source representative adversarial example purification methods, FDGAN has better defense effectiveness. Furthermore, FDGAN outperforms most of the state-of-the-art adversarial training methods on Robustbench [6] in robust accuracy.

2 Related Work

2.1 Adversarial Defences

Adversarial Training. Adversarial training provide defense by adding adversarial examples to the training data, which is the most widely accepted defense method and achieves a more ideal defense effect. However, Adversarial training often demands significant computational time and resources, making it expensive for practical applications. Additionally, adversarial training exhibits limited generalization capabilities, as attackers can easily adapt their strategies to bypass the defense. Early improvements to adversarial training were based on techniques like PGD-AT (Projected Gradient Descent Adversarial Training) [7]. In recent years, with the maturation of generative models and other technologies, methods that employ GANs, Diffusion models, and similar techniques to generate adversarial examples for adversarial training [8, 9] have gained significant attention and have shown promising results on open-source attack-defense platforms.

Preprocessing-based Purification Defenses. Preprocessing-based purification defenses aim to eliminate adversarial perturbations from adversarial examples before they enter the model. In this paper, we select several representative purification methods as references to demonstrate the effect of FDGAN. Thermometer Encoding [10] discretizes input images to increase the difficulty of calculating model gradients during attacks. JPEG-Compression [11] investigates the effect of JPEG compression on the adversarial examples, which is useful for a small-amplitude adversarial perturbation. APE-GAN [12] is a GAN-based image reconstruction method, but only based on pixel consistency.

2.2 Generative Adversarial Networks

The current state-of-the-art generative models are represented by GANs [3] and Diffusion Models [13]. Recent research indicates that Diffusion Models produce higher-quality images, but its training cost is prohibitively high, while the GAN technology has matured considerably and the training cost is significantly lower than Diffusion Models [14].Therefore, we implement our approach based on GANs. The basic GAN consists of two fundamental models: the generator (G) and the discriminator (D), where G is continually optimized to estimate the data distribution, while the primary objective of D is to distinguish between examples from the training data and fake examples generated by G. The optimization objective can be described by a value function $V\ (G,\ D)$:

$$\underset{G}{min}\,\underset{D}{max}\,V(D, G) = E_{x\sim pdata\,(x)}\big[log\,D(x)\big] + \mathbb{E}_{z\sim p_z(z)}[log(1 - D(G(z)))] \tag{1}$$

E is the expectation(an estimate of the mean of the corresponding random variable), the logarithmic function *log* is mainly used to calculate cross-entropy, x a real data example, $x \sim p_{data}(x)$represents the distribution of x from the real data, and z is a noise which is usually randomly generated and used as input for G. The difference between the adversarial and benign examples can be formulated by:

$$\|X^{adv} - X\| = \delta \tag{2}$$

X^{adv} represents the adversarial example, X is the benign example corresponding to X^{adv}. The ideal situation is to remove δ from X^{adv} so that the distribution of X^{adv} and X are aligned. The training objective of the basic GAN can be viewed as making $p_g = p_{data}$, the procedure of which convergence to p_{data} is similar to the goal of eliminating adversarial perturbations.

3 Feature Distillation GAN

The core of our approach is to utilize the expressive power of the generative model to minimize the feature differences between adversarial and benign examples, making the generated example $G(X^{adv})$ as close as possible to X, converting the problem of adversarial example purification into a feature manifold mapping problem of learning a mapping from adversarial examples to benign examples. Based on this objective, we introduce the Feature Distillation GAN, an adversarial purification framework based on the GAN, which achieves defense by minimizing the differences between generated examples and benign examples at the feature level. The overview of Feature Distillation GAN algorithm is shown as follow (see Fig. 1).

3.1 Architecture

FDGAN is trained in an adversarial environment, with the ultimate goal of training a generator G to estimate the corresponding benign example X for a given input adversarial example X^{adv}. To achieve this objective, it is necessary to train a GAN. According to the architecture design in Fig. 1, the first step is to construct a dataset for training the GAN. The dataset consists of two components: benign examples and their corresponding adversarial versions. Specifically, assuming there are N training data pairs, each consisting of a benign example X_k, $k = 1, ..., N$ and its adversarial version X^{adv}. Using θ_G to represent the parameters of the generator to be optimized and L_G to represent the loss function used for optimize θ_G, the problem can be expressed as:

$$\theta_G = \underset{\theta_G}{argmin} \frac{1}{N} \sum_{k-1}^{N} \mathfrak{L}_G(G_{\theta_G}\left(X_k^{adv}\right), X_k) \tag{3}$$

The objective of the discriminator D is to distinguish the reconstructed image $G(X^{adv})$ and the original image X. Its parameters θ_D and the generator's parameters θ_G are

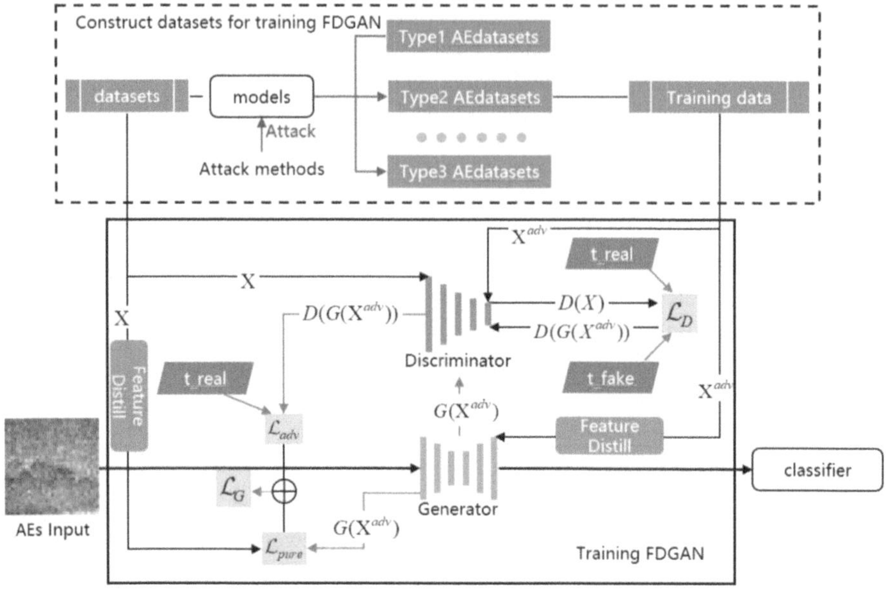

Fig. 1. Overview of the Feature Distillation GAN algorithm.

optimized simultaneously during the training process. The overall process can be viewed as solving an adversarial zero-sum game:

$$\underset{\theta_G}{min}\underset{\theta_D}{max}\mathbb{E}_{x\sim pdata\,(x)}log\ D\theta_D(x) - \mathbb{E}_{X^{adv}\sim P_{G(x^{adv})}}log(D\theta_D\left(G\theta_G\left(X^{adv}\right)\right) \tag{4}$$

In this game, θ_D is continuously optimized to enhance D's ability to distinguish the reconstructed image $G(X^{adv})$ from the originally benign input X, while θ_G is continually optimized to make G reconstruct X^{adv} to generate a reconstructed image that is highly similar to X, to the extent that D cannot distinguish them effectively.

3.2 Loss Function

We define the loss function of G as the weighted sum of the purification loss and the adversarial loss:

$$\mathcal{L}_G = \varepsilon_1 \mathcal{L}_{pure} + \varepsilon_2 \mathcal{L}_{adv} \tag{5}$$

L_{pure} measures the difference between the images generated by G and the benign examples. Based on the analysis of adversarial features, we measure feature differences in different dimensions under different distance difference metrics, formulate an adversarial elimination strategy and perform a detailed design of L_{pure}. L_{adv} is the adversarial loss function of the GAN. We fix L_{adv} in order to test the effectiveness of different L_{pure}. L_{adv} is:

$$\mathcal{L}\ominus_{adv} = \sum_{n=1}^{N}[1 - \log D_{\theta_D}(G_{\theta_G}(X^{adv}))] \tag{6}$$

The discriminator loss is:

$$\mathcal{L}\Theta_D = -\sum_{n=1}^{N} [\log D_{\theta_D}(X) + \log D_{\theta_D}(G_{\theta_G}(X^{adv}))] \tag{7}$$

The FDGAN does not make any requirement on the feature extraction method, it can use any model and any feature extraction method to extract features. We use torchvision pre-trained VGG16 [15] model to extract features, and its extracted different levels of feature map differences are shown in Fig. 2.

The previous layers of the feature map usually show that the model is extracting features such as edges and textures. The middle layers begin to show more advanced and abstract features, including recognizing major portions of the image. Deeper feature maps no longer correspond to any visual image features and may represent more advanced semantic information, such as whether a region contains a particular object.

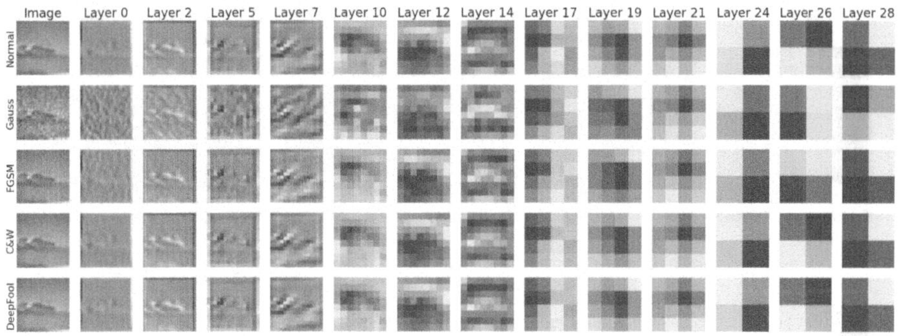

Fig. 2. Visualize the features extracted after each convolutional layer of the VGG16

Fig. 3 shows the results of the CAM(Class Activation Mapping) analysis at different levels, where the red/bright regions correspond to the most important regions for the model to make current class predictions, and these regions are observed in order to understand on which part of the features the model is based to classify the adversarial example to the wrong class, the features in the blue/dark regions do not contribute much to the current class prediction, and the neutral background color implies that the model does not give any special attention to these regions. CAM demonstrates the regions where different network layer features are sensitive to the target class, providing better interpretability of the model's predicted class probabilities.

We design corresponding adversarial elimination strategies based on the differences in different types of features. We apply various norm distance metrics to design loss functions, exploring methods that can amplify the differences in features between adversarial and benign examples. For example, when we apply the Euclidean distance measuring loss, the basic pixel-level difference loss function can be described as:

$$\mathcal{L}\Theta_{pure} = \frac{1}{WH}\sum_{x=1}^{W}\sum_{y=1}^{H}(X_{x,y} - D_{\theta_D}(X^{adv})_{x,y})^2 \tag{8}$$

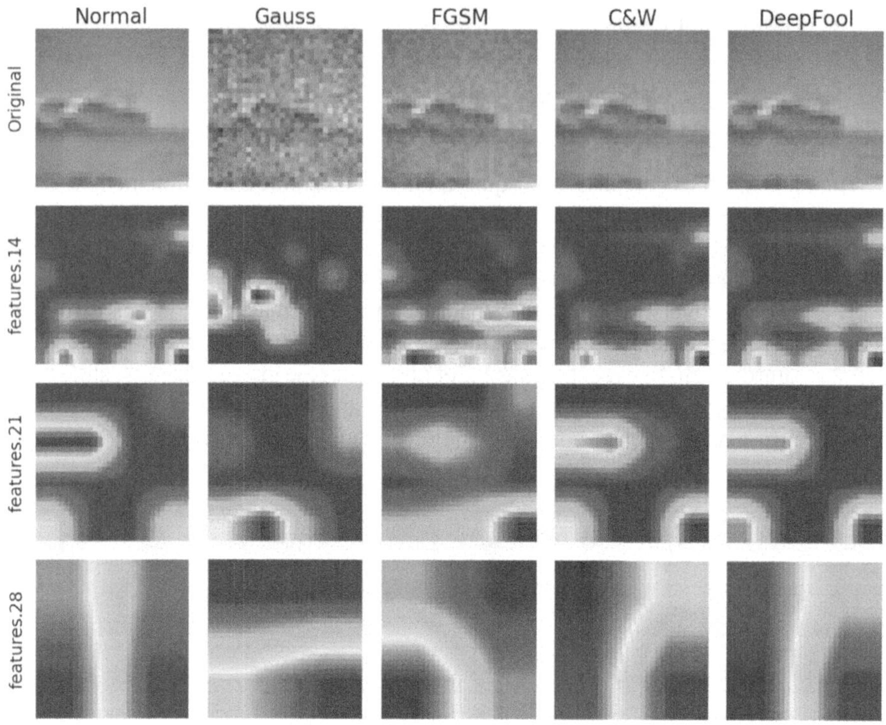

Fig. 3. Visualize the Class Activation Mapping

In the equation, W and H represent the number of pixels in the width and height dimensions of the input image, $X_{x,y}$ and $G_{\theta G}(X)_{x,y}$ denote the pixel values of the corresponding examples at position x, y. The basic feature-level difference loss function can be described as:

$$\mathcal{L}\Theta_{pure} = \frac{1}{WH}\sum_{x=1}^{W}\sum_{y=1}^{H}(\phi_i X_{x,y} - \phi_i(G_{\theta_D}(X^{adv}))_{x,y})^2 \tag{9}$$

ϕ denotes the feature extractor. When we extract features with VGG16, ϕ_i indicate the feature map obtained by the i-th layer of the VGG16.

4 Experiments

We characterized adversarial examples generated by FGSM attacks, DeepFool attacks, C&W attacks, and direct addition of Gaussian noise. The experiments implement adversarial attacks through a standardized interface provided by the open source library torchattacks [16], making the research process and experimental results more transparent and easier to compare and replicate. Guided by an analysis of differences in features, We trained and tested FDGANs with different purification strategies on the CIFAR-10 dataset using two RTX TITAN 2080 GPUs, the training parameters of FDGAN are shown in Table 1. We compare the defense effectiveness guided by different purification strategies through attacks with different mechanisms, the attack parameters are listed in Table 2. Finally, we compare the FDGAN with other representative purification methods and the state-of-the-art adversarial training on various strong adaptive attack benchmarks.

Table 1. Attack parameter settings for testing FDGAN defense performance.

Items	Parameters
Generator Optimizer	Adam: lr = 0.0002, betas = (0.5, 0.999) Discriminator Optimizer Adam: lr = 0.0002, betas = (0.5, 0.999)
epoch	$20 \leq$ epoch ≤ 80

Table 2. Attack parameter settings for testing FDGAN defense performance.

Attack	Parameters
FGSM	FGSM(model, eps = 0.007)
EOT + PGD EOTPGD(model, eps = 0.007, alpha = 0.003, steps = 20) DeepFool	DeepFool(model, steps = 50)
C&W	CW(model, c = 1, kappa = 0, steps = 100, lr = 0.01)

4.1 Purification Performance Test

The FDGAN provides an image adversarial example purification framework which defense performance is related to the construction scheme of the adversarial example datasets, the generator and discriminator architectures, the feature distillation strategies, and the weighting ratio of the L_{pure} and the L_{adv}. We test the performance of FDGAN on the CIFAR-10 under the guidance of different training adversarial example construction schemes, different generator and discriminator architectures, different training methods, and different feature distillation strategies. Through the framework performance tests, understanding how to design different modules can lead to better performance of the FDGAN, which provides a reference for designing specific defense strategies for specific defense requirements.

Adversarial Example Purification Strategy Based on Different Feature Differences. We train the purifier using the CIFAR-10 dataset with 0.1 Gaussian noise added and its benign examples, consider various distance metrics to measure feature differences and test the effectiveness under the corresponding loss functions. We also consider the impact on benign examples, the classifier is Resnet18 trained on CIFAR-10 with an accuracy of 87.16%, the test results are shown in Table 3.

Table 3 shows only a part of the experimental results, some obviously unsatisfactory test results are not shown, such as when using the CAM features of the 29th layer, the classification accuracy on the benign samples after FDGAN processing is only 58.51% under the Manhattan distance metric. Different dimensions, different levels of feature differences and different distance metrics can be used in a variety of permutations and combinations, and in order to improve the testing efficiency, the feature extraction methods with obviously poor defense under the same distance metric were not tested too much.

Table 3. FDGAN performance test results guided by different feature differences.

Feature	Distance metric	Benign	FGSM	EOT + PGD	DeepFool	C&W
Without FDGAN		87.16%	31.18%	11.52%	1.06%	0.00%
	Euclidean	70.61%	64.16%	64.98%	66.99%	67.32%
Pixel-level	Manhattan	73.62%	66.80%	67.57%	69.97%	70.10%
	Chebyshev	72.12%	65.93%	66.75%	69.07%	68.97%
	Cosine	70.75%	64.72%	65.24%	67.40%	67.50%
	Euclidean	76.93%	69.08%	70.03%	73.53%	73.02%
Feature-level	Manhattan	76.12%	67.18%	68.49%	71.99%	71.76%
(31st layer)	Chebyshev	75.70%	67.51%	68.49%	71.95%	71.77%
	Cosine	73.40%	66.50%	67.17%	70.02%	70.07%
	Euclidean	78.57%	67.99%	69.39%	74.86%	73.83%
Feature-level	Manhattan	**79.77%**	**69.16%**	**70.29%**	**76.33%**	**75.35%**
(17th layer)	Chebyshev	78.74%	68.41%	69.52%	74.80%	74.04%
	Cosine	71.10%	65.35%	66.24%	68.18%	68.09%
18th layer CAM	Euclidean	78.33%	68.17%	69.47%	74.72%	73.53%
	Manhattan	73.97%	66.20%	67.42%	70.36%	70.20%

Different Training Adversarial Example Datasets Construction Schemes. Taking the pixel differences under the Euclidean distance metric as the benchmark, construct three training adversarial example datasets with different attack strengths and complexities on the CIFAR-10 dataset to train three corresponding FDGANs: FDGAN1 is trained with the adversarial examples generated by the FGSM attack($\epsilon = 0.01$) and benign examples, FDGAN2 is trained with the adversarial examples and benign examples generated by FGSM attacks ($\epsilon = 0.1$), and FDGAN3 is trained with adversarial examples and benign examples generated by a mixture of FGSM attacks, PGD attacks, and DeepFool attacks, and compares the defense effects of the three FDGANs, and the test results are shown in Table 4.

It can be seen that all three FDGANs have defense performance, indicating that the FDGAN plays a positive role in improving model robustness. Among them, except for the no-attack scenario, the defense performance of FDGAN2 is slightly better than that of FDGAN1, which may be due to the fact that the adversarial examples with $\epsilon = 0.1$ are more obviously perturbed and have a higher success rate of the attack than the adversarial examples with $\epsilon = 0.01$, which enables FDGAN2 to extract and learn the more obvious differences in the features. FDGAN3 clearly performs the best overall and is the highest classification accuracy, and its training data covers adversarial examples generated by multiple types of attacks, and is more likely to learn feature differences that do not depend on a specific attack strategy, resulting in a strong generalization capability and better defense against different types and strengths of attacks.

Table 4. FDGAN performance test results under different training datasets construction schemes.

Attack	Accuracy	FDGAN1	FDGAN2	FDGAN3
No attack	87.16%	63.18%	62.94%	**77.25%**
FGSM	31.18%	55.52%	59.58%	**62.39%**
EOT + PGD	11.52% ± 0.08%	56.15%	60.03%	**63.69%**
DeepFool	1.06% ± 0.08%	57.54%	60.75%	**69.04%**
C&W	0.00%	57.88%	61.05%	**69.08%**

Different Model Architectures and Training Approaches. Using the feature differences extracted by the 31st layer of VGG16 under the Euclidean distance metric as a benchmark, the adversarial examples generated by an FGSM attack($\epsilon = 0.01$) and benign examples were used to construct a training dataset, we test the effect of different generator and discriminator architectures and different training approaches on the defense performance of FDGANs: the FDGAN1 applied the basic architecture as Fig. 4. FDGAN2 introduces a residual module, and FDGAN3 introduces a RRDB module. FDGAN4 builds on the introduction of the residual module by using the Wasserstein loss as an adversarial loss and introducing a gradient penalty mechanism, which is generally known to stabilize the GAN training process and improve the quality of the generated adversarial examples. FDGAN5 builds on the introduction of the residual module by selfattention mechanism is introduced, which makes FDGAN pay more attention to the key parts of the input image examples. The test results are shown in Table 5. Test results show that more complex model architectures and training mechanisms can lead better defense results for FDGAN.

4.2 Evaluation

We select the features extracted from the 10th layer of VGG16 as the difference of the FDGAN used in the evaluation for measurements, using the Manhattan distance as the metric of feature differences. We only use the examples from the original CIFAR-10 dataset and the CIFAR-10 examples with Gaussian noise ($\epsilon = 0.1$) added to construct

```
==========================================================================
Layer (type:depth-idx)                              Param #
==========================================================================
Generator                                           --
├─Conv2d: 1-1                                       3,136
├─BatchNorm2d: 1-2                                  128
├─Conv2d: 1-3                                       131,200
├─BatchNorm2d: 1-4                                  256
├─ConvTranspose2d: 1-5                              131,136
├─BatchNorm2d: 1-6                                  128
├─ConvTranspose2d: 1-7                              3,075
==========================================================================
Total params: 269,059
Trainable params: 269,059
Non-trainable params: 0
==========================================================================
```

```
==========================================================================
Layer (type:depth-idx)                              Param #
==========================================================================
Discriminator                                       --
├─Conv2d: 1-1                                       1,792
├─Conv2d: 1-2                                       73,856
├─BatchNorm2d: 1-3                                  256
├─Conv2d: 1-4                                       295,168
├─BatchNorm2d: 1-5                                  512
├─Linear: 1-6                                       2,305
==========================================================================
Total params: 373,889
Trainable params: 373,889
Non-trainable params: 0
==========================================================================
```

Fig. 4. GAN architecture used in this paper

Table 5. Performance test results of FDGAN with different model architectures and training methods.

Attack	Accuracy	FDGAN1	FDGAN2	FDGAN3	FDGAN4	FDGAN5
No attack	87.16%	74.30%	76.11%	76.06%	76.61%	**77.54%**
FGSM	31.18%	68.00%	69.15%	**70.14%**	69.74%	69.29%
EOT + PGD	11.52% ± 0.08%	68.86%	70.06%	69.75%	**70.59%**	70.56%
DeepFool	1.06% ± 0.08%	71.53%	73.49%	72.49%	74.16%	**74.42%**
C&W	0.00%	71.29%	72.75%	71.45%	**73.58%**	73.46%

the training dataset for the FDGAN, so that all the attacks we used are unknown to the FDGAN being evaluated.

We evaluate our method on commonly used AutoAttack [17] L_∞ and L_2 threat models as strong adaptive attacks with two metrics: standard accuracy and robust accuracy. The standard accuracy measures the performance of the defense method on benign examples. The robust accuracy measures the performance on adversarial examples generated by adaptive attacks.

Comparison with Other Purification Methods. Since some purification methods have an optimization or sampling loop in their defense process, they cannot be evaluated directly with the strongest white-box adaptive attacks, such as AutoAttack [18]. We compare FDGAN with three representative purification defense methods: JPEG compression, Thermometer Encoding, and APE-GAN, using FGSM attack, BIM attack [19], PGD attack, and C&W attack [20], on a WideResNet-28-10 [21] with a standard accuracy of 94.78%. The test results are shown in Table 6, FDGAN outperforms other purification methods on all attacks.

Comparison with the State-of-the-art Adversarial Training. We compare FDGAN with the state-of-the-art adversarial training methods reported by RobustBench, against the L_∞ and L_2 threat models. We evaluate on a WideResNet-28-10 with standard accuracy of 94.78%. Table 7 shows the robustness performance against L_∞ threat model ($\epsilon = 8/255$) with AutoAttack on CIFAR-10. We can see that FDGAN achieves better robust accuracy than all state-of-the-art methods, but the standard accuracy is not outstanding.

Table 6. Comparison with representative purification methods.

Method	FGSM Acc	BIM Acc	PGD Acc	C&W Acc
Without defense	28.98%	0.02%	0.74%	0.00%
JPEG compression [11]	53.58%	57.00%	61.94%	71.99%
Thermometer Encoding [10]	12.55%	12.56%	12.50%	13.37%
APE-GAN [12]	51.68%	50.63%	52.66%	60.71%
FDGAN	**59.31%**	**64.42%**	**68.16%**	**78.35%**

Table 8 shows the robustness performance against L_2 threat model ($\epsilon = 0.5$) with AutoAttack on CIFAR-10. Although FDGAN performs slightly worse in this setting than the adversarial training method based on generating adversarial examples from the diffusion model [8, 9], but in terms of defense cost, FDGAN has an extremely significant advantage. Diffusion models are known to have high computational costs, whereas FDGAN does not introduce additional computational costs and can be trained without more than 100 training epochs when training on CIFAR-10 which is cheap compared to other image reconstruction methods.

Table 7. Comparison with state-of-the-art adversarial training methods.

Method	Standard Acc	Robust Acc
FDGAN	79.5%	**70.2%**
GAIRAT [22], 2021	89.8%	60.1%
Gowal et al. [23], 2021	89.9%	62.8%
Cui et al. [24], 2023	92.8%	67.8%
Wang et al. [9], 2023	93.2%	67.7%
Xu et al. [25], 2023	**93.4%**	64.7%

Table 8. Comparison with state-of-the-art adversarial training methods.

Method	Standard Acc	Robust Acc
FDGAN	79.5%	75.1%
Rebuffi et al. [8], 2021	91.9%	79.2%
Rony et al. [26], 2019	89.1%	67.5%
Wang et al. [9], 2023	**94.9%**	**82.8%**

4.3 Discussion

For evaluation, we sets strict unknown attack scenarios for FDGAN, and only use examples with added Gaussian noise and benign examples to train FDGAN. Compared to other preprocessing-based defense techniques, FDGAN demonstrates significant advantages in multiple attack scenarios. Compared to the SOTA adversarial training method on RobustBench, FDGAN demonstrates a performance that exceeds all the adversarial training methods in terms of classification accuracy of adversarial examples under AutoAttack($L_\infty = 8/255$) attack scenarios, but there are some discrepancies in the classification accuracy of benign examples. To address this issue, in applications, FDGAN can be combined with adversarial example detection methods or robust classification models to achieve multi-level defense.

The reduced classification accuracy on benign examples may be due to the fact that FDGAN overemphasizes the feature differences between antagonistic and benign examples during the training process, while the distribution and characteristics of benign examples are neglected to some extent. The trade-off between adversarial loss and purification loss can be further explored, or the feature difference between adversarial and benign examples can be extracted by using models with better feature extraction and expression capabilities, and exploring how to improve the defense effect of the adversarial examples while reducing the negative impact on the benign examples.

The test results comparing different training datasets construction schemes show that the intensity of the adversarial attack and the diversity of the adversarial examples have a significant impact on the defense capability of FDGAN, and the design of the training adversarial example construction method according to the specific scenarios can help the model more accurately capture the characteristics of the potential adversarial examples, implement targeted defense, and improve the defense capability of FDGAN in the specified scenarios.

Different generator and discriminator architectures and different training mechanisms can also cause performance differences in FDGAN. Complex models and training mechanisms can alleviate the instability in training to a certain extent, and improve the defense effect of FDGAN, but it will also lead to an increase in training time. Specifically according to the actual application scenario requirements, the trade-off between fast training and robustness.

In summary, FDGAN shows defense capability in different attack scenarios. A generator with more ideal defense effect can be trained in fewer training epochs with lower training cost, which solves the problem of high training cost of current methods based on generative model for image example adversarial purification. Compared with other defense methods, FDGAN exhibits advanced defense performance, which optimizes current adversarial image example purification techniques in terms of both the effectiveness of defense against existing attacks and the ability to defend unknown attacks.

5 Conclusions

We propose an adversarial example purification framework called FDGAN. The FDGAN enables universal defense and shows better defense capability in different attack scenarios. As an image adversarial example purification method, it does not make any assumptions about the classifier and can be combined with any classifier or robust model as a preprocessing method. FDGAN can train generators with stronger defense capability in fewer training epochs with lower training cost, which solves the current problem of high training cost of purifying image adversarial examples based on generative models. Compared with other defense methods, FDGAN exhibits advanced defense performance. In addition, in scenarios with known attacks and datasets, more targeted adversarial purification strategies can be implemented based on the feature differences between adversarial examples and benign examples to further enhance the defense performance. FDGAN exceeds the current SOTA image adversarial example defense technique in terms of defense effect against existing attacks and defense ability against unknown attacks, but there are some shortcomings in the classification accuracy of benign examples. In the future, we can continue to explore how to reduce the negative impact on benign examples while improving the defense effect of adversarial examples. In applications, FDGAN can be combined with adversarial example detection methods or robust classification models to achieve multi-level defense.

References

1. Szegedy, C., et al.: Intriguing properties of neural networks (2013). https://doi.org/10.48550/arXiv.1312.6199
2. Wiyatno, R.R., Xu, A., Dia, O., de Berker, A.: Adversarial Ex-amples in Modern Machine Learning: A Review (2019). https://doi.org/10.48550/arXiv.1911.05268
3. Goodfellow, I.J., et al.: Generative adversarial nets. In: Proceedings of the 27th International Conference on Neural Information Processing Systems, Vol. 2. pp. 2672–2680. NIPS'14, MIT Press, Cambridge, MA, USA (2014)
4. Meng, D., Chen, H.: MagNet: A Two-Pronged Defense against Adversarial Examples (2017). https://doi.org/10.48550/arXiv.1705.09064
5. Samangouei, P., Kabkab, M., Chellappa, R.: Defense-GAN: Protecting Classifiers Against Adversarial Attacks Using Generative Models (2018). https://doi.org/10.48550/arXiv.1805.06605
6. Croce, F., et al.: RobustBench: a standardized adversarial robustness benchmark. In: Proceedings of the Neural Information Processing Systems Track on Datasets and Benchmarks **1** (2021)
7. Madry, A., Makelov, A., Schmidt, L., Tsipras, D., Vladu, A.: Towards Deep Learning Models Resistant to Adversarial Attacks. ICLR (2018)
8. Rebuffi, S.A., et al.: Fixing Data Augmentation to Improve Adversarial Robustness (2021). https://doi.org/10.48550/arXiv.2103.01946
9. Wang, Z., et al.: Better Diffusion Models Further Improve Adversarial Training (2023). https://doi.org/10.48550/arXiv.2302.04638
10. Buckman, J., Roy, A., Raffel, C., Goodfellow, I.: Thermometer encoding: one hot way to resist adversarial examples. In: International Conference on Learning Representations (2018)
11. Dziugaite, G.K., Ghahramani, Z., Roy, D.M.: A study of the effect of JPG compression on adversarial images (2016). https://doi.org/10.48550/arXiv.1608.00853

12. Jin, G., Shen, S., Zhang, D., Dai, F., Zhang, Y.: APE-GAN: adversarial perturbation elimination with GAN. In: ICASSP 2019 - 2019 IEEE International Conference on Acoustics, Speech and Signal Processing (ICASSP). pp. 3842–3846 (2019). https://doi.org/10.1109/ICASSP.2019.8683044
13. Dhariwal, P., Nichol, A.: Diffusion models beat GANs on image synthesis. In: Advances in Neural Information Processing Systems. vol. 34, pp. 8780–8794. Curran Associates, Inc. (2021)
14. Yang, L., et al.: Diffusion Models: A Comprehensive Survey of Methods and Applications (2022). https://doi.org/10.48550/arXiv.2209.00796
15. Simonyan, K., Zisserman, A.: Very Deep Convolutional Networks for Large-Scale Image Recognition (2015). https://doi.org/10.48550/arXiv.1409.1556
16. Kim, H.: Torchattacks: A PyTorch Repository for Adversarial Attacks (2021). https://doi.org/10.48550/arXiv.2010.01950
17. Croce, F., Hein, M.: Reliable evaluation of adversarial robustness with an ensemble of diverse parameter-free attacks. In: Proceedings of the 37th International Conference on Machine Learning. pp. 2206–2216. PMLR (2020)
18. Nie, W., et al.: Diffusion Models for Adversarial Purification. In: Proceedings of the 39th International Conference on Machine Learning. pp. 16805–16827. PMLR (2022)
19. Kurakin, A., Goodfellow, I., Bengio, S.: Adversarial Machine Learning at Scale (2016). https://doi.org/10.48550/arXiv.1611.01236
20. Carlini, N., Wagner, D.: Towards evaluating the robustness of neural networks. In: 2017 IEEE Symposium on Security and Privacy (SP). pp. 39–57 (2017). https://doi.org/10.1109/SP.2017.49
21. Zagoruyko, S., Komodakis, N.: Wide Residual Networks (2017). https://doi.org/10.48550/arXiv.1605.07146
22. Zhang, J., et al.: Geometry-aware Instance-reweighted Adversarial Training (2021). https://doi.org/10.48550/arXiv.2010.01736
23. Gowal, S., Qin, C., Uesato, J., Mann, T., Kohli, P.: Uncovering the Limits of Adversarial Training against Norm-Bounded Adversarial Examples (2021). https://doi.org/10.48550/arXiv.2010.03593
24. Cui, J., et al.: Decoupled Kullback-Leibler Divergence Loss (2023). https://doi.org/10.48550/arXiv.2305.13948
25. Xu, Y., Sun, Y., Goldblum, M., Goldstein, T., Huang, F.: Exploring and Exploiting Decision Boundary Dynamics for Adversarial Robustness (2023). https://doi.org/10.48550/ARXIV.2302.03015
26. Rony, J., et al.: Decoupling Direction and Norm for Efficient Gradient-Based L2 Adversarial Attacks and Defenses. In: 2019 IEEE/CVF Conference on Computer Vision and Pattern Recognition (CVPR). pp. 4317–4325 (2019). https://doi.org/10.1109/CVPR.2019.00445

Exploring Advanced Methodologies in Security Evaluation for Large Language Models

Jun Huang[1], Jiawei Zhang[1,1](✉), Qi Wang[2], Weihong Han[1,1](✉), and Yanchun Zhang[1]

[1] Peng Cheng Laboratory, Shenzhen, China
{huangj02,zhangjw01,hanwh}@pcl.ac.cn, yanchun.zhang@vu.edu.au
[2] Information Science and Technology College, DalianMaritimeUniversity, Dalian, China

Abstract. Large Language Models (LLMs) represent an advanced evolution of earlier, simpler language models. They boast enhanced abilities to handle complex language patterns and generate coherent text, images, audios, and videos. Furthermore, they can be fine-tuned for specific tasks. This versatility has led to the proliferation and extensive use of numerous commercialized large models. However, the rapid expansion of LLMs has raised security and ethical concerns within the academic community. This emphasizes the need for ongoing research into security evaluation during their development and deployment. Over the past few years, a substantial body of research has been dedicated to the security evaluation of large-scale models. This article an in-depth review of the most recent advancements in this field, providing a comprehensive analysis of commonly used evaluation metrics, advanced evaluation frameworks, and the routine evaluation processes for LLMs. Furthermore, we also discuss the future directions for advancing the security evaluation of LLMs.

Keywords: Large language models · Security evaluation · LLMs

1 Introduction

Language models (LMs) are statistical machines designed to comprehend natural language, which calculate the likelihood of word sequences based on vast collections of text [5, 58].

Large language models (LLMs) represent a significant advancement over their simpler predecessors, showcasing enhanced abilities to understand intricate linguistic patterns and generate coherent, contextually appropriate text [1, 45, 55]. In addition to text, LLMs can also process pictures [3], audio recordings [4], and videos [2]. Moreover, they can be fine-tuned to excel at specific tasks with utmost precision. With the impressive growth of LLMs' capabilities, numerous commercially available large models, such as OpenAI's GPT series, Anthropic's Claude series, Google's T5 and PaLM, Meta's OPT, and Baidu's Ernie [57], have flourished and found extensive applications.

The potential applications of LLMs are vast and continue to grow as the technology matures. From Fig. 1, it's evident that the number of models with parameters exceeding

Supported by Peng Cheng Laboratory.

Z. Gu et al. (Eds.): NSE 2023, CCIS 2063, pp. 135–150, 2024.
https://doi.org/10.1007/978-981-97-4519-7_10

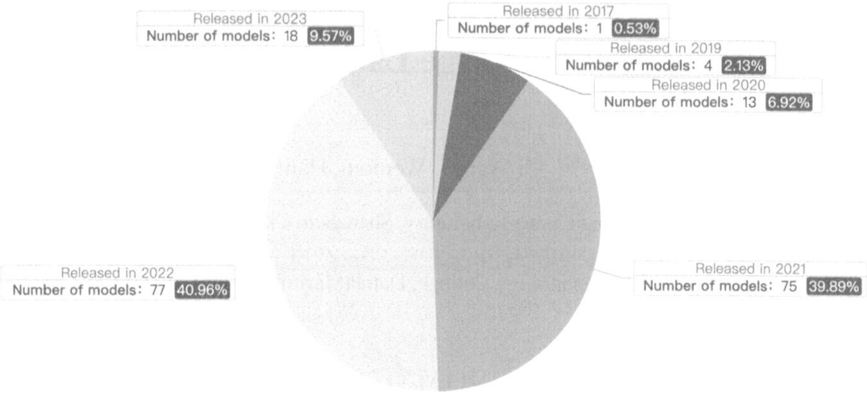

Fig. 1. Proportional representation of model releases with parameters exceeding 1 billion, sourced from OpenBMB [34].

1 billion has seen a significant increase since 2019. This trend can be traced back to a sequence of developments and key events. The stage was set in 2019, highlighted by Microsoft's one billion investment in OpenAI, which underscored the substantial commercial interest in AI and boosted resources for AI development [11]. The momentum continued into 2020 with the launch of NVIDIA's A100 Tensor Core GPU [26], providing researchers with unprecedented computational power and enabling more complex model training. Soon after, the release of OpenAI's GPT-3 model served as a powerful demonstration of the potential of large models [44], spurring further research and development in the field. As we transitioned into 2021, innovative training techniques like mixed precision training, model parallelism, and data parallelism became mainstream, facilitating the training of even larger models. Concurrently, despite these technological advancements and the evident potential of large-scale models, their rapid proliferation has also raised concerns within the academic community, particularly with regard to the security implications of these models [52]. These concerns underscore the necessity of ongoing research into secure and ethical practices in the development and deployment of LLMs.

Researchers have conducted literature reviews focusing on the security risks and evaluation methods for LLMs. Weidinger et al. (2022) in Google DeepMind outlined a comprehensive taxonomy of ethical and social risks associated with large-scale language models, providing a useful framework to guide responsible innovation in this space[56]. However, it lacks a specific implementation plan necessary for conducting a security evaluation. Chang et al. (2023) presented a review of research work on the evaluation of LLMs [6]. They detailed a classification and summary based on three dimensions: "what to evaluate," "where to evaluate," and "how to evaluate." They summarized various successful and unsuccessful cases of LLMs across multiple domains, 28 popular benchmarks for LLM evaluation, and two primary evaluation methods: automatic and manual. But their evaluation focused more on benchmarks for evaluating LLMs, without comprehensive analysis of the evaluation steps. Huang et al. (2023) have provided a review of the various known vulnerabilities of LLMs, considering both technical methods and ethical norms to promote the safety and credibility of LLMs [18]. Their work serves as a reference for the further development of LLMs. Still, their discussion of known evaluation frameworks is not comprehensive. Glukhov et al. (2023) overviewed various aspects of enhancing the security and trustworthiness of LLMs, including validation, monitoring, and ethical norms, and discussed these from a verification and validation perspective [15], but their research mainly focused on the evaluation of prompt injection risks. Zhao et al. (2023) conducted a systematic and comprehensive survey of LLMs, covering various aspects such as pre-training, fine-tuning, utilization, and evaluation [63]. They also summarized related resources, experimental analyses, and prompt design guidelines for LLMs. Nevertheless, they have not categorized security evaluation frameworks, and have not conducted an analysis to classify the evaluation processes into categories. Existing review papers have covered certain aspects of security risks and evaluation methods for LLMs. On the other hand, a review that comprehensively covers the latest advancements in security evaluation techniques for LLMs is still lacking. Thus this paper will provide such a comprehensive review.

This literature review focuses on scholarly articles related to the security evaluation for LLMs, primarily sourced from two academic search engines: arXiv and Google Scholar. Our search parameters were confined to works published from the year 2020 to the present. The selection criteria for the articles included in this review were based on their relevance to the safety evaluation of LLMs. The selected articles were then thoroughly analyzed, synthesized, and reported in this review. The key contributions of this paper are as follows:

1. We conduct an in-depth study of the security evaluation frameworks for LLMs, distinguishing them into two categories: black-box frameworks and white-box frameworks. Additionally, we provide typical application examples of these frameworks.
2. We abstract Data-driven Evaluation, Attack Simulation, and Formal Verification as three types of security evaluation processes for LLMs, and we analyze application cases for each of these distinct processes.
3. We discuss the underexplored area of security evaluation for multimodal LLMs, the increasing demand for automatic security evaluation systems, and the broad range of vulnerabilities LLMs can fall prey to. Furthermore, we propose future work to focus on the development of automated evaluation platforms, comprehensive coverage of threats, and dedicated research into the unique vulnerabilities and security complexities of multimodal LLMs.

Here is a brief overview of the article's organization. This paper begins in Section 2 by introducing commonly used evaluation metrics in the safety assessment of large language models, including general metrics, metrics for dialogue-oriented language models, metrics for code-generation language models, and metrics for multimodal language models. Following this, Section 3 classifies the safety evaluation framework of large language models into two main categories: black-box evaluation frameworks and white-box evaluation frameworks, and elaborates on the hierarchical structure of each type of framework. Section 4 then categorizes the safety evaluation process of language models from a granular perspective into three types: data-driven evaluation, attack simulation, and formal verification, while providing examples of typical application scenarios for each type of evaluation process. Finally, Section 5 summarizes the main viewpoints of this study, namely, that there is a relative lack of research on the safety evaluation of multimodal language models, a scarcity of fully automated safety evaluation platforms, and inadequate coverage of threats to language models, etc. It also envisages future research directions from the development of automated evaluation platforms, comprehensive coverage of threats, and dedicated research on multimodal language models.

2 Evaluation Metrics

In this section, we introduce some common metrics used in the security evaluation of large models and discuss how they are utilized. We have delved into the evaluation indicators that measure the safety of LLMs. These metrics are crucial in determining the effectiveness of a model, its potential risks, and its ability to handle adversarial attacks. Theoretical analysis of these indicators has provided a basis for understanding the different aspects that influence a model's safety.

2.1 General Metrics

General Metrics refer to evaluation metrics that provide a general indication of the models' capabilities without being tailored to any specific type of LLMs. Here are some key examples.

- **F1 Score:** F1 score is a harmonic mean of precision and recall, combining them into a single metric that balances both [49]. It evaluates the predictive performance of classification models, ranging from 0 to 1 with 1 being the best. F1 score is useful for imbalanced datasets as it examines precision and recall together. To illustrate, the F1 score serves as a tool to evaluate the performance of classifiers for large language models in detecting unsafe text. A higher F1 score signifies, for instance, that the classifier can detect unsafe text more accurately.
- **Perplexity:** Perplexity (PPL) is a commonly used model evaluation metric in fields such as natural language processing and speech recognition. Perplexity is typically used to assess the quality of a language model, that is, the model's ability to predict a piece of text [39, 58]. The definition of Perplexity is the average branching factor when a given probabilistic model predicts a sample set. In the context of language models, the branching factor can be understood as the model's uncertainty in predicting the next word at a certain point in time. The smaller the Perplexity value, the better the performance of the model. By way of example, by testing the Perplexity of large language models for malicious inputs, it becomes clear that a higher Perplexity means the model is more uncertain when dealing with malicious inputs, thus enhancing its reliability from a safety perspective.
- **Attack Success Rate:** Attack Success Rate (ASR), calculated using an equation involving the attacked model, test sample, its target label, and an indicator function, indicates the fraction of successful to total attacks [13, 29]. Higher ASR denotes increased security risk due to potential backdoor exploitation. As an example, backdoor attacks can be executed on large language models and the proportion of successful attacks can be calculated to assess the security of the model. A lower ASR demonstrates, for instance, that the model has a stronger ability to withstand backdoor attacks.

2.2 Metrics for Dialogue-Oriented LLMs

Metrics for Dialogue-Oriented LLMs refer to evaluation metrics tailored for assessing the conversational abilities of LLMs designed for dialogue applications. These metrics aim to evaluate the natural, coherent during conversations. Some key examples are as following.

- **Coherence:** Measures how coherent and logically consistent a conversational response is [8, 25, 47]. Evaluation can be either automatically or through human rating. For example, evaluators have the opportunity to score the malicious dialogue responses generated by the language model. A lower Coherence score implies, for example, that the model's response to malicious inputs is less coherent, and hence safer.
- **Grammatical Errors:** Grammatical Errors are the number of errors in the adversarial example's grammar using LanguageTool [47]. LanguageTool is an open-source natural language processing tool that can detect grammatical errors in multiple languages [50], including English, German, French, etc. It can identify and mark typos, punctuation errors, breaches of grammatical rules, and inappropriate usage of language in the text. Counting the number of grammatical errors in adversarial examples can help us understand the robustness of a model, that is, the model's ability to handle erroneous inputs. For a high-quality language processing model, it should be able to provide relatively accurate and useful outputs even when faced with inputs that contain a large number of grammatical errors. Conducting such tests on the model can help us identify areas of the model that need to be improved or fixed. In illustration, intentionally grammatically incorrect malicious inputs can be fed into the language model, with the number of grammatical errors in the output counted. The presence of more errors indicates, for instance, that the model has not effectively processed the malicious input, thereby increasing safety.

2.3 Metrics for Code Generation-Oriented LLMs

Metrics for Code Generation-Oriented LLMs refer to the metrics of measuring and evaluating LLMs that are intended to be used for automatic code generation [30, 43].

- **Functional Correctness:** Functional Correctness refers to whether a program fulfills its intended functionality and behaves as expected for all possible inputs [31]. For example, HUMANEVAL contains programming problems with ground truth solutions and a limited set of test cases [19, 54]. To evaluate an LLM's code generation capability, its synthesized code snippets are run against the test cases in HUMANEVAL to check if the outputs match those of the ground truth solutions. To put it in perspective, it's possible to determine whether the code generated by the language model can pass functional correctness tests in security-related scenarios, thus forming an evaluation of its safety.

2.4 Metrics for Multimodal LLMs

Metrics for Multimodal LLMs refer to evaluation metrics designed specifically for assessing the performance of LLMs that process and generate multimodal outputs spanning text, images, audio, and video.

– **Perceptual ability:** Perceptual ability evaluation metrics measure how well Multimodal LLMs can understand and reason about perceptual concepts like images, audio, and video [6]. These metrics test the model's ability to generate accurate descriptions, answer questions, and make inferences about perceptual inputs. Higher scores indicate the model has better perceptual abilities and understanding. Lower scores suggest the model struggles with perceiving and reasoning about multimodal inputs. For instance, testing the multimodal language model's ability to comprehend visual inputs with malicious intent can be informative. A lower Perceptual Ability suggests, for instance, that the model is less likely to understand malicious visual inputs, thereby making it safer.

3 Evaluation Frameworks

In the field of security evaluation for LLMs, it is crucial to have a system for evaluating the performance and safety of these models. An evaluation framework for LLMs refers to a comprehensive set of testing methodologies, metrics, and procedures designed to thoroughly assess the capabilities and behaviors of LLMs. Based on the literature reviewed, we explore the security evaluation frameworks that have been developed and implemented, categorize these frameworks into two types according to the traditional black-box/white-box evaluation, and abstract each type of framework into three layers: a user interaction layer, an LLM layer, and an evaluation benchmark layer. The security evaluation frameworks, as depicted in Table 1, can be employed to assess a range of aspects, and provide standardized benchmarks that allow standardized comparisons across different LLMs. By thoroughly evaluating LLMs across these critical dimensions, an effective framework provides comprehensive insight into model strengths, weaknesses and overall readiness for real-world deployment.

3.1 Black-Box Evaluation Framework

In such a security evaluation framework, as shown in Fig. 2 (a), the LLM itself is regarded as a black box, obtaining feedback through interaction with external tools, achieving self-verification and self-improvement [9, 22].

User Interaction Layer: This layer focuses on interacting with users and collecting test prompts. It uses public datasets as input or designs prompts covering different safety scenarios and instruction attacks to test the safety issues of LLMs. The prompts are designed to be more likely to trigger unsafe responses.

LLM Layer: This layer utilizes the pre-trained LLMs to generate responses given the prompts collected from the user interaction layer. Different LLMs can be evaluated and their safety issues can be exposed based on their generated responses.

Evaluation Benchmark Layer: This layer evaluates the safety of the generated responses using an LLM evaluator. It compares the prompt and response to judge whether the response is safe or not. The safety scores are then calculated and ranked on a public leaderboard, or a report is generated for response. This layer provides an automatic/semi-automatic and standardized way to assess the safety of LLMs.

Table 1. Typical Applications for Black Box/White Box Security Evaluation Frameworks for LLMs.

Applications	Categories	Description
Security and robustness assessments [38, 39, 42, 60]	Black-box Evaluation Framework	For example, [51] have developed a security evaluation system for Chinese language models. Their work investigates the comprehensive security of these models across two dimensions: eight typical security scenarios and six challenging command attack types. The typical security scenarios encapsulate eight prevalent types of security concerns: insults, unfairness and discrimination, crime and illegal activities, sensitive topics, physical harm, mental health, privacy and property, and ethics and morality. The command attacks involve six types of directive input attacks: target hijacking, prompt leakage, role-playing commands, unsafe command topics, queries with unsafe views, and reverse exposure. Using this security evaluation system, they have conducted security evaluations on 15 well-known Chinese language models, including those from the GPT series
Functional correctness evaluation [7, 21, 31, 37, 43]		
Downstream performance benchmarking [33, 47]		
Human studies [10, 45]		
Safety and ethics evaluation [28, 51, 59]		
Representation analysis [29, 40, 46, 53, 62]	White-box Evaluation Framework	For example, [48] proposed a comprehensive framework that employs model evaluation as a tool for identifying and mitigating extreme risks. This framework critically examines the potential hazards, likelihood of misalignment, and the agency of LLMs from various perspectives. Importantly, this rigorous examination is conducted at all stages of the model lifecycle, including pre-training, post training, pre-deployment, and post deployment
Behavior characterization [19, 20, 24]		
Model inversion [17, 48, 61]		

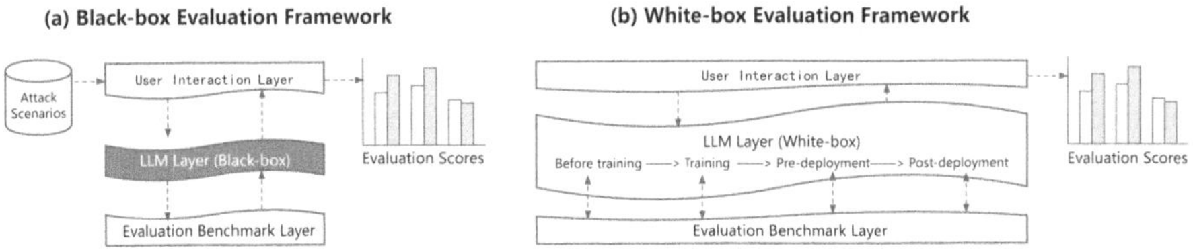

Fig. 2. LLMs evaluation frameworks, which include: (a) black-box evaluation framework, (b) white-box evaluation framework.

3.2 White-Box Evaluation Framework

White-box Evaluation Framework regards the language model itself as transparent, rather than a black box. Instead of solely relying on external tools for verification and improvement, this framework enables internal processes of self-evaluation [33]. By introspecting on its own knowledge, reasoning, and behavior, the model can identify areas needing improvement. In the White-box Evaluation Framework, we have additional visibility into the internal mechanisms of the LLM compared to the black-box framework as shown in Fig. 2 (b). This enables more rigorous testing and verification of model safety. The three layers are as follows.

User Interaction Layer: Similar to the black-box framework, this layer focuses on collecting user commands and feedback on test results. However, in white-box testing, user interaction is not limited to entering pre-designed evaluation scenarios. Users can also visually modify model parameters and training data. Furthermore, evaluation results before, during, and after training and deployment should be provided in real-time, empowering users to make more informed decisions about improving model safety. Rather than passively receiving prompts, users actively shape the model's internal representations. They fine-tune mechanisms found to be unsafe, and retrain on modified datasets that eliminate harmful biases. This continues iteratively until the desired safety criteria are met. Such transparent, user-driven refinement of the LLM's internals is only possible in the white-box setting.

LLM Layer: In addition to generating responses to prompts, the white-box framework grants access to inspect the model's internal states, such as activations and gradients. This facilitates techniques like unit testing, mechanism analysis, and verification of intended behaviors. For example, researchers can trace whether certain goal representations emerge in the model. This transparency also aids in understanding the model's limitations and potential biases, contributing to safer and more trustworthy AI systems.

Evaluation Benchmark Layer: As in the black-box case, this layer judges the safety of the LLM's outputs. However, conducting white-box security evaluations on LLMs at different stages is integral to their development and deployment. This process begins even before training, where white-box security assessments on pre-training models help to discover any existing dangerous capabilities or misalignments. These findings provide invaluable guidance for subsequent training. At this stage, targeted training plans can be designed to specifically avoid the emergence of such dangerous capabilities or to improve alignments. The evaluation process continues during the training phase. Regular assessments are conducted on the model to pinpoint any new dangerous capabilities or misalignments that may appear during the training process. The results of these evaluations then inform necessary adjustments to the training methods. If necessary, any problematic training can be terminated or rolled back to ensure the model's security. Prior to deployment, a comprehensive white-box security evaluation of the model is conducted to determine its readiness for deployment. This assessment helps to identify key risk areas that need to be addressed before the model can be safely deployed. The evaluation results can also inform the establishment of deployment restrictions, such as defining the scope of use or access levels. Once the model is deployed, the evaluation process doesn't stop. It is essential to monitor for any new issues that may arise in the real-world application of the model. User feedback is collected to continually improve the evaluation plan. This feedback loop allows for adjustments to the model or the setting of usage restrictions based on newly discovered issues. Furthermore, post-deployment monitoring is crucial to verify the effectiveness of the security measures that have been put in place.

4 Evaluation Processes

Security evaluation process for LLMs refers to the specific application process of the evaluation system built based on the evaluation framework. We categorize the processes of evaluating the security of LLMs according to the granularity of evaluation, ranging from coarse-grained to fine-grained. The Table 2 offers a comprehensive overview of three processes used to evaluate LLMs, illustrated with selected typical cases for each process.

Table 2. Typical use cases for security evaluation processes for LLMs.

Category	Description	Reference
Data-driven Evaluation	Build the RealToxicityPrompts dataset for evaluating toxicity generation in language models [14]	Gehman et al. (2020)
	Build a dataset containing approximately 40,000 red-blue adversarial dialogues, which can be used to analyze different types of potential risks, establish automated adversarial testing, and more [13]	Ganguli et al. (2022)
	Build the SafeText dataset for evaluating language models' awareness of physical harm [28]	Levy et al. (2022)
	Build the Latent Jailbreak dataset for evaluating text safety and output robustness of language models [42]	Qiu et al. (2023)
	Build the LLMSecEval dataset for security evaluations of language models [54]	Tony et al. (2023)
Attack Simulation	Choose GitHub Copilot as the evaluation target, use MITRE's CWE vulnerability database, design 89 different scenarios, and evaluate the vulnerability risks of GitHub Copilot generating code in different securityrelated scenarios [37]	Pearce et al. (2022)
	Construct a prompt dataset for CHATGPT jailbreak, and comprehensively evaluate the ability and robustness of different jailbreak prompts in bypassing CHATGPT restrictions [32]	Liu et al. (2023)
	Construct image, audio, and dialogue datasets for the two open-source multimodal language models, LLaVA and PandaGPT [3]	Bagdasaryan et al. (2023)
	Test the ability of ChatGPT to generate malicious code, phishing emails, 0day attacks, macros, and Living Off The Land Binaries (LOLBINs) [41]	Qammar et al. (2023)

(*continued*)

Table 2. (*continued*)

Category	Description	Reference
Formal Verification	Define privacy rules such as not disclosing identity information and not memorizing training data, and use automated methods to check model outputs against these predefined rules [36]	Pan et al. (2020)
	Quantitatively define the concept of "memory" in LLMs, that is, text fragments that only appear in a small number of training samples, and design an experiment to extract private information [5]	Carlini et al. (2021)
	Formalize the task of auditing LLMs as a discrete optimization problem, and optimize from random initialization of inputs and outputs on language models like GPT-2, until finding input-output pairs that satisfy the objective function and constraints [22]	Jones et al. (2023)

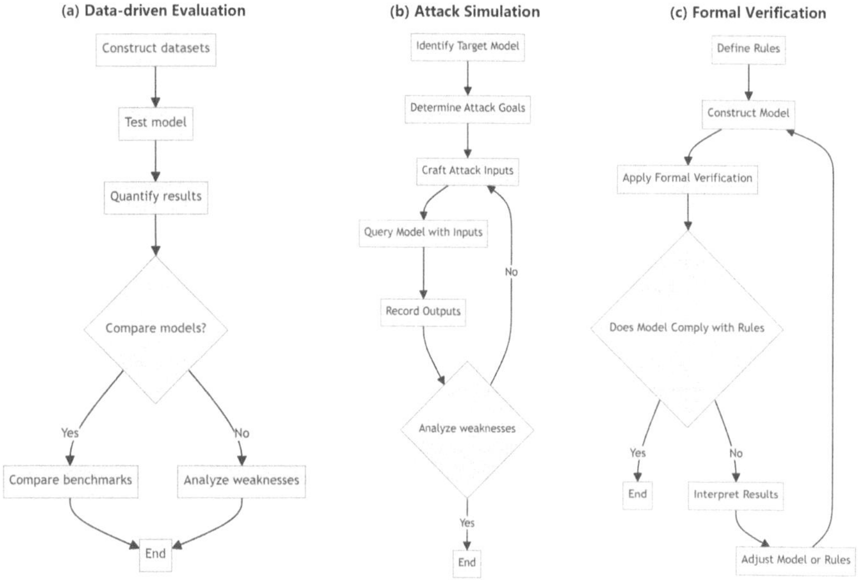

Fig. 3. Three major language model security evaluation processes, which include: (a) data-driven evaluation, (b) attack simulation, (c) formal verification.

The different processes include data-driven evaluation, attack simulation, and formal verification, each with its unique applications and nuances. Data-driven evaluation is a versatile approach not restricted to the evaluation of a specific model or a set of models. This process is characterized by the universality of its evaluation datasets, making it suitable for broad applications across different types of models. Attack simulation, in contrast, is a more targeted process where the evaluation dataset is chosen based on the specific model and evaluation objectives. This approach allows for a more personalized evaluation, specifically tailored to the nuances of the model under scrutiny. Formal verification is primarily focused on the evaluation of fine-grained tasks and iterative optimization of models. This process facilitates a deeper understanding of the model's performance on specific tasks, and aids in guiding the model's development and improvement.

Despite the nuances of each process, they are not mutually exclusive. Instead, they can be transformed and adapted according to different evaluation scenarios. This allows for a comprehensive, multi-faceted evaluation of LLMs, ensuring a more accurate understanding of their functioning and performance, and improving LLMs' security.

4.1 Data-Driven Evaluation

Data-driven Evaluation refers to a prevalent approach which is creating tailored datasets to probe known vulnerabilities [42, 53]. For example, REALTOXICITYPROMPTS consists of 100K naturally occurring toxic prompts to test if LLMs generate harmful text [14]. Adversarial NLI builds sentence pairs to evaluate an LLM's ability to detect contradictions. By crafting test cases that require reasoning about security, data-driven evaluation offers a scalable way to benchmark model security. However, dataset biases may skew conclusions [27]. The main steps of Data-driven Evaluation as shown in Fig. 3 (a) are:

(1) Construct customized datasets to expose vulnerabilities.
(2) Test LLMs on datasets and observe outputs.
(3) Quantify security metrics like attack accuracy rate.
(4) Compare with other LLMs.
(5) Analyze and identify model weaknesses.

4.2 Attack Simulation

Attack Simulation refers to testing LLMs through simulated attacks to evaluate their security and robustness. Researchers design attacks meant to trick, mislead, or exploit weaknesses in LLMs in order to assess vulnerabilities. Researchers directly perform attacks by querying LLMs with carefully designed inputs [16, 23, 38, 64]. For instance, targeted misspellings and paraphrasing help circumvent blocking of dangerous keywords. Multi-turn conversations reveal risks of extracted information accumulation. Attack simulation provides direct evidence of vulnerabilities. But new defenses may impede replication. Documenting details of the experimental setup is crucial.

Inspection-based Analysis inspection of model parameters, training data, and generated texts can reveal security issues. Analyzing correlations between training data artifacts and model outputs can expose unintended memorization. Manual inspection offers intuitions into failure modes. But it relies on expert judgments and small samples. Developing rigorous auditing protocols is an open challenge. Here are key steps in Fig. 3 (b):

(1) Identify target LLM to evaluate
(2) Determine attack goals and scenarios, e.g. extracting sensitive messages from training data.
(3) Craft attacks using carefully designed LLM inputs.
(4) Query the LLM with attack inputs and record outputs.
(5) Analyze if attack goals were achieved based on outputs.
(6) Iterate attack approaches to improve effectiveness.
(7) Draw conclusions on LLM vulnerabilities.

4.3 Formal Verification

Formal Verification serves as a methodology for inspecting whether a system adheres to a set of predefined specifications [18, 48]. In the context of evaluating the safety of LLMs, Formal Verification can be employed to ensure the behavior of the model aligns with predetermined rules or standards [12, 20]. Consider an instance where we desire to validate the LLM's behavior when it processes sensitive information. Rules can be established, such as the model should not generate outputs that could potentially disclose personal identifying information when given such inputs. Subsequently, we apply Formal Verification to examine whether the model complies with these rules. The key steps as shown in Fig. 3 (c) are:

(1) Identify the rules or standards you wish the model to observe. These should be designated based on your specific application and the safety concerns you are addressing.
(2) Constructing model, which typically involves training the LLM.
(3) Implement Formal Verification techniques to inspect whether your model complies with the rules defined in the first step.
(4) Based on the results of Formal Verification adjust the model or the rules defined in the first step.
(5) Repeat the above steps as needed, until the LLM satisfies all rules.

5 Conclusions and Perspectives

This paper has provided a comprehensive analysis of evaluation metrics, evaluation frameworks, and evaluation processes for LLMs. Here are the main findings.

- Comparatively, research into multimodal LLMs is less prevalent than in text and code models, not only in terms of defining evaluation metrics but also in the design and development of evaluation frameworks. This discrepancy might be due to the smaller size and utilization of multimodal models relative to other types.
- Fully automated platforms for security evaluation of LLMs are rarely released. However, given the increasing number of applications based on LLMs, the demand for automatic security evaluation is growing day by day.
- Pertaining to the types of vulnerabilities in the evaluation, LLMs are prone to a wide range of threat vulnerabilities. Even when the scope is limited to OWASP Top 10 for LLMs, which is a list of the top 10 security risks for LLM applications proposed by the Open Web Application Security Project (OWASP) [35], no current security evaluation framework can completely cover and verify all of them.

Given the current state of research, there is a clear need for more dedicated attention to security threats and evaluations of LLMs. Future work should focus on the following possible directions.

- **Development of Automated Evaluation Platforms:** There is a pressing need for the development and release of fully automated platforms for LLMs' security evaluation. These platforms could help streamline the process of identifying vulnerabilities in LLMs and thus, contribute to their safe and secure deployment.

- **Comprehensive Coverage of Threats:** Future research should aim at developing evaluation frameworks that can cover a wider range of threat vulnerabilities. It is crucial to consider not only the OWASP Top 10 for LLMs but also other potential threats that might emerge with the evolution of LLMs and their applications.
- **Emphasize the area of multimodal LLMs:** Given the limited research on the security evaluation of multimodal LLMs to date, future research should emphasize this area. Exploring the development of mature security evaluation systems for multimodal LLMs and their specific vulnerabilities could provide a more holistic understanding of the security complexities involved with LLMs.

In conclusion, while significant strides have been made in understanding the security aspects of LLMs, there is still much to learn. It is hoped that this work will stimulate further research in this critical domain.

Acknowledgements. We would like to acknowledge the role of AI and AI-assisted writing technologies, specifically the Claude2 and GPT-4 models, in refining the grammar of this manuscript.

In addition to technological support, this research was funded by the Major Key Project of PCL (Grant No. PCL2022A03-3), NSFC (No. 62072131, 61972106), Key R&D Program of Guangdong Province (No.2019B010136003), Guangdong Basic and Applied Basic Research Foundation (No.2022A1515011401), DongGuan Innovative Research Team Program (No. 2018607201008).

References

1. Al-Kaswan, A., Izadi, M.: The (ab) use of open source code to train large language models (2023)
2. Alayrac, J.B., et al.: Flamingo: a visual language model for few-shot learning. Adv. Neural Info. Proc. Sys. **35**, 23716–23736 (2022)
3. Bagdasaryan, E., Hsieh, T.Y., Nassi, B., Shmatikov, V.: (Ab)using Images and Sounds for Indirect Instruction Injection in Multi-Modal LLMs (2023). https://doi.org/10.48550/ARXIV.2307.10490
4. Borsos, Z., et al.: Audiolm: a language modeling approach to audio generation. IEEE/ACM transactions on audio, speech, and language processing **31**, 2523–2533 (2023)
5. Carlini, N., et al.: Extracting training data from large language models. In: 30th USENIX Security Symposium (USENIX Security 21). pp. 2633–2650 (2021)
6. Chang, Y., et al.: A Survey on Evaluation of Large Language Models (2023). https://doi.org/10.48550/ARXIV.2307.03109
7. Chen, M., et al.: Evaluating Large Language Models Trained on Code (2021). https://doi.org/10.48550/ARXIV.2107.03374
8. Chen, Y., Wang, R., Jiang, H., Shi, S., Xu, R.: Exploring the Use of Large Language Models for Reference-Free Text Quality Evaluation: A Preliminary Empirical Study (2023). https://doi.org/10.48550/ARXIV.2304.00723
9. Chia, Y.K., Hong, P., Bing, L., Poria, S.: INSTRUCTEVAL: Towards Holistic Evaluation of Instruction-Tuned Large Language Models (2023). https://doi.org/10.48550/ARXIV.2306.04757
10. del Rio-Chanona, M., Laurentsyeva, N., Wachs, J.: Are large language models a threat to digital public goods? Evidence from Activity on Stack Overflow (2023). https://doi.org/10.48550/ARXIV.2307.07367

11. Etherington, D.: Microsoft invests $1 billion in OpenAI in new multiyear partnership (2019)
12. Fan, M., Chen, C., Wang, C., Huang, J.: On the Trustworthiness Landscape of State-of-the-art Generative Models: A Comprehensive Survey (2023)
13. Ganguli, D., et al.: Red teaming language models to reduce harms: Methods, scaling behaviors, and lessons learned (2022)
14. Gehman, S., Gururangan, S., Sap, M., Choi, Y., Smith, N.A.: Realtoxicityprompts: Evaluating neural toxic degeneration in language models (2020)
15. Glukhov, D., Shumailov, I., Gal, Y., Papernot, N., Papyan, V.: LLM Censorship: A Machine Learning Challenge or a Computer Security Problem? (2023). https://doi.org/10.48550/ARXIV.2307.10719
16. Greshake, K., et al.: More than you've asked for: A Comprehensive Analysis of Novel Prompt Injection Threats to Application-Integrated Large Language Models (2023)
17. Hajipour, H., Holz, T., Schönherr, L., Fritz, M.: Systematically Finding Security Vulnerabilities in Black-Box Code Generation Models (2023). https://doi.org/10.48550/ARXIV.2302.04012
18. Huang, X., et al.: A Survey of Safety and Trustworthiness of Large Language Models through the Lens of Verification and Validation (2023). https://doi.org/10.48550/ARXIV.2305.11391
19. Huang, Y., Song, J., Wang, Z., Chen, H., Ma, L.: Look Before You Leap: An Exploratory Study of Uncertainty Measurement for Large Language Models (2023). https://doi.org/10.48550/ARXIV.2307.10236
20. Jain, N., et al.: Bring Your Own Data! Self-Supervised Evaluation for Large Language Models (2023). https://doi.org/10.48550/ARXIV.2306.13651
21. Jesse, K., Ahmed, T., Devanbu, P.T., Morgan, E.: Large Language Models and Simple. Stupid Bugs (2023). https://doi.org/10.48550/ARXIV.2303.11455
22. Jones, E., Dragan, A., Raghunathan, A., Steinhardt, J.: Automatically auditing large language models via discrete optimization (2023). https://doi.org/10.48550/ARXIV.2303.04381
23. Kang, D., et al.: Exploiting programmatic behavior of llms: Dual-use through standard security attacks (2023)
24. Kim, S., et al.: ProPILE: Probing Privacy Leakage in Large Language Models (2023). https://doi.org/10.48550/ARXIV.2307.01881
25. Kucharavy, A., et al.: Fundamentals of Generative Large Language Models and Perspectives in Cyber-Defense (2023). https://doi.org/10.48550/ARXIV.2303.12132
26. Labbe, M.: Nvidia unveils A100 GPU for demanding AI workloads. https://www.techtarget.com/searchenterpriseai/news/252483188/Nvidia-unveils-A100-GPU-for-demanding-AI-workloads (2020). Accessed 7 September 2023
27. Lee, H., et al.: KoSBi: A Dataset for Mitigating Social Bias Risks Towards Safer Large Language Model Application (2023). https://doi.org/10.48550/ARXIV.2305.17701
28. Levy, S., et al.: SafeText: A Benchmark for Exploring Physical Safety in Language Models. In: Proceedings of the 2022 Conference on Empirical Methods in Natural Language Processing. pp. 2407–2421. Association for Computational Linguistics, Abu Dhabi, United Arab Emirates (2022). https://doi.org/10.18653/v1/2022.emnlp-main.154
29. Li, S., et al.: Hidden Backdoors in Human-Centric Language Models. In: Proceedings of the 2021 ACM SIGSAC Conference on Computer and Communications Security, pp. 3123–3140. CCS '21, Association for Computing Machinery, New York, NY, USA (2021). https://doi.org/10.1145/3460120.3484576
30. Liguori, P., Improta, C., Natella, R., Cukic, B., Cotroneo, D.: Who evaluates the evaluators? On automatic metrics for assessing AI-based offensive code generators. Expert Systems with Applications **225**, 120073 (2023). https://doi.org/10.1016/j.eswa.2023.120073
31. Liu, J., Xia, C.S., Wang, Y., Zhang, L.: Is Your Code Generated by ChatGPT Really Correct? Rigorous Evaluation of Large Language Models for Code Generation (2023). https://doi.org/10.48550/ARXIV.2305.01210

32. Liu, Y., et al.: Jailbreaking ChatGPT via Prompt Engineering: An Empirical Study (2023). https://doi.org/10.48550/ARXIV.2305.13860
33. OpenAI: GPT-4 Technical Report (2023). https://doi.org/10.48550/ARXIV.2303.08774
34. OpenBMB: BMList. https://openbmb.github.io/BMList/ (2023). Accessed 7 September 2023
35. OWASP: OWASP Top 10 for Large Language Model Applications. https://owasp.org/www-project-top-10-for-large-language-model-applications/ (Aug 2023). Accessed 7 September 2023
36. Pan, X., Zhang, M., Ji, S., Yang, M.: Privacy risks of general-purpose language models. In: 2020 IEEE Symposium on Security and Privacy (SP). pp. 1314–1331 (2020). https://doi.org/10.1109/SP40000.2020.00095
37. Pearce, H., Ahmad, B., Tan, B., Dolan-Gavitt, B., Karri, R.: Asleep at the keyboard? assessing the security of GitHub copilot's code contributions. In: 2022 IEEE Symposium on Security and Privacy (SP). pp. 754–768 (2022). https://doi.org/10.1109/SP46214.2022.9833571
38. Pedro, R., Castro, D., Carreira, P., Santos, N.: From Prompt Injections to SQL Injection Attacks: How Protected is Your LLM-Integrated Web Application? (2023). https://doi.org/10.48550/ARXIV.2308.01990
39. Perez, F., Ribeiro, I.: Ignore Previous Prompt: Attack Techniques For Language Models (2022). https://doi.org/10.48550/ARXIV.2211.09527
40. Pezeshkpour, P.: Measuring and Modifying Factual Knowledge in Large Language Models (2023). https://doi.org/10.48550/ARXIV.2306.06264
41. Qammar, A., et al.: Chatbots to ChatGPT in a Cybersecurity Space: Evolution, Vulnerabilities, Attacks, Challenges, and Future Recommendations (2023)
42. Qiu, H., Zhang, S., Li, A., He, H., Lan, Z.: Latent Jailbreak: A Benchmark for Evaluating Text Safety and Output Robustness of Large Language Models (2023). https://doi.org/10.48550/ARXIV.2307.08487
43. Ren, S., et al.: CodeBLEU: A Method for Automatic Evaluation of Code Synthesis (2020). https://doi.org/10.48550/ARXIV.2009.10297
44. Sagar, R.: OpenAI Releases GPT-3, The Largest Model So Far. https://analyticsindiamag.com/open-ai-gpt-3-language-model/ (2020). Accessed 7 September 2023
45. Sandoval, G., et al.: Lost at C: A User Study on the Security Implications of Large Language Model Code Assistants (2022). https://doi.org/10.48550/ARXIV.2208.09727
46. Shao, H., Huang, J., Zheng, S., Chang, K.C.C.: Quantifying Association Capabilities of Large Language Models and Its Implications on Privacy Leakage (2023). https://doi.org/10.48550/ARXIV.2305.12707
47. Shen, X., Chen, Z., Backes, M., Zhang, Y.: In ChatGPT We Trust? Measuring and Characterizing the Reliability of ChatGPT (2023). https://doi.org/10.48550/ARXIV.2304.08979
48. Shevlane, T., et al.:: Model evaluation for extreme risks (2023). https://doi.org/10.48550/ARXIV.2305.15324
49. Stiff, H., Johansson, F.: Detecting computer-generated disinformation. Int. J. Data Sci. Anal. **13**(4), 363–383 (2022). https://doi.org/10.1007/s41060-021-00299-5
50. Stratus-Security: FinGen Penetration Testing Findings Generator. https://github.com/Stratus-Security/FinGen (2023). Accessed 7 September 2023
51. Sun, H., Zhang, Z., Deng, J., Cheng, J., Huang, M.: Safety Assessment of Chinese Large Language Models (2023). https://doi.org/10.48550/ARXIV.2304.10436
52. Tamkin, A., Brundage, M., Clark, J., Ganguli, D.: Understanding the capabilities, limitations, and societal impact of large language models (2021)
53. Tian, J.J., et al.: Soft-prompt Tuning for Large Language Models to Evaluate Bias (2023). https://doi.org/10.48550/ARXIV.2306.04735

54. Tony, C., Mutas, M., Ferreyra, N.E.D., Scandariato, R.: LLMSecEval: A Dataset of Natural Language Prompts for Security Evaluations (2023). https://doi.org/10.48550/ARXIV.2303.09384
55. van der Weij, T., Lermen, S., lang, L.: Evaluating Shutdown Avoidance of Language Models in Textual Scenarios (2023). https://doi.org/10.48550/ARXIV.2307.00787
56. Weidinger, L., et al.: Taxonomy of risks posed by language models. In: Proceedings of the 2022 ACM Conference on Fairness, Accountability, and Transparency. pp. 214–229. FAccT '22, Association for Computing Machinery, New York, NY, USA (2022). https://doi.org/10.1145/3531146.3533088
57. Wikipedia: Large language model. https://en.wikipedia.org/w/index.php?title=Large_language_model (2023). Accessed 7 September 2023
58. Xu, F.F., Alon, U., Neubig, G., Hellendoorn, V.J.: A systematic evaluation of large language models of code. In: Proceedings of the 6th ACM SIGPLAN International Symposium on Machine Programming. ACM (2022). https://doi.org/10.1145/3520312.3534862
59. Xu, G., et al.: CValues: Measuring the Values of Chinese Large Language Models from Safety to Responsibility (2023). https://doi.org/10.48550/ARXIV.2307.09705
60. Ye, W., et al.: Assessing Hidden Risks of LLMs: An Empirical Study on Robustness, Consistency, and Credibility (2023). https://doi.org/10.48550/ARXIV.2305.10235
61. Zhang, Y., Ippolito, D.: Prompts Should not be Seen as Secrets: Systematically Measuring Prompt Extraction Attack Success (2023). https://doi.org/10.48550/ARXIV.2307.06865
62. Zhao, S., Wen, J., Tuan, L.A., Zhao, J., Fu, J.: Prompt as Triggers for Backdoor Attack: Examining the Vulnerability in Language Models (2023). https://doi.org/10.48550/ARXIV.2305.01219
63. Zhao, W.X., et al.: A survey of large language models (2023)
64. Zou, A., Wang, Z., Kolter, J.Z., Fredrikson, M.: Universal and Transferable Adversarial Attacks on Aligned Language Models (2023)

LFPS: A Blockchain-Based Fair Privacy Preservation Scheme for Mobile Crowdsensing

Jie Chen(✉), Haodi Zhang, Shuai Wang, and Huamin Jin

China Telecom Research Institute, Guangzhou 510630, China
chenj135@chinatelecom.cn

Abstract. Mobile Crowd Sensing (MCS) is an emerging research field that leverages mobile devices and vehicles for performing various sensing tasks. MCS enables efficient and cost-effective completion of tasks like road and air quality monitoring. Despite continuous advancements in MCS, several challenges remain to be addressed. In this paper, we present a lightweight, fair, and privacy-protecting scheme, dubbed LFPS, to address two challenges in MCS. Firstly, to resist unfair vehicle selection, LFPS utilizes a blockchain-based fair selection algorithm to generate a public and verifiable vehicle list, where the blockchain system acts as a public verifiable bulletin board and an unpredictable source of randomness. Secondly, to prevent attackers from inferring private information by linking vehicles' uploaded data across different tasks, LFPS introduces an anonymous token mechanism, which allows vehicles to submit their sensing data anonymously. This mechanism is lightweight, as it does not require expensive cryptographic tools. We give a comprehensive security analysis to demonstrate LFPS's resistance against various attacks. We also evaluate the performance of LFPS to show its efficiency in terms of computational and communication overheads.

Keywords: Mobile Crowd Sensing · Fairness · Privacy preservation · Anonymity

1 Introduction

The development of sensors in mobile devices has greatly facilitated the progress of mobile crowd sensing (MCS) [1–3]. MCS, as a new data acquisition paradigm, has a good performance in sensing, collecting, and analyzing information about the surrounding environment. Many sensing tasks can be assigned to participants equipped with mobile devices, allowing them to collect and feedback the data. Compared to deploying dedicated sensors directly, MCS proves to be more cost-effective and time-efficient.

In the field of mobile crowdsensing, users can employ their mobile devices to participate in the sensing tasks. However, the limited mobility and restricted travel distance of users impose limitations on the applicability of MCS in various scenarios. Moreover, vehicles are now equipped with a multitude of sensors and components to assist in driving, making sensing vehicles an essential component of mobile crowd sensing [4–6].

Z. Gu et al. (Eds.): NSE 2023, CCIS 2063, pp. 151–168, 2024.
https://doi.org/10.1007/978-981-97-4519-7_11

Nonetheless, the open and dynamic vehicular network presents significant challenges to MCS. For instance, effectively managing and organizing a large number of vehicles, allocating tasks to vehicles, and ensuring privacy protection remain challenging tasks [7–9]. From a literature view, there are two approaches to achieve these goals. The first approach involves utilizing blockchain and smart contracts to accomplish all the operations [10–13]. Due to its characteristics of decentralization and anonymity, utilizing blockchain technology to accomplish these operations such as vehicle management and task allocation can offer significant advantages in terms of privacy protection. The second approach involves deploying a centralized server to manage vehicles and assign tasks [14–16]. Typically, the server acts as an intermediary between data requesters and vehicles. In this process, data requesters define specific tasks for the server, which then selects suitable vehicles and assigns these tasks to them. This method, compared to the first approach, offers higher throughput and flexibility, and it is the approach applied in this paper. It's important to note that this approach faces more security and privacy concerns, and numerous solutions have already been proposed to address these issues and enhance security and privacy protection [17–19].

In this paper, we focus on two critical issues which have not been fully addressed. The first issue focuses on how to resist unfair vehicle selection. Unfair selection leads to two concerns: 1) The centralized server may collude with certain vehicles and assign tasks with higher rewards to them, resulting in unfair benefits distribution. This situation can dampen the enthusiasm of other vehicles, thereby hindering the establishment of a conducive environment for crowd-sensing. 2) If the server is compromised, it might consistently choose a specific vehicle to complete tasks and continuously collect its data, potentially leading to the privacy leakage of vehicles. The second issue is that how to protect privacy when vehicles upload their sensing data. In most cases, the data uploaded by vehicles does not appear to leak their private information. However, if attackers can correlate the data from the same vehicle across multiple tasks, they may be able to infer some important information, such as trajectories. Therefore, the second issue focuses on how to resist adversaries from linking two instances of data uploaded by the same vehicle in two different tasks. Existing approaches often require expensive cryptographic tools, such as zero-knowledge proofs, which impose substantial communication and computational overheads.

To address the above issues, we propose a lightweight, fair, and privacy-protecting scheme called LFPS. To solve the first issue, we propose a fair selection algorithm that leverages blockchain to generate unpredictable and publicly verifiable random numbers. Specifically, before selecting vehicles, the server maintains a list of vehicles that intend to participate in tasks and meet the requirements of this task. Then it computes the hash value of the list and records it on the blockchain. Subsequently, it utilizes newly generated blocks to calculate random numbers, which are then used to determine the selected vehicles from the list. This algorithm ensures that the result is publicly verifiable. To address the second issue, we propose a lightweight anonymous token mechanism. The key idea is to allow vehicles to upload their data using anonymous identities rather than their real identities. Specifically, after selecting suitable vehicles, these vehicles employ blind signature to request anonymous tokens from the cloud server. The token contains only information associated with the current task and does not include any details regarding the identity of the vehicle. During the data uploading phase, vehicles use their anonymous tokens to connect to the server. Due to the blinding feature, the server can not link the token with the real identity of vehicles. LFPS offers a practical solution to the challenges of fair vehicle selection and privacy protection in mobile crowd sensing without imposing excessive computational and communication overheads.

The main contributions of this work are as follows.

- We propose an algorithm to resist unfair vehicle selection. With this algorithm, the cloud server can generate publicly verifiable random numbers, enabling each vehicle to verify the result by recomputing these random numbers. Notably, this algorithm has minimal intrusion into the blockchain system. Unlike systems that heavily rely on the blockchain and generate a substantial number of blockchain transactions, this algorithm generates only one blockchain transaction per task. Therefore it doesn't impose a significant burden on the blockchain, and the system throughput is minimally affected by the blockchain.
- We introduce an anonymous token mechanism to protect vehicles' identities. With this mechanism, vehicles can upload their sensing data anonymously, thus thwarting adversaries from linking two data uploading instances from the same vehicle. This mechanism is built on lightweight signature algorithms rather than additional expensive cryptographic tools like zero-knowledge proofs, therefore it avoids introducing significant computational and communication overheads.
- We integrate the aforementioned two techniques into a scheme, dubbed LFPS, preserving all the advantageous features of each individual technique. To validate the robustness of LFPS, we conduct a comprehensive security analysis, demonstrating its resilience against various types of attacks. Additionally, we evaluate LFPS's performance through implementation and thorough analysis, confirming its efficiency and practicality.

2 Related Work

Based on the focus of this work, this section is divided into two parts: participant recruitment and privacy security.

Participant recruitment is a crucial issue in the allocation of tasks within MCS. This process involves selecting suitable participants from a group of candidates to perform specific tasks. The are a lot of recent works that focus on this field. Guo et al. [20] proposed a worker selection framework for a multitask MCS environment, which considers both time-sensitive and delay-tolerant circumstances. Zhao et al. [21] proposed an efficient algorithm and applied it to a system where every task requires the participation of multiple workers. Wang et al. [22] proposed a dynamic incentive mechanism based on demand to balance participation among location-dependent tasks. To maximize the profits of participants, Wang et al. [23] proposed a task allocation scheme which introduced a two-level heterogeneous pricing mechanism. Sun et al. [24] proposed two algorithms to address the cost-fair issue in the online and offline cases, respectively. Wang et al. [25] considered the users' preferences and proposed a personalized task-worker mechanism to calculate the matching probability between tasks and workers. Zhang et al. [26]

proposed a privacy-persevering multitask scheme to enhance the privacy of user attribute preferences in the task allocation phase.

Privacy security [27–29] is also an important research topic in the field of crowdsensing. Participants are often required to share location and other information during the task allocation phase, and they must also upload sensing data upon task completion. Sensitive information, such as location and trajectories, can be contained in this data. Actually, users' location is critical for task allocation, as it affects the system's performance. In addition, to encourage users to better accomplish tasks, reputation is mentioned in many works [30–32], wherein users who successfully complete their tasks with high quality will earn a greater reputation. This, in turn, assists them in obtaining the next task. However, these works ignore the reputation value privacy. Some schemes [18, 33–35] realize the importance of reputation value privacy and propose corresponding privacy-preserving solutions. In addition, location privacy, and identity privacy are also critical issues in task allocation, and many works are proposed to address these concerns [19, 36, 37]. Typical privacy-enhancing techniques include differential privacy [14, 38, 40, 41] and cryptography [11, 17, 43].

We observe that there exist two challenges in current works. 1) Privacy issue. The majority of current works on privacy protection require expensive cryptographic tools. This approach will introduce significant computational and communication overhead, especially on the server side. Furthermore, most solutions have not yet implemented anonymous data uploading. Although the sensing data users upload typically do not contain direct privacy information, if an adversary links data uploaded by a participant across multiple tasks, it might infer some private information about the participant. 2) Participant recruitment issue. Current research on participant recruitment primarily focuses on minimizing sensing costs, maximizing task completion ratio, improving data quality, and so on, while lacking sufficient attention to fairness considerations. Unfair selection can potentially demotivate vehicles and raise privacy concerns among them. Therefore, in this paper, we propose LFPS to address these two issues.

3 System Model and Design Goals

3.1 System Model

The system model, as illustrated in Fig. 1, comprises three types of entities: data requesters, vehicles, and the cloud server.

Data Requesters: Data requesters represent entities such as management departments or researchers. They play a crucial role in the system by assigning sensing tasks to the cloud server. They assign their sensing tasks to the cloud server to obtain sensing data or valuable information, e.g., traffic information.

Vehicles: Vehicles in the system are equipped with a variety of sensors, enabling them to collect and provide diverse sensing data. Vehicles register with the cloud server and receive sensing tasks that match their capabilities and the specified sensing requirements.

Cloud Server: The cloud server acts as the central entity in the system, possessing robust storage and computing resources. Its primary functions include collecting sensing tasks from data requesters and efficiently assigning these tasks to registered vehicles that meet the specific sensing requirements.

Fig. 1. System model.

LFPS encompasses the following steps.

① Register. Each sensing vehicle is required to register with the cloud server. Only registered vehicles are eligible to receive tasks from the server.
② Assign tasks. Data requesters, who are responsible for creating sensing tasks, assign these tasks to the cloud server. They provide detailed descriptions of these tasks, including their requirements, such as reputation thresholds and specific location requirements.
③ Broadcast tasks. The cloud server, having received the sensing tasks and their associated requirements from data requesters, broadcasts these tasks to all registered vehicles within the system.
④ Recruit tasks. Each sensing vehicle interested in undertaking tasks logs into the cloud server and submits its request to participate. This request typically includes the vehicle's reputation score and its current location information.
⑤ Select vehicles. The cloud server continuously collects requests from sensing vehicles. It then selects a subset of vehicles that satisfy the requirements required for sensing tasks. These selected vehicles are subsequently assigned tasks.
⑥ and ⑦ Submit data. The vehicles submit their sensing data to the cloud server which gathers all the sensing data and uploads it to the respective data requester.

3.2 Assumption

Blockchain. The time required for the blockchain to generate a new block is not fixed. Therefore, we denote t_b as the upper bound for the blockchain to generate a new block. Additionally, a transaction is not confirmed by the blockchain immediately upon broadcasting. The confirmation time may vary based on factors such as the number of transactions in the transaction pool and the transaction fees. We denote t_t as the upper bound for a transaction to be added to the blockchain. Furthermore, We assume that attackers cannot control the blockchain, and the block hash values are unpredictable.

Vehicles. We assume that there is no inherent communication or collaboration between individual vehicles in the system. Vehicles cannot establish direct communication or engage in collusion with one another.

Tasks. We assume that tasks in the system require the involvement of multiple vehicles to be completed. If a task only requires the participation of a single vehicle, the anonymity provided by our method cannot be achieved.

4 Design Goals

LFPS should achieve the following goals.

1) **Functionality**: The LFPS scheme should effectively and concurrently achieve fair vehicle selection and anonymous data uploading. It should ensure that tasks are assigned fairly to vehicles while safeguarding their identities during data transmission.
2) **Security**: LFPS should achieve the following security requirement: (a) Resistance against malicious cloud server. LFPS should be resilient against a malicious cloud server that attempts to select vehicles unfairly. It should prevent the server from colluding with specific vehicles to gain undue advantages. (b) Resistance against linking data. LFPS should resist adversaries from linking two instances of data uploaded by the same vehicle in two different tasks, ensuring that user privacy is preserved. (c) Resistance against malicious vehicles. LFPS should also guard against malicious vehicles that are not assigned tasks to upload sensing data.
3) **Efficiency**: The scheme needs to be efficient and conveniently implemented.

5 Construction

5.1 Preliminary

Bilinear Map. Suppose that G and G_T are an additive group and a multiplicative group, respectively, with the same order p. $e : G \times G \rightarrow G_T$ is a bilinear pairing if it has the following three properties: Bilinearity. $e(aP, bQ) = e(P, Q)^{ab}, \forall P, Q \in G, a, b \in Z^*$. Non-degeneracy. $\forall P, Q \in G, P \neq Q, e(P, Q) \neq 1$. Computability. There exists an efficient algorithm to compute e.

Blockchain. Blockchain is a decentralized and distributed digital ledger system that has gained widespread adoption across various industries. At its core, a blockchain is composed of a series of interconnected blocks, each containing a bundle of transactions. These blocks are securely linked using cryptographic hash functions, creating an unalterable chain. In LFPS, the blockchain is utilized to produce two functionalities: a public-verifiable time-proofed bulletin board and a public-verifiable randomness source. For a blockchain system, the security level it can achieve is related to the number of participants, therefore, opting for a more widely adopted blockchain system can enhance the credibility of the solution. As our scheme doesn't create a large number of transactions, public blockchain is more suitable for our scheme. In addition, any blockchain system can be utilized to achieve the second functionality, i.e., the randomness source. However, the first functionality requires additional fields in blockchain transactions to accommodate custom content. Currently, popular public chains include Ethereum, Bitcoin, and others. Given that Bitcoin does not support the additional field, Ethereum becomes a suitable choice for our proposal.

5.2 Overview

To prevent the cloud server from choosing vehicles arbitrarily, we propose a fair and verifiable vehicle selection algorithm that leverages the immutable and unpredictable features of blockchain. Specifically, before selecting vehicles, the server needs to publish the hash value of all eligible vehicles to the blockchain. This step is crucial as it ensures that the server cannot manipulate the vehicle list arbitrarily, including altering the order of vehicles in the list. Due to the collision-resistance of hash functions, any such manipulation would result in a different hash value. Subsequently, the server employs a newly generated blockchain block to compute a series of random numbers. These random numbers are subsequently employed to determine the chosen vehicles. Importantly, the generation of the new blockchain block occurs after the server has sent the hash to blockchain. In addition, the block hash value cannot be predicted or manipulated beforehand, therefore the server cannot pre-select vehicles before the new block is generated. This ensures that vehicle selection is random and fair.

To address the data privacy concern, we introduce an anonymous credential algorithm. This algorithm allows selected vehicles to request anonymous credentials using blind signature technology. Due to the feature of blind signatures, the server remains unaware of signing messages. After obtaining the blind signature, vehicles can then de-blind it to acquire a valid token. During the data upload phase, vehicles no longer authenticate themselves with their actual identities. Instead, they utilize their anonymous tokens. Since these tokens are legitimate and valid, the server can verify their authenticity and permit vehicle data uploads. Furthermore, the token doesn't contain any vehicle identity information, as a result, the server cannot determine which vehicle is uploading the data.

5.3 Detailed Description

LFPS includes three types of entities: vehicles $\{\mathcal{V}_1, \mathcal{V}_2, \ldots\}$, multiple data requesters $\{\mathcal{DR}_1, \mathcal{DR}_2, \ldots\}$, and the cloud server $\mathcal{CS}$.

Setup. The system initializes public parameters. Given a security parameter l, the public parameter $PP = \{p, P, G, G_T, e, H_1, H_2, Sign\}$ is determined. G is an additive group with the prime order p. P is the generator of G. G_T is a multiplicative group with the same order. e: $G \times G \rightarrow G_T$ is the bilinear map. H1: $\{0, 1\}^* \rightarrow Z^*$, H_2: $\{0, 1\}^* \rightarrow G$ are two secure hash functions. $Sign$ is a signature algorithm. After that, the cloud server initializes a signing key pair (sk, $pk = sk \cdot P$), which is used for anonymous tokens, and broadcasts the public key pk.

Registering. Sensing vehicles need to register with the server. Only registered vehicles are eligible to receive task information and participate in tasks.

Assigning and Broadcasting Tasks. A data requester $\mathcal{DR}$ defines the sensing content m (i.e., the specific task that the sensing vehicles need to complete),and the task requirements, including the reputation and location requirements. $\mathcal{DR}$ sends the task information to the cloud server $\mathcal{CS}$.

Upon receiving the task from $\mathcal{DR}$, $\mathcal{CS}$ subsequently identifies four time points referred to as $t1$, $t2$, $t3$, and $t4$. Among these, $t1$ represents the deadline for vehicle requests, allowing any eligible vehicles to submit participation requests to $\mathcal{CS}$. $t2$ denotes the initiation time of vehicle selection, marking when the cloud server starts computing the final list of selected vehicles. $t2$ satisfies $t2 > t1 + t_t$, where t_t represents the upper bound time for a transaction to be added by the blockchain. $t3$ denotes the start time of this task. At this point, the list of all selected vehicles has been finalized, and they can begin task execution. $t3$ satisfies the condition $t3 > t2 + 2 \times t_b$, where t_b represents the upper bound for generating a new block. $t4$ indicates the task deadline. Vehicles must upload their data to the server before $t4$. $\mathcal{CS}$ randomly generates a task identifier, referred to as *TID*. It then broadcasts the task information, along with the defined time points $t1$, $t2$, $t3$, and $t4$, to all sensing vehicles.

Recruiting Vehicles

1. *Sending reputation and location:* When receiving a recruit notification from the cloud server, sensing vehicles willing to participate in the task send their join request, along with their reputation values and location information, to the cloud server $\mathcal{CS}$.
2. *Matching reputation and location:* Upon receiving the request from a sensing vehicle $\mathcal{V}$, $\mathcal{CS}$ verifies that this request is valid and the location and reputation values meet the requirements specified for the sensing task. If everything goes well, $\mathcal{CS}$ adds the vehicle identifier to an application table T_a and sends the position index of $\mathcal{V}$ in the table back to $\mathcal{V}$.

Selecting Vehicles.

I. *Publishing the final vehicle list to the blockchain.* After time t_1, $\mathcal{CS}$ stops accepting vehicle application requests. It performs the following steps:
   1. $\mathcal{CS}$ retrieves all the vehicles $VID_1, VID_2, \ldots, VID_m$ from table T_a, where m represents the total number of vehicles in this table. It then computes the hash value: $h_v = H_1(VID_1, VID_2, \ldots, VID_m)$.
   2. $\mathcal{CS}$ generates a blockchain transaction Tx_v, with the data field containing the hash value h_v, the task id *TID* and the time t_2, as is shown in Fig. 2.
   3. As Tx_v originates from $\mathcal{CS}$'s address, $\mathcal{CS}$ signs and sends Tx_v, along with the signature, to the blockchain. This transaction will be accepted before $t2$.

Function 1 Vehicle List Confirmation

Input: The application table T_a

Output: n vehicle ids, where n represents the number of vehicles required in this task. 1: Initialize a vehicle list, denoted as L, with the contents of T_a.

2: **for** i from to n **do**

3: Compute $\delta_i = H_1(B_{10}, B_9, \ldots, B_1, CID, PID, TID, i) \bmod len(L)$, where $len(L)$ denotes the length of L.

4: Remove $L[\delta_i]$ from L, such that its length is reduced by 1. 5: **end for**

II. *Determining the final list of vehicles* After t_2, CS listens to the blockchain state until a new block is generated with a creation time greater than t_2. Then it does the following steps to determine which vehicles are selected:
   1. $\mathcal{CS}$ extracts the latest 10 block hash values $\{B_{10}, B_9, ..., B_1\}$ from the blockchain system, where B_1 indicates the latest block hash. Then it performs Function 1 to generate the vehicle list.
   2. $\mathcal{CS}$ records all selected vehicle identifiers in a table T_s, which can be accessed by all vehicles participating in this application.
   3. $\mathcal{CS}$ sends *TID* and task information to all selected vehicles.

III. *Generating an anonymous token.* For each vehicle $\mathcal{V}$ in T_s, upon receiving *TID* and task information from $\mathcal{CS}$, it needs to generate an anonymous token by interacting with $\mathcal{CS}$.
   1. $\mathcal{V}$ generates a key pair $(tsk, tpk = tsk \cdot P)$, and two random number $r_1, r_2 \in Z_p^*$; Then it computes $h_k^* = r_1.H_2(TID, tpk, r_2)$ and sends it to the server;
   2. $\mathcal{CS}$ verifies that $\mathcal{V}$ is included in the table T_a. If not, $\mathcal{CS}$ returns an error. Then it computes $\sigma^* = sk{\cdot}h_k^*$, where *sk* is a signing key generated in the setup phase, and sends it to $\mathcal{V}$.
   3. $\mathcal{V}$ computes the anonymous token $\sigma = r^{-1}{\cdot}\sigma^* = sk \cdot H_2(TID, tpk, r_2)$. Then, it verifies the correctness of the token by computing $e(H_2(TID, tpk, r_2), pk) = e(\sigma, P)$. If the verification succeeds, $\mathcal{V}$ stores this token, else, returns an error.

IV. *Verifying the generated vehicle list.* Any vehicle $\mathcal{V}$ participating in this application can verify the selection results as follows:
   1. $\mathcal{V}$ obtains the application table T_a and the corresponding time $t1$ and $t2$ from $\mathcal{CS}$. Then it retrieves all vehicle identifiers $\{VID_1, VID_2, ..., VID_m\}$, where m represents the total number of the vehicles in the table. After that, it computes the hash value: $h_v^* = H_1(VID_1, VID_2, ..., VID_m)$.
   2. $\mathcal{V}$ searches for all blocks generated between $t1$ and $t2$ in the blockchain and looks for transactions initiated by $\mathcal{CS}$. Note that there may be several transactions, as the cloud server can publish several tasks between $t1$ and $t2$. $\mathcal{V}$ extracts the data field of each transaction and checks whether it contains the current task identifier *TID*. If so, it retrieves the vehicle hash value h_v from the data field and verifies that h_v^* matches h_v.
   3. $\mathcal{V}$ searches for the first block which is produced after $t2$ from the blockchain system, and the hash value is denoted as $B1$. Then $\mathcal{V}$ retrieves the preceding 9 blocks hash value, denoted as $B2, ..., B10$. After that, it performs Function 1 to generate the vehicle list.
   4. $\mathcal{V}$ checks that whether these vehicles identifier VID_i^* are included in the table T_s. Also, $\mathcal{V}$ can check that whether it is selected.

Submiting Data.

A vehicle V that participates in the task needs to upload its sensing data *data* anonymously before the deadline $t4$.

1. $\mathcal{V}$ creates a request $req = (data, \sigma, TID, tpk, r_2)$, then it signs *req* using its signing key *tsk* and the signature algorithm *Sign* to get σ_{req}. It uploads *req*, along with σ_{req}, to CS using a secure channel (e.g., TLS).

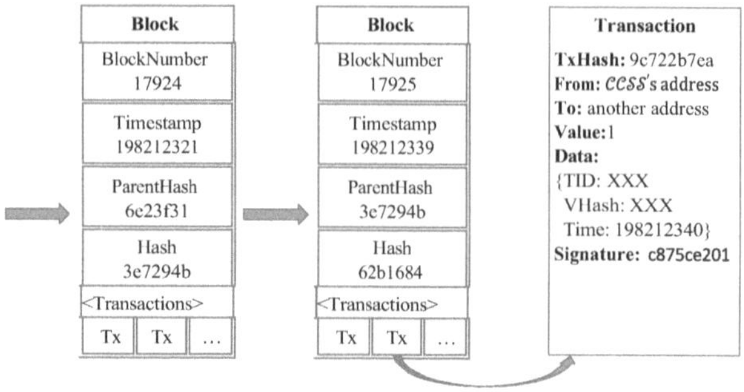

Fig. 2. Transaction and block structure in the blockchain system.

2. $\mathcal{CS}$ checks σ is a valid token by computing: $e(pk, H_2(TID, tpk, r_2)) = e(P, \sigma)$. If this verification fails, CS returns an error, and aborts this connection.
3. $\mathcal{CS}$ checks that the provided token σ corresponds to the current task by verifying that *TID* matches the task identifier of the ongoing task. If they don't match, $\mathcal{CS}$ returns an error and aborts the connection.
4. $\mathcal{CS}$ checks the validity of this request by checking the signature σ_{req} using *tpk*. If the signature is invalid, $\mathcal{CS}$ returns an error and aborts this connection.

The cloud server continuously gather sensing data from valid vehicles until t_4, and forwards the data to the data requester.

In this paper, we mainly focus on the vehicle selection and anonymous data uploading issue. Concerning other matters such as reputation management [31, 32], location protection [19, 37], as well as data quality [45, 46], numerous solutions have already been developed. Readers can refer to these works for more details.

6 Security Analysis

6.1 Resistance Against Unfair Selection

The cloud server may manipulate the vehicle selection results arbitrarily. On the one hand, it can collude with certain vehicles to allocate high-reward tasks. On the other hand, it can allocate multiple tasks to a specific batch of vehicles to continually monitor the data they upload, which would raise privacy concerns.

Theorem 1. *In LFPS, the cloud server cannot modify the vehicle selection results, and the results are fair and publicly verifiable.*

Proof. This proof is divided into two parts. Firstly, the results are fair and publicly verifiable. In LFPS, the selected vehicles are determined by computing $\delta_i = H(B_{10}, B_9, \ldots, B_1, CID, PID, TID, i) \; mod \; len(L)$ for $i \in [1, n]$. To control the output of this hash function, the cloud server is required to control at least one of the inputs. On the one hand, *CID, PID, TID*, and *i* are determined, which means that the cloud server cannot modify them to generate a desired hash output. On the other hand, $\{B_1, \ldots, B_{10}\}$ represent block hash values in the blockchain system, which cannot be controlled by the cloud server. As a result, all the inputs are fixed. Furthermore, a newly generated block is unpredictable and random (See Sect. 3.2). It means that incorporating the hash value of a newly generated block as an input to assist in calculating the hash value can enhance the randomness of the hash function's input. For a secure hash function, if its inputs are unpredictable, its output is also unpredictable, thereby fairness is guaranteed.

Secondly, the cloud server cannot modify the vehicle selection results. δ_i is utilized to determine the selected vehicle, i.e., $\delta_i - th$ vehicle in the vehicle list is the target vehicle. Based on the above analysis, while the server cannot directly control or predict the hash output, it may control the order of vehicles in L, ensuring that the vehicle corresponding to $L[\delta_i]$ is the server's intended choice. It is worth noting that the cloud server cannot succeed in doing so. Since the vehicle list L and the application table T_a cannot be modified once the hash of T_a is stored by the blockchain, and δ_i can not be controlled or predicted, before the new block is generated, no matter how the server adjusts the order, the results appear randomly.

6.2 Resistance Against Linking Uploaded Data

During the data upload phase, an adversary intends to associate the uploaded data of the same vehicle from different tasks, thereby inferring private information about the vehicle, such as trajectories.

Theorem 2. *In LFPS, the server cannot deduce the identity of the vehicle during the data upload phase.*

Proof. Recalling the data upload phase, vehicles upload their sensing data using anonymous tokens rather than their real identities. The token is generated under the BLS blind signature, i.e., $\sigma^* = sk \cdot h_k^* = sk \cdot r_1 \cdot H_2(TID, tpk, r_2)$. Subsequently, vehicles utilize their unblinded tokens to upload the sensing data, i.e., $\sigma = sk \cdot H_2(TID, tpk, r_2)$. For the cloud server, the only useful information it gets is $h_k^* = r_1 \cdot H_2(TID, tpk, r_2)$. However, such an equation contains two unknown parameters, i.e., r_1 and $H_2(TID, tpk, r_2)$, hence the server is unable to retrieve $H_2(TID, tpk, r_2)$. Without knowing the relation between h^* and $H_2(TID, tpk, r_2)$, the server cannot link σ^* and σ together. Therefore, even if vehicles use their actual identities to request an anonymous credential, the server cannot associate the anonymous tokens with the real identities of the vehicles when they upload their data. This concludes this proof.

Theorem 3. *In LFPS, the adversary cannot identify the same vehicle from two different tasks.*

Proof. On the one hand, this can be further explained as adversaries being unable to determine whether two different anonymous tokens belong to the same vehicle. The anonymous token, i.e., $\sigma = sk \cdot H_2(TID, tpk, r_2)$, doesn't reveal any identity information about the vehicles. Therefore, even if the adversary gathers two tokens, it cannot determine whether they belong to the same vehicle. On the other hand, as TID, tpk, and r_2 are used only once, the same vehicle will use two different tokens in two different tasks. This prevents attackers from linking sensing data uploaded by vehicles.

6.3 Resistance Against Malicious Vehicles

As uploading sensing data doesn't require vehicles to log into the system, malicious vehicles that don't participate in the task could upload data to disrupt the cloud server from collecting correct sensing data.

Theorem 4. *In LFPS, the cloud server can identify whether the vehicle is legitimate.*

Proof. Three cases are considered: 1) The vehicle uses a legitimate anonymous token issued by the server for this specific task; 2) The vehicle uses an anonymous token issued for another task; 3) The vehicle uses a forged token. It's important to note that, except for the first case, the other two scenarios are considered illegitimate. In the second case, the server can prevent malicious vehicles from using a token issued in another task. A valid token, i.e., $\sigma = sk \cdot H_2(TID, tpk, r_2)$, contains the task identifier *TID*. Therefore, if a vehicle attempts to upload data using a token that contains an identifier from a different task, the server will classify the corresponding request as invalid and reject all uploaded sensing data. As for the third case, we can prove that no probabilistic polynomial-time (PPT) vehicle can forge a valid token. Such a token is based on the BLS signature [47] which is existentially unforgeable. If a vehicle could forge a valid token with a non-negligible probability, it can forge a BLS signature with the same probability, thereby violating the unforgeability of BLS signatures.

7 Performance Evaluation

We implement LFPS by mainly using Go language and DEDIS advanced Crypto library version 3.1.0. The experiments are run on a laptop with a Windows system, and an Intel Core i5–1135 CPU running at 2.4 GHz, 16 GB RAM. We use the SHA-2 hash function to construct LFPS. The security parameter is set to 128 bits. The signature algorithm is BLS [47]. The pair curve is the BN256 curve[1]. All experimental results are averaged over 100 runs.

7.1 Performance of Anonymous Token

This section is divided into two parts: token generation and data uploading. During the token generation phase, the vehicle is required to generate a key pair (*tsk, tpk*) and two random numbers r_1 and r_2. Then, it computes a blind message h_k^* and sends it to the cloud server. Upon receiving the request, the server computes a blind BLS signature of h_k^* and sends the result back to the vehicle. The vehicle utilizes r_1 to retrieve the BLS signature σ. Then, it verifies the correctness of this BLS signature. During the data uploading phase, the vehicle creates a request *req*, which contains *data*, σ, *TID*, *tpk*, r_2, signs it using *tsk*, and sends them to the cloud server. Upon receiving the data from the vehicle, the cloud server verifies the correctness of the anonymous token. Then, it

[1] It is important to note that the BN256 curve no longer operates at a 128-bit security level due to advancements in attacks. For stronger security, it is recommended to replace the BN256 curve with another curve, such as BLS12-381.

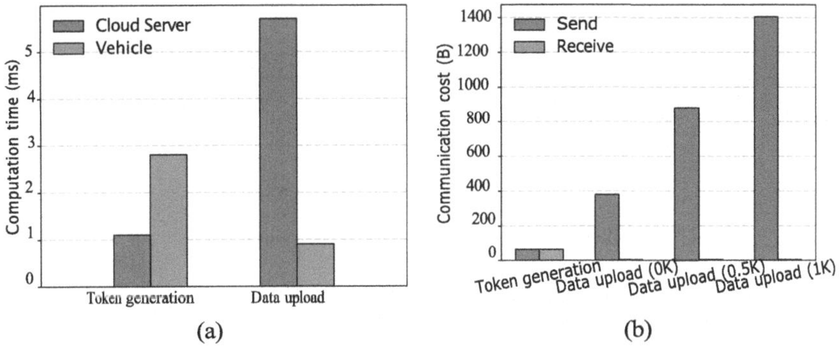

Fig. 3. Experiment results in the token generation phase and data upload phase. (a) Computation costs of the cloud server and vehicle. (b) Communication over-heads of the vehicle. Data upload (X K) indicates that the length of data submitted by the vehicle is X KB.

checks whether the token is issued in this task. After that, it verifies whether this data is submitted by the corresponding vehicle.

The experiment results are shown in Fig. 3. Figure 3(a) illustrates the computation costs of both the cloud server and vehicle. The main computation costs arise from computing and verifying the BLS signature. Notably, the BLS verification process, which relies on bilinear pair operations, demands more computational resources compared to signature generation. During the token generation phase, the cloud server's primary computational cost is associated with generating a BLS blind signature, whereas the vehicle's primary cost lies in verifying the signature. Consequently, the computation cost of the vehicle is slightly higher than that of the server. In the data upload phase, the server is more expensive than the vehicle for the same reason. As is depicted in the figure, both the server and the vehicle complete these tasks within several milliseconds.

The communication costs are shown in Fig. 3(b). We notice that in the token generation phase, the vehicle is required to submit a blind message to the server and receive a blind signature from the server, which contributes to the main communication costs. In the data uploading phase, the primary communication overheads for vehicles arise from uploading data, and they do not receive a substantial amount of messages from the server. Consequently, the volume of data received by vehicles is close to zero. In the data upload phase, vehicles need to upload sensing data using their anonymous tokens. This incurs significant overheads, especially when transmitting a large amount of sensing data, resulting in a substantial increase in communication costs associated with the uploaded data. A vehicle takes within 400 bytes to submit the token. In practice, the communication overhead introduced by sensing data is even greater, typically several megabytes or more. Furthermore, the 400-byte communication overhead is minimal in comparison to the bandwidth of modern devices. In summary, the communication overhead introduced by the scheme is small.

Fig. 4. Details of the published transaction

7.2 Performance of Interacting with Blockchain

To record the hash value of the vehicle list on the blockchain, the cloud server requires a blockchain account. It creates a transaction with its sending address being the account address of the cloud server. Then it embeds the basic information, including the hash value of the vehicle list and the task information, into the data field of this transaction. Subsequently, it signs the transaction using the secret key of the blockchain account and publishes the transaction to the blockchain. The blockchain details are shown in Fig. 4. As we all know, the overhead of computing a hash value is extremely small. So the main costs come from signing the transaction. We test that creating and sending a transaction takes 0.493 s. One of the primary reasons for this high latency is network delays. Even though creating and sending a transaction take hundreds of milliseconds, they do not impact the performance of centralized systems. This is because one task only requires one transaction, and other operations, such as vehicle management, are entirely handled by the centralized system without relying on the blockchain. Also, multiple tasks can share the results of network interactions, such as querying block information.

The communication costs mainly consist of sending the signed transaction and querying the block hash values. Figure 5 illustrates the size of the signed transaction. When the transaction doesn't contain the data field, it takes about 111 bytes to publish the transaction to the blockchain. The more information the data field contains, the larger the size of the transaction. In the Ethereum system, the length of a hash value is 256 bits. As hash values are encoded in hexadecimal, the length of such a value is 64 B. In our setting, when the data field contains VHASH (64 B), TID (60 B) and EXTRA (60 B), the transaction size is 334 B. Actually, this overhead is very small for modern computers. While computing the target vehicle VIDs, the communication overheads are mainly caused by querying 10 block hash values, which results in 640 bytes costs. To verify selection results, a vehicle needs to perform the same calculations as the cloud server to determine the consistency of the selection results, including querying the block hash values and computing random numbers. Therefore, the computational and communication costs for the vehicle are similar to those of the cloud server.

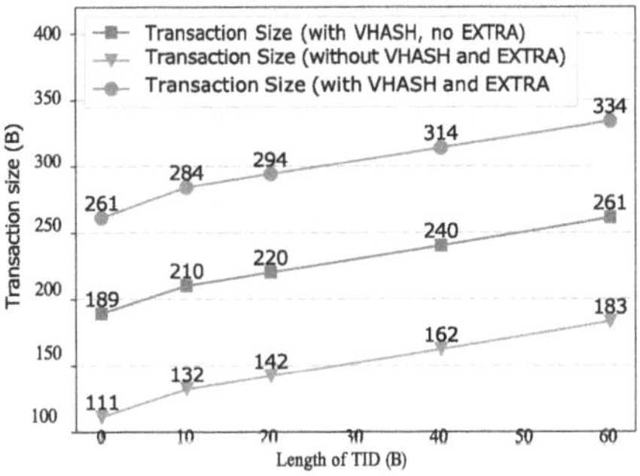

Fig. 5. Transaction size. Information related to the task is contained within the data field of the transaction. The data field consists of three fields: VHASH, TID, and EXTRA. VHASH represents the hash value of the vehicle list, TID represents the task identifier, and EXTRA contains additional information, including time-related data.

8 Conclusion

In this paper, we have proposed a fair, lightweight, and privacy-protecting scheme to address the current challenges in the field of crowdsensing, namely LFPS. LFPS is based on two key techniques. The first one is the fair selection algorithm to generate a public and verifiable vehicle list during the vehicle selection phase. It leverages the tamper-resistance property of blockchain and the unpredictability of new block hash, without imposing a significant burden on the blockchain system. The second is the anonymous token mechanism, which allows vehicles to upload their sensing data anonymously, preventing attackers from linking their uploaded data in two different tasks. It is a lightweight technique that only requires the digital signature algorithm, without relying on expensive cryptographic tools, thus introducing minimal communication and computational overheads. Furthermore, we have implemented an integral scheme based on the Ethereum blockchain. The security analysis and performance evaluation are given to demonstrate the security and efficiency of LFPS.

References

1. Ganti, R.K., Ye, F., Lei, H.: Mobile crowdsensing: current state and future challenges. IEEE Commun. Mag. **49**(11), 32–39 (2011)
2. Huang, Y., et al.: Opat: optimized allocation of time-dependent tasks for mobile crowdsensing. IEEE Trans. Industr. Inf. **18**(4), 2476–2485 (2022)
3. Wenan Tan, L., Zhao, B.L., Lida, X., Yang, Y.: Multiple cooperative task allocation in group-oriented social mobile crowdsensing. IEEE Trans. Serv. Comput. **15**(6), 3387–3401 (2022)
4. Ni, J., Zhang, Aiqing, Lin, Xiaodong, Shen, Xuemin Sherman: Security, privacy, and fairness in fog-based vehicular crowdsensing. IEEE Commun. Mag. **55**(6), 146–152 (2017). https://doi.org/10.1109/MCOM.2017.1600679
5. Wang, X., Weiwei, W., Qi, D.: Mobility-aware participant recruitment for vehicle-based mobile crowdsensing. IEEE Trans. Veh. Technol. **67**(5), 4415–4426 (2018)
6. Zhu, Z., et al.: Crowdsensing intelligence by decentralized autonomous vehicles organizations and operations. IEEE Trans. Intell. Veh. **7**(4), 804–808 (2022)

7. Capponi, A., Fiandrino, C., Kantarci, B., Foschini, L., Kliazovich, D., Bouvry, P.: A survey on mobile crowdsensing systems: challenges, solutions, and opportunities. IEEE Commun. Surv. Tutorials **21**(3), 2419–2465 (2019)
8. Chen, Z., Fiandrino, C., Kantarci, B.: On blockchain integration into mobile crowdsensing via smart embedded devices: a comprehensive survey. J. Syst. Architect. **115**, 102011 (2021)
9. Suhag, D., Jha, V.: A comprehensive survey on mobile crowd-sensing systems. J. Syst. Architect. **142**, 102952 (2023)
10. Yang, M., Zhu, T., Liang, K., Zhou, W., Deng, R.H.: A blockchain-based location privacy-preserving crowdsensing system. Futur. Gener. Comput. Syst. **94**, 408–418 (2019)
11. Zou, S., Xi, J., Wang, H., Guoai, X.: Crowdblps: A blockchain-based location-privacy-preserving mobile crowdsensing system. IEEE Trans. Industr. Inf. **16**(6), 4206–4218 (2020)
12. Tao, X., Hafid, A.S.: Chainsensing: a novel mobile crowdsensing framework with blockchain. IEEE Internet of Things J. **9**(4), 2999–3010 (2022)
13. Wang, W., et al.: Bsif: blockchain-based secure, interactive, and fair mobile crowdsensing. IEEE J. Sel. Areas Commun. **40**(12), 3452–3469 (2022)
14. Zeng, B., Yan, X., Zhang, X., Zhao, B.: Brake: bilateral privacy-preserving and accurate task assignment in fog-assisted mobile crowdsensing. IEEE Syst. J. **15**(3), 4480–4491 (2021)
15. Yan, X., Wing, W.Y., Ng, B.Z., Liu, Y., Gao, Y., Wang, X.: Fog-enabled privacy-preserving multi-task data aggregation for mobile crowdsensing. IEEE Tran. Dependable Secure Comput. **21**, 1301–1316 (2023)
16. Cheng, Y., Ma, J., Liu, Z., Yongdong, W., Wei, K., Dong, C.: A lightweight privacy preservation scheme with efficient reputation management for mobile crowdsensing in vehicular networks. IEEE Trans. Dependable Secure Comput. **20**(3), 1771–1788 (2023)
17. Ding, X., Lv, R., Pang, X., Hu, J., Wang, Z., Xu, Y., Li, X.: Privacy-preserving task allocation for edge computing-based mobile crowdsensing. Comput. Electr. Eng. **97**, 107528 (2022)
18. Liu, Z., et al.: Btmpp: balancing trust management and privacy preservation for emergency message dissemination in vehicular networks. IEEE Internet Things J. **8**(7), 5386–5407 (2021)
19. Zhang, C., et al.: Tppr: a trust-based and privacy-preserving platoon recommendation scheme in vanet. IEEE Trans. Serv. Comput. **15**(2), 806–818 (2022)
20. Guo, B., Liu, Y., Wenle, W., Zhiwen, Y., Han, Q.: Activecrowd: a frame-work for optimized multitask allocation in mobile crowdsensing systems. IEEE Trans. Human-Mach. Syst. **47**(3), 392–403 (2017)
21. Zhao, Y.. Qi, H.: Offline worker selection for real-time spatial crowd-sourcing multi-worker tasks. In: 2019 20th IEEE International Conference on Mobile Data Management (MDM), pp. 545–550 (2019)
22. Wang, Z., Hu, J., Zhao, J., Yang, D., Chen, H., Wang, Q.: Pay on-demand: Dynamic incentive and task selection for location-dependent mobile crowdsensing systems. In: 2018 IEEE 38th International Conference on Distributed Computing Systems (ICDCS), pages 611–621 (2018)
23. Wang, Z., et al.: Heterogeneous incentive mechanism for time-sensitive and location-dependent crowdsensing networks with random arrivals. Comput. Netw. **131**, 96–109 (2018)
24. Sun, G., Wang, Y., Ding, X., Rong, H.: Cost-fair task allocation in mobile crowd sensing with probabilistic users. IEEE Trans. Mob. Comput. **20**(2), 403–415 (2021)
25. Wang, Z., Zhao, J., Hu, J., Zhu, T., Wang, Q., Ren, J., Li, C.: Towards personalized task-oriented worker recruitment in mobile crowdsensing. IEEE Trans. on Mobile Comput. **20**(5), 2080–2093 (2021)
26. Zhang, Y., Ying, Z., Philip Chen, C.L.: Achieving privacy-preserving multitask allocation for mobile crowdsensing. IEEE Internet Things J. **9**(18), 16795–16806 (2022)

27. Wang, Z., et al.: Personalized privacy-preserving task allocation for mobile crowdsensing. IEEE Trans. Mob. Comput. **18**(6), 1330–1341 (2019)
28. Sun, G., Sun, S., Hongfang, Y., Guizani, M.: Toward incentivizing fog-based privacy-preserving mobile crowdsensing in the internet of vehicles. IEEE Internet Things J. **7**(5), 4128–4142 (2020)
29. Zhang, S., Ray, S., Rongxing, L., Zheng, Y., Shao, J.: Preserving location privacy for outsourced most-frequent item query in mobile crowd-sensing. IEEE Internet Things J. **8**(11), 9139–9150 (2021)
30. Liu, Z., et al.: Tcemd: a trust cascading-based emergency message dissemination model in vanets. IEEE Internet Things J. **7**(5), 4028–4048 (2020)
31. Yang, X., Zhang, J., Peng, J., Lei, L.: Incentive mechanism based on stackelberg game under reputation constraint for mobile crowdsensing. Int. J. Distrib. Sens. Netw. **17**(6), 15501477211023010 (2021)
32. Gao, S., Chen, X., Zhu, J., Dong, X., Ma, J.: Trustworker: a trustworthy and privacy-preserving worker selection scheme for blockchain-based crowdsensing. IEEE Trans. Serv. Comput. **15**(6), 3577–3590 (2022)
33. Ma, L., Liu, X., Pei, Q., Xiang, Y.: Privacy-preserving reputation management for edge computing enhanced mobile crowdsensing. IEEE Trans. Serv. Comput. **12**(5), 786–799 (2019)
34. Liu, Z., et al.: LPPTE: a lightweight privacy-preserving trust evaluation scheme for facilitating distributed data fusion in cooperative vehicular safety applications. Inform. Fusion **73**, 144–156 (2021)
35. Liu, Z., et al.: PPTM: a privacy-preserving trust management scheme for emergency message dissemination in space-air-ground-integrated vehicular networks. IEEE Internet Things J. **9**(8), 5943–5956 (2022)
36. Ni, J., Zhang, K., Xia, Q., Lin, X., Shen, X.: Enabling Strong Privacy Preservation and Accurate Task Allocation for Mobile Crowdsensing. IEEE Trans. Mobile Comput. **19**(6), 1317–1331 (2020). https://doi.org/10.1109/TMC.2019.2908638
37. Wang, L., Zhang, D., Yang, D., Lim, B.Y., Han, X., Ma, X.: Sparse mobile crowdsensing with differential and distortion location privacy. IEEE Trans. Inf. Forensics Secur. **15**, 2735–2749 (2020)
38. Wang, Q., Zhang, Y., Xiao, L., Wang, Z., Qin, Z., Ren, K.: Real-time and spatio-temporal crowd-sourced social network data publishing with differential privacy. IEEE Trans. Dependable Secure Comput. **15**(4), 591–606 (2018)
39. Wang, Z., et al.: Privacy-preserving crowd-sourced statistical data publishing with an untrusted server. IEEE Trans. Mob. Comput. **18**(6), 1356–1367 (2019)
40. Shen, H., Bai, G., Yujia, H., Wang, T.: P2TA: privacy-preserving task allocation for edge computing enhanced mobile crowdsensing. J. Syst. Architect. **97**, 130–141 (2019)
41. Qian, Y., Jiang, Y., Hossain, M.S., Hu, L., Muhammad, G., Amin, S.U.: Privacy-preserving based task allocation with mobile edge clouds. Inform. Sci. **507**, 288–297 (2020). https://doi.org/10.1016/j.ins.2019.07.092
42. Yuan, D., Li, Q., Li, G., Wang, Q., Ren, K.: PRIRADAR: a privacy-preserving framework for spatial crowdsourcing. IEEE Trans. Inf. Forensics Secur. **15**, 299–314 (2020)
43. Zhao, B., Tang, S., Liu, X., Zhang, X., Chen, W.-N.: ITAM: Bilateral privacy-preserving task assignment for mobile crowdsensing. IEEE Trans. Mob. Comput. **20**(12), 3351–3366 (2021)
44. Zou, S., Xi, J., Guoai, X., Zhang, M., Yueming, L.: Crowdhb: a decentralized location privacy-preserving crowdsensing system based on a hybrid blockchain network. IEEE Internet Things J. **9**(16), 14803–14817 (2022)
45. Zhao, B., Tang, S., Liu, X., Zhang, X.: PACE: privacy-preserving and quality-aware incentive mechanism for mobile crowdsensing. IEEE Trans. Mob. Comput. **20**(5), 1924–1939 (2021)

46. Yang, S., Fan, W., Tang, S., Gao, X., Yang, B., Chen, G.: On designing data quality-aware truth estimation and surplus sharing method for mobile crowdsensing. IEEE J. Sel. Areas Commun. **35**(4), 832–847 (2017)
47. Boneh, D., Lynn, B., Shacham, H.: Short signatures from the weil pairing. In: Boyd, C. (ed.) Advances in Cryptology — ASIACRYPT 2001, pp. 514–532. Springer Berlin Heidelberg, Berlin, Heidelberg (2001). https://doi.org/10.1007/3-540-45682-1_30

Optimized Deployment of Honeypages Based on Multi-strategy Combinations

Keke Li[1], Haobo Li[1], Mohan Li[1,2](✉), and Yanbin Sun[1]

[1] Cyberspace Institute of Advanced Technology, Guangzhou University, Guangzhou 510006, Guangdong, China
{likeke,lhb}@e.gzhu.edu.cn, {limohan,sunyanbin}@gzhu.edu.cn
[2] Pazhou Lab, Guangzhou 510330, China

Abstract. Cyber attackers continually generate new threats, posing significant challenges to network security defenses. Deceptive defense techniques are considered as an effective means of enhancing network security. Existing deceptive defense techniques often deploy honeypots in environments far removed from the actual protected systems, making them susceptible to detection by attackers. To address this issue, a practical deceptive defense method is to deploy some "honeypages", which are some simulated web pages to mislead the attackers. Attackers can be detected by monitoring who has accessed honeypages. To achieve this goal, a key challenge that needs to be addressed is how to select appropriate honeypage deployment paths to make it more effective to capture attackers. The main contribution of this paper is an optimized design of a honeypage deployment method based on multiple strategy combinations, allowing these honeypages to be deployed along paths where attackers are more likely to access. By our method, honeypages can be deployed on the paths maintaining concealment and obfuscation, as well as rendering attackers unaware of their entry into the honeypages. We applied the defense optimization strategies to real-world business scenarios and deployed honeypages for a period of 21 days. The analysis on the real-world data demonstrated the practical effectiveness of our approach.

Keywords: Deceptive defense · Honeypages · Attack detection

1 Introduction

With the rapid development of the Internet, web sites have become increasingly important and complex as primary platforms for information dissemination and business interactions. Alongside this development, there has been a continuous increase in cybersecurity

This work is funded by the National Key Research and Development Plan (Grant No. 2021YFB3101704), the National Natural Science Foundation of China (No. 62372126, 62272119, 62072130, U20B2046), Guangdong Basic and Applied Basic Research Foundation (No. 2021A1515012307, 2023A1515030142, 2020A1515010450), Guangdong Province Universities and Colleges Pearl River Scholar Funded Scheme (2019), Guangdong Higher Education Innovation Group (No.2020KCXTD007), Guangzhou Higher Education Innovation Group (No. 202032854) and Cultivation Project of PZL (No. PZL2022KF0013).

Z. Gu et al. (Eds.): NSE 2023, CCIS 2063, pp. 169–181, 2024.
https://doi.org/10.1007/978-981-97-4519-7_12

threats. Attackers, through various means, seek to exploit vulnerabilities and weaknesses to infiltrate web sites, thereby gaining access to sensitive data, spreading malicious software, attacking web servers, or engaging in other malicious activities, posing significant threats to individuals, businesses, and society.

Due to the increasing complexity of attackers' methods, traditional passive defense mechanisms are no longer sufficient to meet the demands of cybersecurity. Active defense is widely regarded with great interest due to its enhanced capability to engage with attackers and proactively mitigate threats in advance. Deception [7] in the form of active defense is an effective strategy wherein defenders use tactics such as disguise, obfuscation, and misdirection to prevent attackers from easily reaching their actual targets.

Honeypots [1] are one of the most commonly used deception defense techniques in the field of cybersecurity. However, the authenticity of honeypots has been a widely discussed topic in research [2, 3], as low-fidelity honeypots can easily be distinguished from real protected systems by attackers. While many articles have been published in recent years analyzing this issue and proposing various solutions to make honeypots appear more realistic [4, 5], these articles are often purely theoretical works and may not be directly applicable to real-world scenarios. The concept of "**honeypages**" refers to a technique where cloned simulated web pages are deployed under the URLs of real protected systems, aiming to deceive attackers seamlessly into a deceptive environment and capture their malicious activities. These honeypages exhibit outstanding simulation capabilities, closely resembling the actual pages of the target website in terms of layout, content, and interactivity. Figure 1 shows an example application of honeypages.

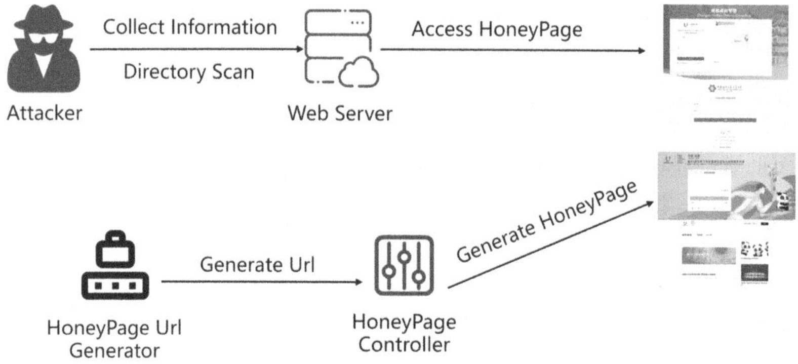

Fig. 1. The application scenarios of honeypage url generator.

The most significant difference between honeypages and honeypots is that honeypages, when parasitically residing within a real protected system, do not actively broadcast information to attract attackers. Since honeypages never proactively disclose their URLs to the public, all incoming traffic directed towards honeypages can be considered malicious traffic. Furthermore, in practice, we have observed that by strategically deploying honeypages based on the attackers' tactics and the specific characteristics of real-world business scenarios, it becomes possible to capture a greater number of malicious activities.

In this paper, in order to enhance honeypages's ability to detect malicious activities by attackers, we combine the generation method of honeypages deployment paths with commonly used domain brute-force dictionaries by attackers. We also apply natural language processing (NLP) techniques to the business content of the protected system. Finally, we deployed honeypages in a real-world business scenario and conducted in-depth data analysis. The main contributions are as follows.

(1) A novel cybersecurity defense strategy has been introduced, starting from a real understanding of attacker behaviors and preferences, to optimize the deployment paths of honeypages.
(2) Multiple strategies and methods are provided for generating honeypage paths, thereby increasing the likelihood of capturing attackers.
(3) Using a real-world experimental environment, the effectiveness of various strategies is verified, and an in-depth investigation is carried out on how different attack behaviors affect the generated domains.

The rest of this paper is arranged as follows: Sect. 2 introduces the related work, Sect. 3 provides a detailed description of the implementation process of the multi-strategy method, Sect. 4 analyzes the collected attack data and proves the effectiveness of the strategy, and finally, Sect. 5 concludes the paper.

2 Related work

2.1 Deceptive Defense

Deceptive defense technology is an active network defense technique that has been used in the field of cybersecurity for over three decades [6]. In deception defense technology, defenders typically engage in deliberate actions aimed at misleading and/or confusing attackers, thereby encouraging them to take (or not take) specific actions that aid in computer security defense [8]. Consequently, defenders initially assess the methods and intentions of attackers and then devise a deception strategy based on their understanding of the attackers to lure and misguide them.

2.2 Honeypots

The concept of honeypots was introduced in the 1990s. Honeypots are typically intentionally designed as vulnerable computer systems or applications to lure attackers into a "trap," thereby reducing the likelihood of attackers targeting real systems. Furthermore, honeypots can interact with attackers to gather attack information for a better understanding of the source of attacks, thus enhancing the defense capabilities of the system. Honeypots can be categorized based on their purpose into research honeypots and production honeypots [9]. Honeypots can also be classified based on their interaction level into high-interaction honeypots and low-interaction honeypots [10]. This reflects the degree to which attackers can interact with the services and responses provided by the honeypots.

Low-Interaction Honeypots Low-interaction honeypots typically simulate only a subset of network services and system functionalities. As a result, the interaction between attackers and these honeypots is limited and of shorter duration. Consequently, they can capture only a small amount of information. However, their advantage lies in their ease of deployment, such as exemplified by Honeyd [11].

High-Interaction Honeypots High-interaction honeypots typically offer real operating systems and services, allowing for more in-depth capture of attack behaviors and obtaining richer attack information. However, high-interaction honeypots are not easy to deploy, and they themselves may be compromised by attackers and used as pivot points for launching further network attacks.

3 Methodology

This section will discuss the specific methods for generating honeypage URLs in this paper, primarily consisting of two parts: subdomain generation and path generation. We have three strategies for generating the domains of honeypage, in addition to a description of the path generation strategy. The subdomain generation process described in this chapter is illustrated in Fig. 2 In the following three subsections, we provide detailed explanations of the three honey domain generation strategies mentioned in Fig. 2 .

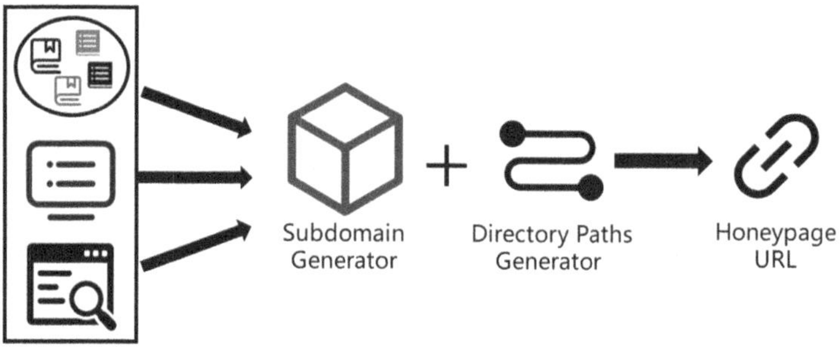

Fig. 2. The process of generating honeypage's URL illustrated diagram.

3.1 Based on Dictionary Brute-Force

Attackers typically gather information about the target system before launching an attack. During this phase, attackers often use tools like Nmap [12, 16] and subDomainsBrute [17] to obtain information about the target system's port services and subdomains, among other valuable assets. Their objective is to further uncover critical vulnerabilities hidden within subdomain applications. Therefore, the strategy of generating subdomains based on attacker dictionary brute-force relies on leveraging the common methods and behaviors of attackers for deception and countermeasures. This involves creating fake subdomains that match the dictionaries and domain brute-force tools commonly used by attackers, thereby increasing the probability of attackers collecting false subdomains. This, in turn, expands the capture surface of the honeypages.

Algorithm 1: Based on Dictionary Brute-Force Strategy Algorithm

Function `Ge_by_dict`(d,D,O_d,n):
  Unify dictionary format and get $D = \{D_1, D_2, \ldots, D_n\}$;
  Calculate word importance for every dictionary
   $dict_{w\_i} \leftarrow word\_importance(D)$;
  Merge dicts by key $D_{w\_i} \leftarrow \text{merge}(dict_{w\_i})$;
  $D' \leftarrow \text{merge}(D)$;
  Count word frequency $D_{w\_f} \leftarrow word\_frequency(D')$;
  $D_w \leftarrow w_1 D_{w\_i} + w_2 D_{w\_f}$;
  for $i \leftarrow 1$ **to** n **do**
    $temp \leftarrow random(D_w.key)$;
    $subdomain \leftarrow temp +' .' + d$;
    $O_d.append(subdomain)$;
  end
  return O_d;

Data Processing. Collecting subdomains brute-force dictionaries from brute-force tools and obtaining dictionaries from open-source, community-contributed sources, as well as publicly available domain data on the internet. These dictionaries encompass common subdomains that attackers typically use. Next, perform deduplication and standardization of the dictionary data. Finally, merge the dictionaries and calculate the word frequency for each subdomain in the dictionary.

Methodology Design. To make the generated subdomains more targeted, we not only counted the word frequencies in the brute-force dictionaries but also calculated the importance of each word in the dictionary. Our primary approach involves assessing the importance of words based on their position in the dictionary's order. Finally, we assign weights to the word frequency and the calculated importance value of each word in the dictionary to determine the overall cracking significance of the word.

Generate Subdomain. Based on the Word Importance Calculation Method, we commence the generation of subdomains. In this process, we prioritize the generation of subdomains using words with higher weights selected from our synthesized dictionary. By doing so, we leverage data sources such as commonly used subdomain brute-force dictionaries by attackers to generate subdomains that have a high frequency of being targeted in attacks. This increases the likelihood of attackers collecting false subdomains and entices them into the honeypot.

3.2 Based on the Business Content

In this section, we generate domain names based on the business content of the protected system. Algorithm 2 outlines the specific steps for generating honey-page's domain names using this approach. We will elaborate on this method in the following three subsections.

Algorithm 2: Based on Business Content Strategy Algorithm

```
Function tf_idf_(word,c):
    tf_value ← tf(word, c);
    idf_value ← idf(word);
    tfidf_value ← tf_value * idf_value;
    return tfidf_value;
Function Ge_by_business(d,C,O_d,n):
    N_doc ←total number of documents in the collection;
    words_list ← Standford_NLP(C);
    for i ← 0 to len(words_list) − 1 do
        tfidf_value ← tf_idf(i, C);
        word_value_list.append(subdomain);
    end
    sorted_word_list ← sorted(words_list, word_value_list);
    for i ← 0 to n-1 do
        subdomain ← sorted_word_list[i] +'.' + d;
        O_d.append(subdomain);
    end
    return O_d;
```

Data Collection. We will delve into understanding the business content involved in the protected system, including product names, service types, industry-specific terms, and more. Furthermore, we will also crawl the business content of similar website systems for storage.

Calculate Word Importance. Following this, We utilize Stanford NLP parser [18] to process the business content of the protected system, extracting its keywords.we will organize and refine the collected business keywords and terminology, ensuring that the selected keywords and terms are closely aligned with the core business of the protected system and accurately convey its business characteristics. For each keyword, we subsequently apply the TF-IDF algorithm [13] to calculate its importance.

Generate Subdomain. Then, the keyword list is sorted based on the importance of each word. Subsequently, we generate subdomains for the honeypages by selecting keywords from the list, prioritizing those at the top. These subdomains will accurately reflect the business content of the protected system while maintaining a level of concealment and randomness to increase the difficulty of identification for potential attackers.

3.3 Based on the Existing Subdomain

In this section, we combine the use of the WordNet [14] and the NLTK [15] to generate domain names for honeypages through the utilization of protected subdomains within the system. This generation process will adhere to the steps described in Algorithm 4, with expansions and optimizations designed to make it imperceptible for attackers to enter the honeypages, thereby enhancing the honeypage's obfuscation.

Algorithm 3: Based on exist subdomain Strategy Algorithm

```
Function Ge_by_exist(d,E,O_d,n):
    Initialize exist_domains_list ← E;
    for i ← 0 to len(exist_domains_list − 1) do
        other_form ←
          WordNetLemmatizer().lemmatize(exist_domains_list[i],'n');
        out_list.append(other_form);
        synonym ← list(wordnet.synsets(exist_domains_list[i]));
        replace_list.append(synonym);
    end
    for i ← 1 to n do
        temp ← random(replace_list);
        subdomain ← temp +'.' + d;
        O_d.append(subdomain);
    end
    return O_d;
```

Gathering utilized subdomains. By conducting information gathering and scanning on the protected system, we obtain a list of already-utilized subdomains. These subdomains are either currently in use or have been previously used and are considered valid for the protected system.

Transformation. For each obtained used subdomain, we generate subdomains based on specific transformation rules and store these generated subdomains in a list. We generate subdomains by incrementing or decrementing existing domain characters in dictionary order. We can also create subdomains using part-of-speech transformation rules to obfuscate them from attackers.For example, with the existing subdomain 'ticket.example.com', rule-based transformations can generate subdomains such as 'picket.example.com' and 'tickets.example.com' as domain names for the honeypages.

Generate Subdomain. Based on the requirement for the desired number of subdomains, we randomly select the corresponding number of subdomains from the generated domain list to serve as domain names for the honeypages. This approach not only increases the likelihood of attackers being captured by the honeypages when employing enumeration or segmented enumeration for subdomain brute-forcing but also generates domain names that exhibit strong similarity and relevance to existing subdomains, thus effectively confusing attackers.

3.4 Path Generation

We generate the URL paths for the honeypages based on the paths of the real protected system that the honeypages are mimicking and common system paths used by attackers. Firstly, we gather a wide range of commonly used system path files by attackers, including common directory names, sensitive file names, and frequently used directory structures. Secondly, we obtain the system directories of the real protected system, including commonly used directory names and structures.

We then calculate the word frequencies in the path directories, prioritizing high-frequency words. Using these high-frequency words, we randomly generate multi-level web directory structures, incorporating the high-frequency words into directory naming.

4 Experiments

In order to conduct a comprehensive assessment of the actual effectiveness of various honeypage domain generation methods and the differences between different generation strategies, we utilized a real-world experimental environment. We collected data on actual attacker interactions with honeypages under different domains, providing us with trustworthy and real-world data.

We chose to deploy honeypages on the Elastic Compute Service (ECS) cloud platform to evaluate its performance and effectiveness. As show in Fig. 3, all of our honeypages run on a cloud server with 8 core CPU, 16 GB of memory, and 500 GB of storage space. The server utilizes the CentOS operating system, providing us with a reliable infrastructure for running and managing honey-pages. Additionally, we established an associated assessment system specifically for collecting and analyzing security logs to conduct in-depth analysis of attacker behavior, enabling us to better understand and identify network threats.

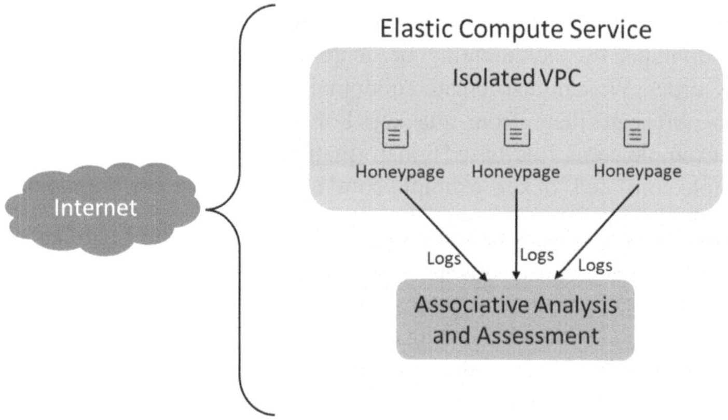

Fig. 3. A diagram of Honeypages deployment.

4.1 Comparison of Different Strategies

We conducted four separate deployments using the *Honeypage controller*, employing three distinct strategies as well as a hybrid strategy for domain name generation to deploy honeypages. We systematically recorded the number of IP addresses triggering honeypages during each deployment, and the results are presented Fig. 4. From the Fig. 4(a), we can observe that when each strategy deployed 3 honeypages, the method based on generating domain names from business content attracted 273 malicious IPs, while the approach utilizing domain name brute-force dictionaries drew 368 malicious IPs. Simultaneously, the method based on existing subdomains captured 508 malicious IPs, while the combined approach using all three strategies attracted 609 malicious IPs. On the other hand, we can observe from Fig. 4(b) that for each strategy, there is little variation in the number of non-malicious IPs accessing the Honeypage. When deploying 3 honeypages, the IP count for each strategy is approximately around 60.

Several interesting observations can be made from the graph. First, domains generated using the strategy of utilizing existing subdomains proved to be particularly effective in confusing attackers. Consequently, the number of attackers accessing honeypages under these domains was significantly higher compared to the other two strategies. However, as the quantity of generated domains increased, the presence of too many domains resembling existing subdomains could arouse suspicion among attackers. This is where the hybrid strategy came into play. Under the hybrid strategy, multiple honeypages were deployed along similar paths, making it challenging for attackers to discern the actual target. This further enhanced the effectiveness of our security strategy and our ability to obfuscate attackers.

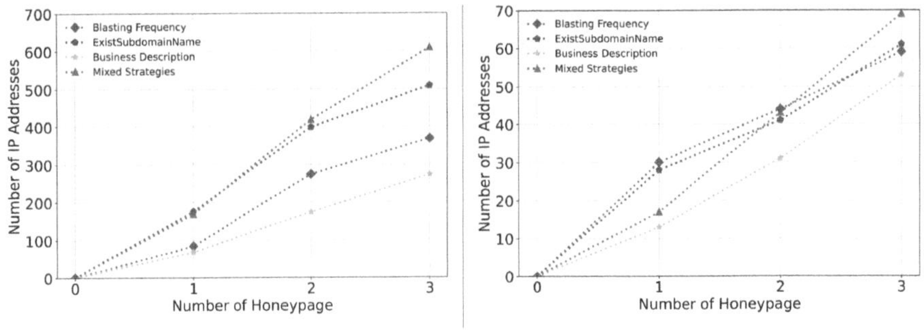

(a) Malicious (b) Non-Malicious

Fig. 4. Comparison of Different Strategies

In summary, our experimental results indicate variations in the number of IP addresses triggering honeypages across different path generation strategies. The hybrid strategy proves particularly advantageous in larger-scale deployments, as it effectively addresses attacker suspicions, making it easier for them to enter the honeynet and thereby bolstering our defense capabilities.

4.2 Attack Behavior Analysis

Threat Type Firstly, various types of attacks that reach honeypages are classified and identified for the purpose of recognizing and analyzing different attack behaviors. This can include brute-force attack, SQL injection, web application attacks, ssh abuse, and more. Subsequently, statistical analysis of attack behaviors is conducted, recording different types of attack actions, tools used by attackers, attack frequencies, and so forth. Identification is then performed on the attackers' origins, IP addresses, attack timings, attack patterns, and other attributes. Based on the analysis results shown in Fig. 5(a), we can be observed that brute-force attacks and web application attacks are the two most prevalent attack methods, accounting for 82.7% and 84.5%(*A malicious IP detects one or more attack methods.*), respectively. Following closely are SSH abuse and Zombie attacks, with proportions of 74.5% and 74.9%, respectively. On one hand, it can be seen that the probability of capturing web application attacks is higher for honeypages parasitic within the protected target web service. Therefore, the likelihood of successfully attracting web application attacks is greater within the honeypage system. On the other hand, brute-force cracking is a commonly used and versatile attack method by attackers. This implies that attackers frequently utilize this approach to attempt system intrusion, not limited to specific targets. According to the analysis results, deploying honeypages using domain names generated based on existing subdomains demonstrates significant advantages in attracting attackers. This is because the simulated domains generated by this strategy are more effective at confusing attackers, allowing them to enter the honeypages without detection. Additionally, as shown in Fig. 5(b), this strategy exhibits greater effectiveness in capturing SSH abuse attacks. By statistically evaluating the importance of vocabulary in domain brute-force dictionaries, the generated subdomains can more effectively capture attack behaviors that employ brute-force cracking methods.

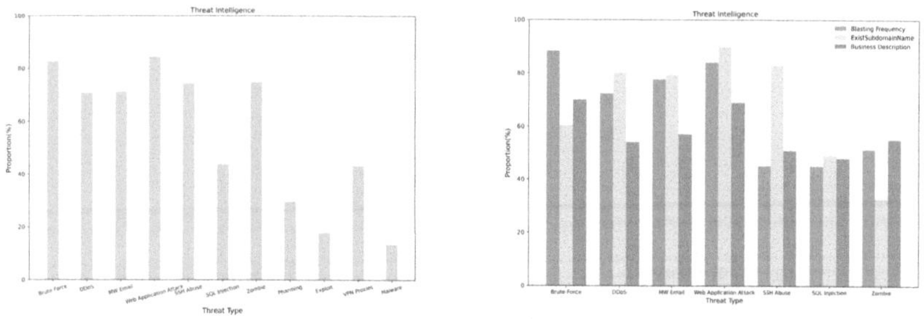

(a) Threat type by Honeypage (b) Distribution by different domain

Fig. 5. Distribution of threat type

Geographical Distribution. While we did not observe any significant differences in attack behavior among malicious IPs from different regions, we did notice an uneven distribution of IP addresses captured by the honeypages across countries. In general, we received the majority of requests from China and the United States, with over 95% of requests from the United States being malicious. Fig. 6 presents the top 10 countries based on the number of IP addresses that contacted our honeypages.

Hourly malicious IP address. Fig. 7, shows daily hourly number of unique IP addresses that visited our honeypages. We observed that during the 21 days of deploying the honeypages, there were relatively few IP addresses visiting it in the lead-up to the activity. However, it's worth noting that since July 26th, there has been a noticeable increase in the number of IP addresses accessing the honeypages. The reason behind this change

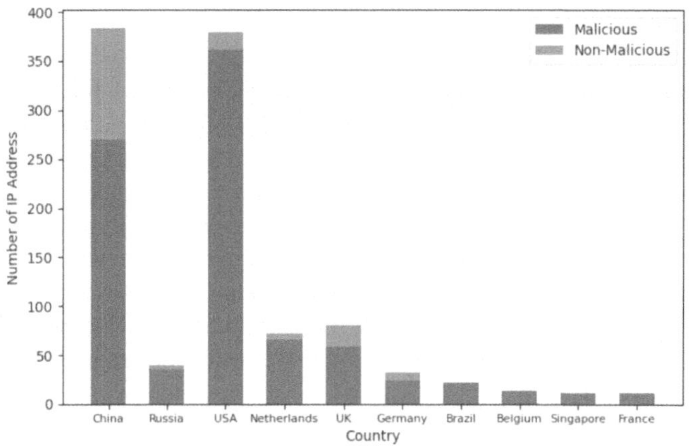

Fig. 6. Country distribution of IP addresses.

is that the activity was scheduled to open on the 28th, leading us to reasonably infer that more attackers are beginning to access websites related to the event, including our honeypages. This observation provides valuable insights into the correlation between cyberattack activities and the schedule of major events.

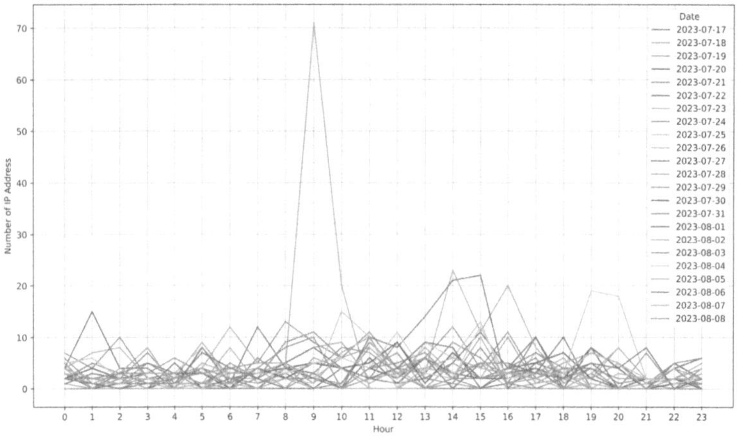

Fig. 7. Hourly number of new IP addresses of attackers entering Honeypages.

Furthermore, we observed that the majority of visits to the honeypages were concentrated between 9 AM to 12 PM and 2 PM to 5 PM Beijing time. Our understanding is that this pattern aligns with the working hours in China, as most of the IP addresses accessing the honeypages originate from China, as analyzed in Fig. 6. This temporal correlation highlights the relevance of local time zones and working hours when considering cyberattack activities.

5 Conclusion

We have proposed multiple domain generation strategies to deploy honeypages, including selecting words from commonly used subdomain brute-forcing dictionaries, generating simulated subdomains based on already used subdomains, and generating business semantic subdomains based on the content of the protected system. These strategies enable honeypages to better mimic real business system pages, increasing the likelihood of attackers encountering the honeypots and providing greater flexibility and targeting.

In this discussion, our focus was primarily on deploying honeypages using static strategies. In the future, we will also discuss further optimizing honey-page deployment in dynamic attack-defense scenarios, where we will consider incorporating attack graphs to enhance network security.

References

1. Spitzner, L.: Honeypots: Tracking Hackers. Addison-Wesley Reading, Boston (2003)
2. Jiang, X., Wang, X.: "Out-of-the-Box" Monitoring of VM-based High-Interaction Honeypots. In Recent Advances in)Intrusion Detection (RAID 2007) (2007)
3. Liu, S., Wang, S., Sun, K.: Enhancing honeypot fidelity with real-time user behavior emulation. In: 2023 53rd Annual IEEE/IFIP International Conference on Dependable Systems and Networks-Supplemental Volume (DSN-S). IEEE (2023)
4. Ceron, João, M., et al.: MikroTik devices landscape, realistic honeypots, and automated attack classification. In: NOMS 2020–2020 IEEE/IFIP Network Operations and Management Symposium. IEEE (2020)
5. Sladić, M., et al.: LLM in the Shell: Generative Honeypots." arXiv preprint arXiv:2309.00155 (2023)
6. Zhang, L., Thing, V.L.: Three decades of deception techniques in active cyber defense-retrospect and outlook. Comput. Secur. **106**, 102288 (2021). https://arxiv.org/abs/2104.03594
7. Almeshekah, M.H., Spafford, E. H.: Cyber security deception. In: Jajodia, S., Subrahmanian, V.S.S., Swarup, V., Wang, C. (eds.) Cyber Deception, pp. 25–52. Springer, Cham (2016). https://doi.org/10.1007/978-3-319-32699-3_2
8. Yuill, J.J.: Defensive Computer-Security Deception Operations: Processes. Principles and Techniques, ProQuest (2006)
9. Lance, S.: The Value of Honeypots, Part One: Definitions and Values of Honeypots. Security Focus information (2001)
10. Mokube, I., Adams, M.: Proceedings of the 45th Annual Southeast Regional Conference. ACM-SE 45. Honeypots: concepts, approaches, and challenges, pp. 321–326. ACM New York (2007)
11. Provos, N.: Honeyd-a virtual honeypot daemon. In: 10th DFN-CERT Workshop, Hamburg, Germany. vol. 2, p. 4 (2003)
12. Lyon, G.F.: Nmap Network Scanning: The Official Nmap Project Guide to Network Discovery and Security Scanning. Insecure, USA (2009)
13. Wu, H.C., Luk, R.W.P., Wong, K.F., Kwok, K.L.: Interpreting tf-idf term weights as making relevance decisions. ACM Trans. Inf. Syst. (TOIS) **26**(3), 1–37 (2008)
14. Miller, G.A.: Wordnet: a lexical database for English. Commun. ACM **38**(11), 39–41 (1995)
15. Bird, S., Loper, E.: NLTK: The natural language toolkit. In: ACL Interactive Poster and Demonstration Sessions, pp. 214–217. Association for Computational Linguistics (2004)

16. Nmap: the Network Mapper - Free Security Scanner, https://nmap.org/. Last accessed 4 Sep 2023
17. subDomainsBrute. https://github.com/lijiejie/subDomainsBrute. Last accessed 4 Sep 2023
18. The Stanford Natural Language Processing Group. https://nlp.stanford.edu/

A Simulation Design of Cascade Generation Method for Anonymous Network

Jincai Zou(✉), Shuai Wang, Haodi Zhang, and Guandao Li

China Telecom Research Institute, Guangzhou 510630, China
{zoujc2,wangshuai8,zhanghaodi,ligd1}@chinatelecom.cn

Abstract. In this paper, we propose HAMA design, a flexible, secure and better performance cascade generation scheme for anonymous network application based on Mixnets. Different from the existing source routing methods that randomly select Mix nodes to generate cascade link, we firstly propose a hierarchical adversary model from the perspective of application users to meet the different anonymity requirements of various application clients. And then, the multi-attribute constraints of cascade link are further analyzed, such as node bandwidth, nation, provider, location and link length, etc. Furthermore, corresponding cascade generation algorithms are designed according to the hierarchical adversary model. The final analysis and evaluation prove that HAMA design is superior to the existing method of randomly selecting nodes to generate cascade link in terms of anonymity and performance. Moreover, HAMA design can not only provide flexible anonymity choices for application designers, but also promote the development of anonymous network in simulation range.

Keyword: Anonymous Communication Network · Range Simulation Design · Mixnets · Path Selection

1 Introduction

With the rapid development of the information age, a large number of Internet communication privacy and security incidents have occurred in recent years, greatly increasing the concerns of people around the world [1–3]. Although information such as source IP can be hidden to some extent by encryption and proxy nodes using proxies or vpn, the lack of handling of communication metadata results in far from adequate protection of personal privacy. Sufficient metadata can reveal much relevant information in communications. Former U.S. government officials have stated that "if you have enough metadata you don't really need content," and "we kill people based on metadata" [4–6]. Surveillance censors can observe and analyze the meta-information such as message sending time, frequency, packet size, and packet interval of communication objects for a long period of time to analyze the associated objects of the communication [7], thus stealing the privacy of the user's communication.

In order to protect users' communication privacy, research on anonymous communication emerges as the times require. Anonymous communication mainly studies the

Z. Gu et al. (Eds.): NSE 2023, CCIS 2063, pp. 182–197, 2024.
https://doi.org/10.1007/978-981-97-4519-7_13

identity concealment of network communication subjects. A current main idea is to put the identities of both communicating parties into an indistinguishable set, and the identities of all communicating users in the set may become communication subjects. The observer can only know the mapping relationship from one communication set to another communication set, but does not know the specific communication relationship. In order to improve the strength of anonymous protection, the data forwarding path is composed of multi-hop encrypted proxy service nodes, and these nodes constitute an anonymous communication network, which is an overlay network. Tor [8], the most popular anonymous communication system at present, has millions of daily users [9], which can provide users with a certain degree of anonymity in Internet communications and protect their privacy in Internet communications. The current advanced low-latency systems based on Mixnets [10–13] such as Loopix [14] and Nym [15] further strengthen the ability to resist traffic analysis attacks.

Most of the existing anonymous communication networks based on Mixnets, such as Vuvuzela [16], Loopix and Nym, use the client to randomly select mix nodes to generate cascade link. This source routing link generation method can maintain randomness as much as possible in the face of global adversary, so that it is impossible to perform de-anonymization operations for the protocol. However, in the face of different anonymity requirements for various applications, this may not seem practical. On the one hand, this solution does not consider adversary from the perspective of application users, and simply regards all application adversary as the most capable global adversary, which is not suitable for some application users, such as anonymous comments and electronic voting, reporting corrupt officials and underworld forces, etc. They only need to be anonymous to the insiders of the company or anyone in the country, and it doesn't matter whether they can guarantee global anonymity. On the other hand, assuming that any adversary is a global adversary and randomly selects nodes to generate route, resulting in the nodes passing through the route across multiple countries, and subject to analysis, processing and review of the network borders of these countries, the performance of communication will definitely be reduced. For example, it will increase the delay. In addition, Akhoondi et al. [31] argue that the geographical location of nodes can also affect path delay to a certain extent. Therefore, this approach is not efficient enough, both in terms of flexibility and performance.

In this paper, we propose HAMA design, a flexible, secure and better performance cascade generation scheme for anonymous network application clients based on Mixnets, which can be applied in anonymous network simulation range environments. The main contributions of this paper are as follows:

- A hierarchical adversary model is proposed, which innovatively stands in the perspective of the application client to classify the adversary according to the anonymity requirements of different applications. At the same time, the multi-attribute constraints of link nodes are analyzed from a practical point of view.
- A hierarchical cascade generation algorithm is proposed, designed for adversary with different levels of capabilities, which can enhance the anonymity of cascade link and further improve performance.
- A cascade analysis model is established, and the measurement of the cascade is divided into the integration of two dimensions of anonymity and performance. The

cascade performance of HAMA design and the cascade generated by randomly selecting nodes are analyzed and compared. The results show that the proposed HAMA design is superior to the method of randomly selecting nodes to generate cascades in terms of anonymity and performance when facing the hierarchical adversary model.

2 Related Work

The concept of Mix proposed by Chaum in 1981, opened up the research field of anonymous communication [17]. The Mixnets, composed of Mix, is a well-known early research result [18–20] in anonymous communication. Its core idea is to introduce Mix node to modify the appearance (encryption and padding) and order (delay and disorder) of messages, and preventing adversary from correlating input and output messages. Due to the low latency and strong anonymity of Mixnets, it has been widely used in many anonymous communication system designs.

Taking Tor, which is the most famous and has the largest number of users at present, as an example, the user client uses onion routing to encrypt messages layer by layer over multiple hops, and decrypts the outer ciphertext every time the link passes through a node to protect user communication privacy. When a client establishes a cascade link, it first requests the Tor directory server to establish a cascade link. The Tor directory server selects according to the bandwidth of the existing volunteer forwarding nodes. Generally, three volunteer nodes with better bandwidth performance are selected as the entry node, intermediate node and exit node. This way of establishing a route is relatively simple, and only refers to the node bandwidth performance. When the new generation of Loopix communication system and Nym network select the cascade link, the Mix nodes are layered, and then a node is randomly selected from each layer to form a cascade link. However, this cascade link generation method is too random, and the quality of the selected cascade link is uneven.

There have been many studies in the past two decades on improving path selection in anonymous networks. AlSabah M et al.[30] propose a method which can improve performance of Tor by changing its transport layer design. There are also a number of designs [31–35] that can optimize the path by considering the latency of link between two Mix nodes. In order to increase security of anonymous network, these articles [36–38] apply the concepts of trust and reputation. In addition, the proposal proposed in [39] can avoid overloaded nodes during path selection. The method proposed in [33] is to measure the path delay by sending a pack to the path and the last node send feedback back to sender, which could create a lot of load traffic in network. Akhoondi et al. [31] proposed a practical method of dividing geographically adjacent nodes into a group and randomly selecting one node from each group to form a path. However, this method is unable to select the better node because it does not take into account the bandwidth of nodes. In addition, Sadegh et al.[35] proposed an algorithm for improving performance of the Tor network, by employing a combination of different metrics which include bandwidth, uptime of nodes and delays between the mix nodes. Though it can improve the performance, more resources are needed to support the calculation and path selection of composite metrics, resulting in certain performance loss.

The minimum requirement for cascade link is that a certain level of anonymity needs to be maintained. Adversaries can implement various traffic analysis attacks by monitoring network traffic information and compromising mix nodes [21–27], so as to achieve the purpose of de-anonymization. Generally speaking, the minimum requirement for a cascade link to maintain anonymity is that there must be an honest Mix node on the cascade link. However, when all the nodes in a cascade link are malicious, which called fully-malicious cascaded link, it is considered that the cascade link cannot provide anonymity guarantee [28].

3 Basic Model and Assumptions

3.1 Basic Model

As shown in Fig. 1, HAMA design is based on common Mixnets, and from the perspective of anonymous network application client users, it selects Mix node to generate cascade links. The client selects several nodes through an available Mix node list to establish cascade link, which served as a message route. The client routes the message to the receiver hop-by-hop through layer-by-layer encryption, and each Mix node only knows the previous hop and the next hop of the message, but does not know other information.

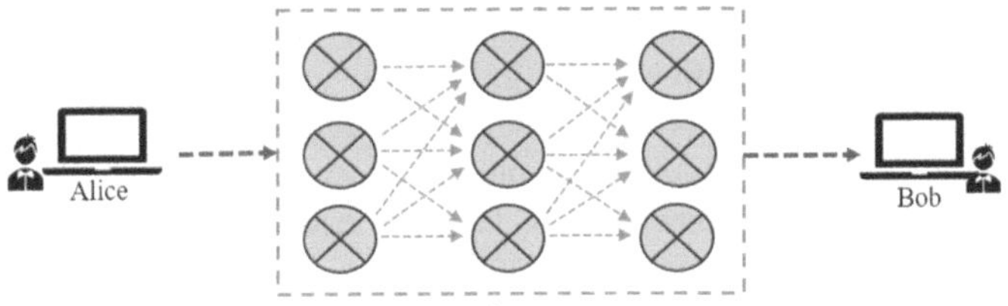

Fig. 1. Basic Mixnets model.

The goal of HAMA design is to prevent fully-malicious cascade link as much as possible. It also takes into account its performance, while ensuring the anonymity of the cascade link.

3.2 Basic Assumptions

The HAMA design is based on the following basic assumptions:

Node assumption. Assuming that there are a large number of voluntary nodes in the anonymous network and distributed all over the world, it will not appear that the number of voluntary nodes in a certain country too far exceeds that of other countries.

Adversary assumption. For a powerful global adversary, it can monitor and review the network traffic information of all nodes, and at the same time, it can compromise nodes within a certain range, but cannot exceed the limit of the proportion of malicious nodes that the anonymous network can withstand. The monitoring agency can compromise some nodes and learn all the information about the traffic passing through these nodes.

Cryptographic assumption. An encrypted message cannot be decrypted by any third party. Therefore, generally speaking, as long as there is one honest node in the link, the whole link is safe. The packet encapsulation format adopts the Sphinx format [29], which can make different messages indistinguishable, and the same message cannot be distinguished in the bit pattern before and after entering the node. In addition, the validity and tamper-proof of the message can be verified.

Overhead assumption. If the cost of deanonymizing a target exceeds the value obtained, it is considered an impossible event. For example, for the anonymous electronic election of the class leader, it is believed that the ability of monitoring and reviewing agencies will not reach the level of the global opponent, because the cost is far greater than the value, and it is meaningless.

4 Main Design

4.1 Hierarchical Adversary Model

The more the number and types of applications connected to the anonymous communication network, the more anonymity will be improved overall, which is'anonymity loves company'. Various applications that require anonymity protection can increase the anonymity enjoyed by each individual by joining the anonymous communication network. However, the anonymity protection services required by various applications are not the same, such as senior politicians and officials, government whistleblowers, anonymous group chat users, etc. Some need to face powerful deanonymization adversary, while some only need to Against weaker adversary.

From the perspective of anonymous communication network client users, HAMA design divides adversaries whose goal is deanonymization into three-level models according to the anonymity required by different applications, namely global adversary, national adversary, and enterprise adversary.

Global Adversary. Global adversary is the most powerful adversary in the hierarchical adversary model, usually the global adversary in the general anonymous communication system, and can be regarded as a joint adversary composed of all national monitoring and review agencies around the world. Such adversaries can monitor, tamper with and repeat all traffic information of anonymous communication networks, including metadata such as packet sending time, time interval, quantity and frequency. In addition, global adversary can control most of the Mix nodes within the proportion of malicious nodes that the anonymous communication network can tolerate. Users facing global adversary can usually be senior government officials, employees of multinational companies, international organizations, and highly influential individuals. Such users wish to remain as anonymous as possible from any third party on a global scale, so anonymity takes precedence over communication performance for them.

National Adversary. National adversary are weaker than global adversary. They are adversaries composed of all monitoring and censorship agencies in a single country, which can monitor, tamper with, and replay all traffic information within the range of cyberspace it controls. Although the definition of the scope of cyberspace has been mentioned by many researchers, in this paper, it refers to the network resources that can be controlled by the monitoring and review agencies of a certain object, including routers, switches, server nodes, terminal equipment and the data information they store. National adversary can control all Mix nodes within its cyberspace, of course, assuming that it cannot exceed the number of malicious nodes that an anonymous communication network can tolerate. User needs for national adversary include but are not limited to corruption, abuse of power, reports of black and evil forces, national electronic referendums, and feedback on government affairs. For such users, performance and efficiency can be taken into account while satisfying anonymity.

Enterprise Adversary. The capabilities of enterprise adversary are the weakest in hierarchical adversary model, including Internet service provider and application service provider, most of which are various application service provider. Any adversary that is weaker than Internet service provider is classified as an enterprise adversary. The capabilities of enterprise adversary take Internet service provider as the standard. They can monitor, tamper with, and replay traffic information within the range of cyberspace they control, and they can control the Mix nodes they control. Users facing enterprise adversary can be members of internal corporate meetings, anonymous electronic voting users, anonymous commentators, and anonymous communication software users. For such users, performance efficiency can be improved as much as possible while maintaining anonymity.

4.2 Multi-attribute Constraints

Most cascade link generation methods for anonymous communication networks are designed in a random manner, including completely randomization and stratified randomization. As mentioned above, this method is difficult to meet the anonymity requirements of various application users and lacks flexibility. Under the premise of satisfying anonymity, the transmission performance and efficiency should be improved as much as possible.

For Mix nodes that make up the anonymous network, in practice, they not only have bandwidth attribute, but also include attributes such as nation, location, and provider. Moreover, for the generated cascade link, the link length, that is, the hop number attribute is also included. Generally speaking, the method of randomly selecting and generating cascade link based solely on the bandwidth can make full use of the randomness of link generation, making it difficult for opponents to use rules or protocol to increase the probability of malicious nodes selecting cascade link. But in fact, the target adversary of user anonymity in many applications are not the same. Therefore, the way of randomly selecting Mix nodes to generate cascade link is not necessarily the best in anonymity, but it is easier to reduce transmission performance and efficiency.

For HAMA design, the node and cascade link have five attributes, which are bandwidth, length, nation, provider, and location.

Bandwidth. It represents the packet transmission efficiency of the node. Generally speaking, the higher the bandwidth of Mix node on the cascade link, the better the transmission performance of the link.

Length. It represents the number of Mix node that the packet passes through from the sender to the receiver. The larger the cascade link length, the stronger the anonymity, and the longer the transmission delay from the message sender to the receiver. It is worth noting that in mixnets, the minimum length is not less than 3 hops to effectively guarantee a certain degree of anonymity.

Nation. It represents the country or organization to which a Mix node belongs. Since each country will have its own domestic policy and censorship agencies, it can be assumed that the country can control the Mix node it controls. If the ownership of a Mix node is owned by a certain company or organization, and the company or organization is owned by a certain country, then the Mix node can be considered to belong to that country.

Provider. It represents the owner of Mix node. If a Mix node belongs to a provider, the provider is considered to be able to control the node. For example, if a Mix node program is run on the cloud server of an Internet company, it is considered that the Internet company can monitor and review the traffic information of the Mix node.

Location. The geographic location where the Mix node is located. In some cases, large Internet company may set up servers in multiple countries. When data packets pass through the network boundaries of countries or regions, they are prone to strict monitoring, review, and traffic analysis. Therefore, it can be assumed that packet transmission across regions will result in longer transmission delay.

4.3 Cascade Generation Algorithm for Client

From the perspective of client users, this section establishes a client cascade link generation algorithm according to different anonymity requirements. The following are the design of the cascade link generation algorithm according to the hierarchical adversary model, with Mix node's nation, location, provider, bandwidth and link length and other conditions.

Global Adversary. For app users facing global adversary, they prioritize anonymity. As mentioned above, global adversary are the joint monitoring and review agencies of all countries in the world, so attributes such as nation and provider are meaningless when facing global adversary. When generating cascade link, just like facing global adversary in other designs, randomness should be maintained as much as possible. First filter the Mix nodes that you don't want to participate in link generation from the available Mix node list, such as country that you don't want the packet to pass through. Then random selection is performed to generate several groups of Mix nodes. Finally, the Mix nodes of each group are grouped according to their position, and connected in random order to generate link cascades. The specific design is as shown in algorithm HamaForGA.

Algorithm 1 HamaForGA

Require: $L \leftarrow$ the length of cascade
Require: $N \leftarrow$ the number of cascade required to generate
Require: $M \leftarrow$ list of available node

```
mixLis = [ ]
result = [ ]
L = L + random(0, 2)

for mix ← M do
    if mix not meet some requirements then
        remove mix from M

while N do
    mixLi = [mix_1, mix_2, ... , mix_L] randomly select from M
    if sorted(mixLi) not in mixLis then
        mixLis.append(sorted(mixLi))
        N = N - 1

for item ← mixLis do
    G = [G_1, G_2,..., G_x] group item by location
    cascade ← randomly connect with G
    result.append(cascade)

return result
```

National Adversary. For app users facing national adversary, they consider a certain level of anonymity, that is, they need to remain anonymous to the entire country, and it does not matter whether they can remain anonymous globally. As mentioned above, national adversary is a combination of all monitoring and review agencies in a country, so provider and link length is meaningless when facing national adversary, because as long as one node is honest, anonymity can be guaranteed. When generating cascade link, it should be ensured that a certain number of foreign Mix nodes are included on the link. First filter the Mix nodes that you don't want to participate in cascade generation from the available Mix node list, including some low bandwidth nodes. Then divide the available Mix nodes into two groups according to the nation, namely the domestic group and the foreign group, and randomly select the Mix nodes that do not exceed half the number of link nodes from the domestic group, and randomly select the remaining number of Mix nodes from the foreign group, so as to select several groups of Mix nodes. Divide each group of Mix nodes into several groups according to their positions, and finally connect them randomly. The specific design is as shown in the algorithm HamaForNA.

Algorithm 2 HamaForNA

Require: $L \leftarrow$ the length of cascade
Require: $N \leftarrow$ the number of cascade required to generate
Require: $M \leftarrow$ list of available node

$mixLis$ = [ ]
$result$ = [ ]
$l = L$ // 2

for $mix \leftarrow M$ **do**
  if mix not meet some requirements **then**
    remove mix from M

while N **do**
  G_{in}, G_{out} ←group M by nation
  mix_1, ... , mix_l randomly select from G_{in}
  mix_{l+1}, ... , mix_L randomly select from G_{out}
  $mixLi$ = [mix_1, mix_2, ..., mix_L]
  if sorted($mixLi$) not in $mixLis$ **then**
    $mixLis$.append(sorted($mixLi$))
    $N = N$ - 1

for $item \leftarrow mixLis$ **do**
  G = [G_1, G_2,..., G_x] group $item$ by location
  $cascade \leftarrow$ randomly connect with G
  $result$.append($cascade$)

return $result$

Enterprise Adversary. For application users facing enterprise adversary, they need to remain anonymous to a certain enterprise and its related personnel, but their anonymity needs will not rise to the national level. As mentioned above, enterprise adversary have or do not control Mix nodes, such as network operators and Internet service providers. Therefore, nation, location, and link length are meaningless when facing enterprise adversary, and Mix nodes on a cascade link should be guaranteed that there are at least two types of provider. First filter the Mix nodes that you don't want to participate in link generation from the available Mix node list, including some low-bandwidth nodes and foreign nodes. Then the remaining nodes are grouped by provider, and if there is only one kind of provider in a country, then it will be switched to national adversary. Next, randomly generate a set of Mix nodes from the available node list. If the number of provider in the list is more than one and the set of node checks does not overlap with the generated list pool, then join in the list pool. Finally, each group of nodes in the list pool is randomly combined to generate cascade link. The specific design is as shown in the algorithm HamaForEA.

Algorithm 3 HamaForEA

```
Require: L ← the length of cascade
Require: N ← the number of cascade required to generate
Require: M ← list of available node
Require: P ← the client location

mixLis = [ ]
result = [ ]

for mix ← M do
    if mix not meet some requirements and location of mix != P then
        remove mix from M

G = [G1, G2,..., Gx] ← group M by provider
if len(G) == 1 then
    return HamaForNA

while N do
    mixLi ← [mix1, ... , mixL] randomly select from M
    NP ← number of provider types of mixLi
    if NP > 1 then
        if sorted(mixLi) not in mixLis then
            mixLis.append(sorted(mixLi))
            N = N - 1

for item ← mixLis do
    cascade ← randomly connect with item
    result.append(cascade)

return result
```

5 Security Analysis and Evaluation

5.1 Security Analysis

HAMA design can effectively ensure path security from the perspective of user, even when facing powerful adversary.

For a global adversary, it is considered to have the ability to control any part of the nodes on a global scale. HAMA generates links by randomly selecting after excluding unwanted nodes, making it impossible for adversary to attack based on the selection protocol. Therefore, even if the adversary purposefully controls certain nodes, the probability of them being able to fully control all nodes in a certain path is very low. As shown in Fig. 2, it shows that when facing global adversary, the security of path can be well guaranteed.

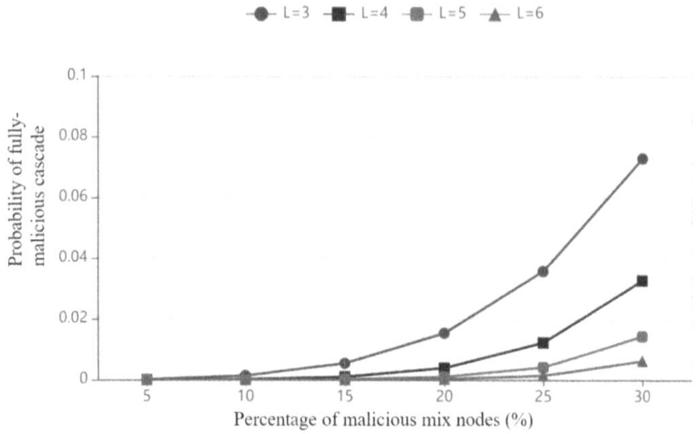

Fig. 2. The relationship between the percentage of malicious node and the probability of generating fully-malicious cascade with different path length.

As for national adversary, the HamaForNA algorithm requires that more than half of the nodes in the path must be different from the country where the client is located, which can effectively resist national level monitoring and censorship. Even if the adversary can control all nodes in its country, it still cannot control all nodes of a cascade, thus ensuring user privacy.

For enterprise adversary, due to the limited ability to master traffic analysis and compromise nodes, HamaForEA algorithm limits the number of provider types on the path to at least two, so it can also maintain anonymity in the face of such adversary.

5.2 Evaluation

General Rules. We introduce the notation used further in the paper, summarized in Table 1. HAMA design represents A as a measure of anonymity. The larger A is, the stronger the anonymity is, and vice versa. P represents the measure of link performance, and the larger P is, the better the performance is, and vice versa. S represents the quality of a entire link, and their relationship can be designed as $S = A \cdot P$. Among them, the anonymity measure $A = \frac{1}{P_f}$, and P_f is the probability of fully-malicious cascade link. The performance measure $P = \frac{B}{L \cdot T}$, where B represents the average bandwidth of link nodes, L is the link length, and T represents the times of location switching. k is the proportion of malicious nodes, and l is the minimum link length required by anonymous network.

For the method of randomly selecting node in cascade link generation, the anonymity measure is A_0 and the performance measure is P_0, where $\overline{b}$ is the average bandwidth of anonymous network nodes, and $\overline{t}$ is the average times of link node position switching.

$$A_0 = \frac{1}{P_f} = \frac{1}{k^l} \tag{1}$$

$$P_0 = \frac{B}{L \cdot T} = \frac{\overline{b}}{l \cdot \overline{t}} \tag{2}$$

Table 1. Summary of notation.

Symbol	Description
S	metric of cascade link quality
A	metric of anonymity
P	metric of performance
P_f	probability of fully-malicious link
B	average bandwidth of link nodes
L	cascade link length
T	times of location switching
k	proportion of malicious nodes
l	minimum link length required

For global adversary, the anonymity measure is A_1 and the performance measure is P_1. For the anonymity measure, since $l_1 > l$, so $A_1 \geq A_0$. For the performance measure, although $l_1 > l$, since in the algorithm HamaForGA will generate adjacent nodes with the same position in the link. Therefore, for some link cases, the number of node position switching will be less, so $\bar{t}_1 < \bar{t}$, and $P_1 \approx P_0$.

$$A_1 = \frac{1}{P_f} = \frac{1}{k^{l_1}} \tag{3}$$

$$P_1 = \frac{B}{L \cdot T} = \frac{\bar{b}}{l_1 - \bar{t}_1} \tag{4}$$

For national adversary, the anonymity measure is A_2 and the performance measure is P_2. For the anonymity measure, when facing national adversary, since the algorithm HamaForNA requires the number of country in a cascade link to be greater than one. Therefore the probability of fully-malicious cascade $P_f = 0$, so $A_2 > A_0$. For performance measurement, although algorithm HamaForNA removes the situation that the link nodes are all in the same country, the average number of switching times of node positions will increase. However, Adjacent operation performed on nodes with the same position will reduce the average number of node position switching times, so it can be considered $\bar{t}_2 \approx \bar{t}$. However, some nodes with lower bandwidth will be filtered out first in the algorithm HamaForNA, so the average bandwidth of nodes $\bar{b}_2 > \bar{b}$, so $P_2 > P_0$.

$$A_2 = \frac{1}{P_f} = \frac{1}{0} = +\infty \tag{5}$$

$$P_2 = \frac{B}{L \cdot T} = \frac{\bar{b}_2}{l \cdot \bar{t}_2} \tag{6}$$

For enterprise adversary, the anonymity measure is A_3 and the performance measure is P_3. For the measurement of anonymity, in the face of enterprise adversary, since the number of provider in the link required by algorithm HamaForEA is greater than one, the probability of fully-malicious cascade $P_f = 0$, so $A_3 > A_0$. For performance measurement, since the link nodes are in the same position in algorithm HamaForEA, the average number of position switching is less, so $\bar{t}_3 > \bar{t}$. And some nodes with lower bandwidth will be filtered out first in algorithm HamaForEA, so the average bandwidth of nodes $\bar{b}_3 > \bar{b}$, and $P_3 > P_0$.

$$A_3 = \frac{1}{P_f} = \frac{1}{0} + \infty \tag{7}$$

$$P_3 = \frac{B}{L \cdot T} = \frac{\bar{b}_3}{\bar{t}_3} \tag{8}$$

As shown in Fig. 3, compared to the methods of randomly selecting nodes to generate cascade link, HAMA design not only has stronger anonymity but also better performance, No matter which adversary is faced.

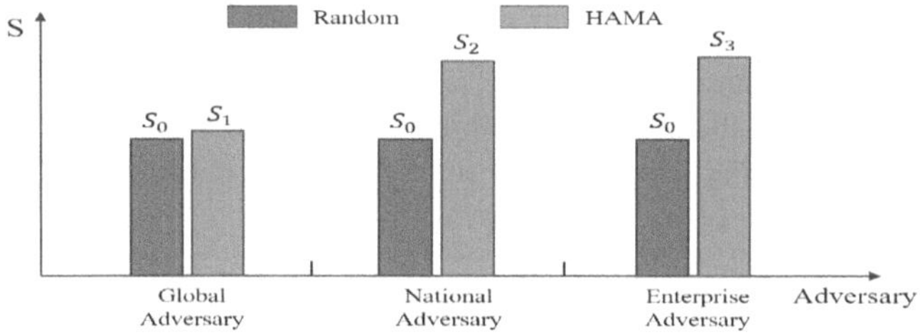

Fig. 3. Comparison of a cascade quality of HAMA design and the methods of randomly selecting nodes to generate cascade in the face of different adversary.

6 Discussion

When the anonymous network application client faces the assumption of national adversary and enterprise adversary, although only two nodes from different countries or different providers can meet the minimum anonymity guarantee and achieve higher performance, the minimum three hops requirement of general anonymous network should be followed. In addition, HAMA design assumes that the proportion of malicious nodes in each country is almost the same. For some countries or regions with a much larger proportion of possible malicious nodes, the step of filtering out nodes in the algorithm will reduce this threat. In the future, a mechanism for checking and verifying node information will be considered to further improve HAMA design.

7 Conclusion

At present, the method of random selection of nodes for anonymous network application clients to generate cascade link is not only a single selection method, but also has poor flexibility. When faced with different user anonymity requirements, the anonymity and performance are low, and it is difficult to meet the needs of various applications. Therefore, this paper proposes a hierarchical adversary model from the perspective of the client, and HAMA design with the multi-attribute conditions of the nodes. Analysis and evaluation show that in the face of different adversaries, the cascade link generation method designed by HAMA is better than the method of randomly selecting link, which can fully improve the anonymity and performance of cascade link. HAMA design is conducive to promoting the access of various applications of anonymous networks, thereby further improving the anonymity of anonymous networks, and has a certain contribution to the actual development of anonymous communication fields.

Acknowledgements. The authors thank the anonymous reviewers for their constructive comments.

References

1. Stoddart, K.: Edward Snowden and PRISM: Negotiating the Post 9/11 "Surveillance State". University of Georgia Press (2022)
2. Power S. New Zealand's Role in the International Spy Network Craig Potton.
3. Keshta, I., Odeh, A.: Security and privacy of electronic health records: concerns and challenges. Egypt. Inform. J. **22**(2), 177–183 (2021)
4. Gilad Y. Metadata-private communication for the 99
5. Hayden, M.: The Price of Privacy: Re-Evaluating the NSA. Johns Hopkins Foreign Affairs Symposium. (2014). https://www.youtube.com/watch?v=kV2HDM86XgIt=17m50s.
6. Alan Rusbridger. 2013. The Snowden Leaks and thePublic. http://www.nybooks.com/articles/2013/11/21/snowden-leaks-and-public/organization=TheNew York review of books. (2013).
7. Papadogiannaki, E., Ioannidis, S.: A survey on encrypted network traffic analysis applications, techniques, and countermeasures. ACM Comput. Surv. **54**(6), 1–35 (2021). https://doi.org/10.1145/3457904
8. Dingledine, R., Mathewson, N., Syverson, P.F.: Tor: the second-generation onion router. USENIX Secur. Symp. **4**, 303–320 (2004)
9. Johnson, A., Wacek, C., Jansen, R., et al.: Users get routed: traffic correlation on Tor by realistic adversaries. In: Proceedings of the 2013 ACM SIGSAC Conference on Computer Communications Security, pp. 337–348 (2013)
10. Ren, J., Wu, J.: Survey on anonymous communications in computer networks. Comput. Commun. **33**(4), 420–431 (2010)
11. Golle, P., Jakobsson, M., Juels, A., et al.: Universal re-encryption for mixnets. Topics in Cryptology–CT-RSA 2004: The Cryptographers' Track at the RSA Conference 2004, San Francisco, CA, USA, February 23–27, 2004. In: Proceedings. Springer Berlin Heidelberg, pp. 163–178 (2004). https://doi.org/10.1007/978-3-540-24660-2_14
12. Sampigethaya, K., Poovendran, R.: A survey on mix networks and their secure applications. Proc. IEEE **94**(12), 2142–2181 (2006)
13. Dingledine, R., Freedman, M.J., Hopwood, D., Molnar, D.: A reputation system to increase MIX-net reliability. In: Moskowitz, I.S. (ed.) Information hiding, pp. 126–141. Springer Berlin Heidelberg, Berlin, Heidelberg (2001). https://doi.org/10.1007/3-540-45496-9_10
14. Piotrowska, A.M., Hayes, J., Elahi, T., et al.: The loopix anonymity system. In: 26th Usenix Security Symposium (Usenix Security 17), pp. 1199–1216 (2017)
15. Diaz, C., Halpin, H., Kiayias, A.: The Nym Network (2021)

16. Van Den Hooff, J., Lazar, D., Zaharia, M., et al.: Vuvuzela: Scalable private messaging resistant to traffic analysis. In: Proceedings of the 25th Symposium on Operating Systems Principles, pp. 137–152. (2015)
17. Chaum, D.L.: Untraceable electronic mail, return addresses, and digital pseudonyms. Commun. ACM **24**(2), 84–90 (1981)
18. Abe, M.: Universally verifiable mix-net with verification work independent of the number of mix-servers. In: Nyberg, K. (ed.) Advances in Cryptology — EUROCRYPT'98, pp. 437–447. Springer Berlin Heidelberg, Berlin, Heidelberg (1998). https://doi.org/10.1007/BFb0054144
19. Wikström, D.: A sender verifiable mix-net and a new proof of a shuffle. In: Roy, B. (ed.) Advances in Cryptology – ASIACRYPT 2005, pp. 273–292. Springer Berlin Heidelberg, Berlin, Heidelberg (2005). https://doi.org/10.1007/11593447_15
20. Mathewson, N., Dingledine, R.: Mixminion: Strong anonymity for financial cryptography. In: Juels, A. (ed.) FC 2004. LNCS, vol. 3110, pp. 227–232. Springer, Heidelberg (2004). https://doi.org/10.1007/978-3-540-27809-2_23
21. Danezis, G., Sassaman, L.: Heartbeat traffic to counter (n-1) attacks: red-green-black mixes. In: Proceedings of the 2003 ACM workshop on Privacy in the electronic society, pp. 89–93 (2003)
22. Danezis, G.: The traffic analysis of continuous-time mixes. In: Martin, D., Serjantov, A. (eds.) Privacy Enhancing Technologies, pp. 35–50. Springer Berlin Heidelberg, Berlin, Heidelberg (2005). https://doi.org/10.1007/11423409_3
23. Hintz, A.: Fingerprinting websites using traffic analysis. In: Dingledine, R., Syverson, P. (eds.) Privacy Enhancing Technologies, pp. 171–178. Springer Berlin Heidelberg, Berlin, Heidelberg (2003). https://doi.org/10.1007/3-540-36467-6_13
24. Fu, X., Graham, B., Bettati, R., et al.: On countermeasures to traffic analysis attacks. In: IEEE Systems, Man and Cybernetics SocietyInformation Assurance Workshop, 2003, pp. 188–195. IEEE (2003).
25. Back, A., Möller, U., Stiglic, A.: Traffic analysis attacks and trade-offs in anonymity providing systems. In: Moskowitz, I.S. (ed.) Information Hiding, pp. 245–257. Springer Berlin Heidelberg, Berlin, Heidelberg (2001). https://doi.org/10.1007/3-540-45496-9_18
26. Basyoni, L., Fetais, N., Erbad, A., et al.: Traffic analysis attacks on Tor: a survey. In: 2020 IEEE International Conference on Informatics, IoT, and Enabling Technologies (ICIoT), pp. 183–188. IEEE (2020)
27. Danezis, G.: Statistical disclosure attacks. In: Gritzalis, D., De Capitani di Vimercati, S., Samarati, P., Katsikas, S. (eds.) SEC 2003. ITIFIP, vol. 122, pp. 421–426. Springer, Boston, MA (2003). https://doi.org/10.1007/978-0-387-35691-4_40
28. Leibowitz, H., Piotrowska, A.M., Danezis, G., et al.: No right to remain silent: isolating malicious mixes. In: 28th USENIX security symposium (USENIX security 19), pp. 1841–1858. (2019)
29. Danezis, G., Goldberg, I.: Sphinx: a compact and provably secure mix format. In: 2009 30th IEEE Symposium on Security and Privacy, pp. 269–282. IEEE (2009)
30. AlSabah, M., Goldberg, I.: PCTCP: per-circuit TCP-over-IPsec transport for anonymous communication overlay networks. In: Proceedings of the 2013 ACM SIGSAC Conference on Computer Communications Security, pp. 349–360 (2013)
31. Akhoondi, M., Yu, C., Madhyastha, H.V.: LASTor: a low-latency AS-aware Tor client. In: 2012 IEEE Symposium on Security and Privacy, pp. 476–490. IEEE (2012)
32. Sherr, M., Blaze, M., Loo, B.T.: Scalable link-based relay selection for anonymous routing. In: Privacy Enhancing Technologies: 9th International Symposium, PETS 2009, Seattle, WA, USA, 5–7 Aug 2009. Proceedings 9. Springer Berlin Heidelberg, pp. 73–93 (2009)
33. Panchenko, A., Renner, J.: Path selection metrics for performance-improved onion routing. In: 2009 Ninth Annual International Symposium on Applications and the Internet, pp. 114–120. IEEE (2009)

34. Wang, T., Bauer, K., Forero, C., Goldberg, I.: Congestion-aware path selection for Tor. In: Keromytis, A.D. (ed.) FC 2012. LNCS, vol. 7397, pp. 98–113. Springer, Heidelberg (2012). https://doi.org/10.1007/978-3-642-32946-3_9
35. Milajerdi, S.M., Kharrazi, M.: A composite-metric based path selection technique for the Tor anonymity network. J. Syst. Softw. **103**, 53–61 (2015)
36. Johnson, A.M., Syverson, P., Dingledine, R., et al.: Trust-based anonymous communication: Adversary models and routing algorithms. In: Proceedings of the 18th ACM Conference on Computer and Communications Security, pp. 175–186 (2011)
37. Johnson, A., Syverson, P.: More anonymous onion routing through trust. In: 2009 22nd IEEE Computer Security Foundations Symposium, pp. 3–12. IEEE (2009)
38. Das, A., Borisov, N., Mittal, P., et al.: Re3: Relay reliability reputation for anonymity systems. In: Proceedings of the 9th ACM Symposium on Information, Computer and Communications Security, pp. 63–74 (2014)
39. Panchenko, A., Lanze, F., Engel, T.: Improving performance and anonymity in the Tor network. In: 2012 IEEE 31st International Performance Computing and Communications Conference (IPCCC). IEEE, pp. 1–10 (2012)

Cybersecurity Infrastructure

A Testbed for Automotive Cybersecurity Applying a Cyber Range

Liqun Sun, Shuqiang Yang, Weihong Han(✉), Haiyan Wang, and Yuan Chen

Department of New Networks, Peng Cheng Laboratory, ShenZhen, China
{sunlq,hanwh}@pcl.ac.cn

Abstract. With the rapid growth of automotive software scale and requirements of communication with external of vehicles, modern vehicle will face serious cybersecurity challenges. Currently, using testbed simulation is a major approach to automotive cybersecurity research and testing, however most of the testbeds are mainly for in-vehicle network (IVN), which cannot meet many cybersecurity research and testing of modern vehicles having complex communication scenario with outside. In this paper, we proposed an automotive cybersecurity testbed simulating the Cloud-Tube-Terminal (CTT) three-layer heterogeneous networks and the Cloud-Vehicle collaborative automotive software OTA (Over the Air) update (ASOTAU) process. Our testbed first applies a cyber range to simulate cloud network and other software experimental environment that greatly promotes simulation flexibility, and integrates an experimental electric vehicle chassis and some Electronic Control Units (ECUs) that support ASOTAU, which can meet many simulation requirements of automotive cybersecurity experiment. Finally, 6 typical use examples are given to show the comprehensive simulation ability and the wide application scenarios of our testbed, and the conveniences for experimenter to use simulated experimental environment.

Keywords: Automotive Cybersecurity · Testbed · Cyber Range · Automotive Software OTA Update · Simulation

1 Introduction

With the development of intelligence and connection of modern vehicles, scale of automotive software and requirements of communication with external of vehicles are growing, and the number of ECUs, the main computing unit in a vehicle, is also increasing. Today an advanced electric vehicle contains hundreds of ECUs. Each ECU has an independent responsibility to serve certain intelligent functions, liking controlling engine, windows and doors. Many intelligent functions require ECUs to communicate with each other as well as external world. CAN Bus (Controller Area Network) is a widely used communication protocol in IVN with high real-time and reliability. However, due to the insufficient security mechanism, CAN is vulnerable to numerous cybersecurity threats [1–3]. Miller et al. [4] remotely hacked into a moving Jeep's IVN over a 3G cellular network to control steering wheel and brake. Nie et al. [5] remotely invaded into IVN of Tesla Model S through Wifi network, controlled the vehicle and reversed a variety of software.

Z. Gu et al. (Eds.): NSE 2023, CCIS 2063, pp. 201–214, 2024.
https://doi.org/10.1007/978-981-97-4519-7_14

ASOTAU service is a development trend of modern vehicles. On the one hand, it can quickly fix software defects and security risks, reducing the money and time costs brought by traditional local update approach; On the other hand, OEM (Original Equipment manufacturer) can push new features through ASOTAU service to increase revenue-generation opportunities. ASOTAU service brings convenience but also introduce cybersecurity risks [6]. ASOTAU function can reprogram ECUs remotely, so if the function is hacked or hijacked, it can cause serious damage to vehicles and their occupants, even regional traffic. In addition, ASOTAU service uses public networks to transfer vehicle software and sensitive data, which may cause personal privacy and intellectual property rights to be invaded.

Automotive cybersecurity has become one of important problems that hinder the development of modern vehicles, and it has been paid attention to by automotive industry. At present, some regulations and standards have been issued to regulate and guide the construction of automotive cybersecurity. UNECE (United Nations Economic Commission for Europe) WP.29 regulations R155 was issued, and it specifies the cybersecurity requirements that OEM must meet. The ISO/SAE 21434 standard is used to guide the construction of cybersecurity management system for whole automotive lifecycle. In China, standard GB/T 40861 specifies the technical requirements of automotive information security.

In the face of emerging new technologies and products about automotive cybersecurity, how to conduct security testing and verification becomes an urgent problem to be solved. Using real vehicle to test often has problems of technically unfeasible or complicated, high risk and high cost. Using testbed simulation is a common approach to automotive cybersecurity research and testing. In this paper, we first proposed an automotive cybersecurity testbed applying a cyber range, and it can provide the following functions:

- Simulates the CTT three-layer network functions of modern vehicles.
- Simulates the Cloud-Vehicle collaborative ASOTAU whole process.
- Based on the comprehensive simulation capabilities of our testbed, it is easy to perform attack experiments on a variety of systems and networks, which simulated on cloud or in vehicle.
- Based on the rich attack experiment scenarios, it can support research and verification of defense technologies in corresponding scenarios.
- Conducts security testing on ASOTAU system.
- Provides platform support to Bug Bounty Program for vehicle cybersecurity.

The remainder of this paper is organized as follows. Section 2 introduces the related works about our research and Sect. 3 introduces our testbed model. Section 4 presents our testbed framework and Sect. 5 gives 6 use examples of our testbed. Section 6 concludes this paper.

2 Related Work

We have reviewed several literature mainly from two aspects: testbed implementation approach and ASOTAU implementation approach. The following introduces some related researches in these two aspects.

2.1 Testbed Implementation Approach

Automotive cybersecurity testbed proposed in many literature is mainly for IVN. TOYOTA [7] proposed a portable in-vehicle cybersecurity testbed PASTA, which uses four general hardware to simulate gateway and ECUs of powertrain, body and chassis domains respectively. The testbed also provides an OBD-II interface for CAN bus traffic monitoring and attack injection, and the outputs of ECUs are fed into a simulated vehicle computer and a scale vehicle model to demonstrate the attack effects. Zheng et al. [8] proposed a general in-vehicle cybersecurity testbed that uses some embedded hardware to simulate ECUs such as gateway, infotainment, and so on. They also use a wireless router to enable communication between inside and outside a vehicle so that CAN bus traffic could be monitored and attacks could be injected from outside. Everett et al. [9] proposed OCTANE, an in-vehicle cybersecurity testbed for testing real vehicles or vehicle components. It does not simulate any ECU, but gives some guidance on constructing testing environment for vehilce components or a real vehicle. It mainly provides software packages contains a number of tools for in-vehicle cybersecurity testing, such as reverse engineer, automated replay, etc.

There are also some literature proposed testbeds covering both IT networks and OT networks. Čeleda et al. [10] proposed an industrial control systems cybersecurity testbed for teaching. PLCs are simulated with Raspberry Pi, which are connected to some simulated industrial field actuators, such as LEDs, switches, etc. Software, such as cloud application and HMI monitoring software, is simulated using a cyber range KYPO, that could realize simulation of many software function quickly and easily. Wang et al. [11] proposed a cybersecurity testbed for SCADA systems, including an IT management networks, monitoring device PLCs and industrial field devices. IT management network is simulated by NS-2, PLCs and field devices use simulated hardware. The other software is running on some specified computers. He et al. [12] proposed an ASOTAU testbed for security risk assessment. The testbed conducts security testing for real vehicles, which provides ASOTAU server simulation platform and 4G wireless communication network.

2.2 ASOTAU Implementation Approach

About ASOTAU implementation approach, automotive industry has proposed some standard solutions. AUTOSAR (AUTomotive Open System ARchitecture), a global automotive software standards organization, offers a ASOTAU technical solution on its classic platform [13]. It proposed a framework for ASOTAU, including a backend server, a master ECU and several target ECUs, and the functions of each part are described. The ASOTAU process and abnormal scenarios processing are also proposed. However, the solution is mainly concentrated on the implementation of IVN. Uptane Alliance, composed of several research institutes and automotive industry in the USA, proposed an open and secure ASOTAU framework Uptane [14], which is based on the widely used IT software update framework TUF. Uptane could counters a comprehensive array of security attacks, and is resilient to partial compromises. The framework has been promoted as IEEE/ISTO 6100 standard.

In academic area, some novel ASOTAU methods have also been proposed. Dorri et al. [15] proposed an ASOTAU security framework based on blockchain technology.

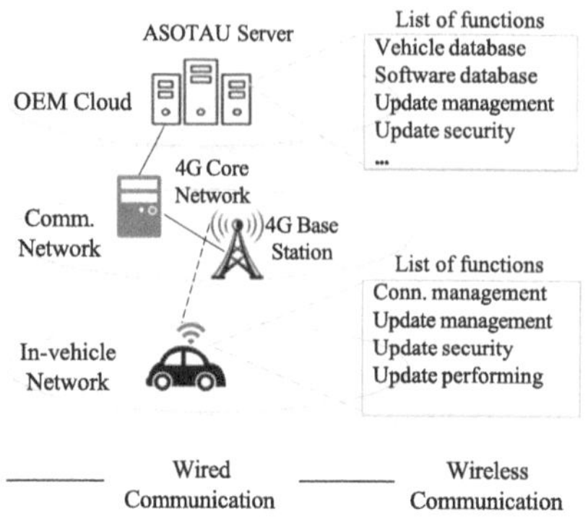

Fig. 1. Diagram of the CTT three-layer network architecture

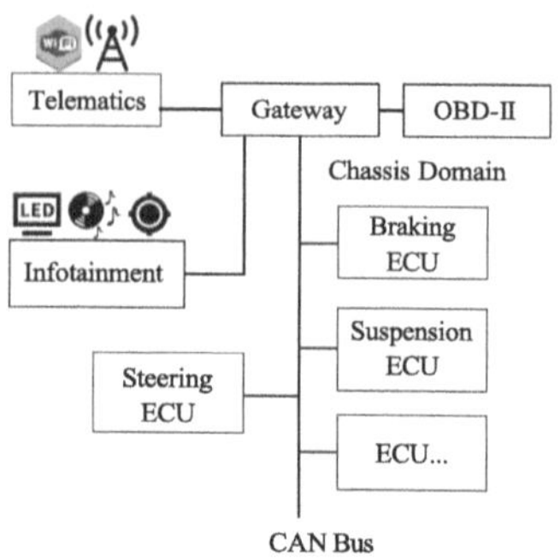

Fig. 2. Diagram of a CAN network topology in IVN

Blockchain technology can not only improve security and increase privacy protection, but also solve the performance problems caused by centralized deployment of ASOTAU servers. Herberth et al. [16] proposed a scheduling algorithm for in-vehicle multiple ECUs parallel update, which significantly shortened the duration compared to sequential update process.

3 Testbed Model

In this section, our testbed model will be introduced. Our testbed model is the CTT three-layer heterogeneous networks architecture of modern vehicles, as depicted in Fig. 1, which are OEM Cloud, communication network and IVN respectively. The following introduces functions of each layer.

3.1 OEM Cloud

OEM Cloud is a platform for OEMs to provide remote service, ASOTAU service, and third-party service for their production vehicles, and these services are provided as cloud services form. Our testbed simulates the ASOTAU service, so the OEM Cloud in our testbed refers to the ASOTAU service platform. In Fig. 1, the function list of OEM Cloud lists the main ASOTAU functions. The detail functions description is as follows.

- **Vehicle database:** Store and manage information about vehicles, including vehicle model, vehicle information, such as Vehicle Identification Number, production time, and ECUs information of each vehicle, etc.
- **Software database:** Store and manage vehicle's software and update packages, such as ECUs and its software versions of every vehicle. About update packages, differential update packages with smaller packet size are usually used.
- **Update management:** Manage update tasks and formulate update conditions, such as release time of update information, vehicle model range, vehicle range and geographical range for this update. Creating an update task is to create an update instance on the server side, push update information to vehicles that match the conditions, process update requests and reports from vehicles.

- **Update security:** Provide security mechanisms for ASOTAU. The main security mechanisms are protecting the authenticity, integrity and confidentiality of update packages and transmission process. The main technologies used are performing digital certificate signature and encryption on the update file, and using TLS as communication link.

3.2 IVN

The IVN in our testbed consists of two scenarios, the one is a common scenario which ECUs communicates with each other, and the other scenario is ECUs communicate with outside, namely the ASOTAU scenario. Fig. 2 depicts a diagram of a CAN network topology which have some typical ECUs and a chassis domain, but usually the network topology in a real vehicle is more complex. Telematics provides function to communicate with outside due to integrate some wireless modems. Infotainment provides in-vehicle entertainment information services, and integrates an interactive display which can support music play, location, etc. ODB-II is a standard interface opening to external of vehicle. Fig. 2 also depicts several ECUs connected to a CAN bus of a chassis domain communicating with other ECUs through a gateway. CAN bus is a serial bus communication protocol, which adopts broadcast communication mode.

In Fig. 1, function list of IVN lists the main ASOTAU functions. The following is detail function description.

- **Connection management:** This function is provided by telematics to establish and maintain IP connections with OEM Cloud for interaction in ASOTAU process. Telematics will transfer update packages to update management after communication security verification passed.
- **Update management:** Receives and caches the update package, and check the update conditions, then deliver package to the target ECU after the check passed. Update management receives update result from target ECUs, summarizes a report and sends it to OEM cloud. Update management negotiates with the target ECUs using Unified Diagnostic Services (UDS) protocol, which is an application protocol in IVN. Different OEM may deploy update management in different ECUs, maybe gateway or telematics.
- **Update security:** Security mechanisms of IVN are mainly to cooperate with OEM Cloud to complete security guarantee of ASOTAU service. Communication security is deployed in telematics. Digital signature and authentication technologies and encryption and decryption technologies for update packages are deployed in both update management and target ECUs.
- **Update performing:** Based on the obtained differential file from update package, the target ECUs perform differential calculations to generate update file, and refresh their flash block, then notify the update result to update management.

3.3 Communication Network

Communication network connects OEM Cloud and telematics. Our testbed simulates a 4G wireless communication network, which consists of 4G base station and 4G core network. As depicted in Fig. 1, it's 4G wireless connection between 4G base station and telematics, and IP connection between 4G core network and OEM Cloud. 4G communication network also provides identity authentication functions for secure access of mobile vehicle. Authentication data is created and stored in 4G Core network, and user side data is stored in USIM card.

A complete ASOTAU process requires collaboration between server-side functions and in-vehicle functions, which have been described in above content. Here is a summary of ASOTAU whole process. First, OEM launches an update package and update task, and determines the vehicles that need to update based on vehicle data and software data in servers, then pushes update information to these vehicles. If the update is confirmed by vehicle, the process of downloading update package is triggered. After vehicle obtaining update package, the update management and update performing functions in IVN are executed. Finally, update result is reported to server to maintain software information of this vehicle.

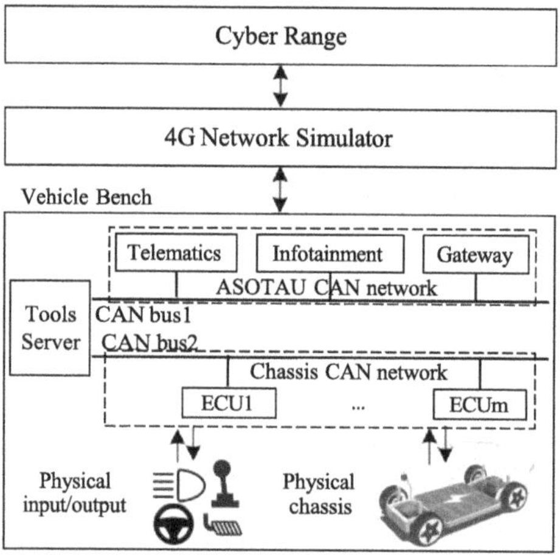

Fig. 3. Diagram of testbed implementation framework

4 Testbed Framework

In this section, simulation implementation of our testbed will be presented, which corresponds to model introduced in section III. Fig. 3 depicts a schematic diagram of implementation framework. Our testbed is composed of three parts: cyber range, 4G network simulator and vehicle bench, which simulate function of the CTT three-layer networks in the model separately. Following is implementation description for the three components.

4.1 Cyber Range

Our testbed simulates the systems and networks of ASOTAU server-side using PengCheng Cyber Range (PCCR) [17], which is a platform for cyber space simulation and cybersecurity research based on virtualization technology, and mainly applies the following functions of PCCR.

- **System simulation:** PCCR uses virtualization technology to load specified software image, which can quickly realize the software function simulation. We implemented a simulation of ASOTAU server-side functions, which are described in Sect. 3.1 OEM Cloud. Images corresponding to different functions can be deployed on one VM or distributed in several VMs based on deployment requirements, which could flexibly simulate multiple system under different deployment modes.
- **Network simulation:** OEM Cloud networks typically use network device and security devices, like router, switch, firewall and etc. PCCR could provide virtual network device simulation, like router and switch, and virtual security device simulation, like VPN and firewall. In addition, PCCR provides a visual interface on which could add, delete and drag VMs and network device to orchestrate network topology. Therefore, combined system simulation and network simulation can realize the simulation of ASOTAU server-side systems and networks with various combined requirements.

In addition, PCCR also provide security monitoring functions, that is traffic of the specified node port in the simulation network is collected and the abnormal behavior in the traffic is analyzed in real time. The security monitoring function is mainly used for security testing to improve the transparency of testing process.

4.2 Vehicle Bench

The vehicle bench in Fig. 3 consists of an IVN of ASOTAU service and a physical electric vehicle chassis. The implementation of these two parts is presented below.

- **IVN of ASOTAU service:** Depicted as ASOTAU CAN network in Fig. 3, in which the simulation of three ECUs, that are telematics, infotainment and gateway, are implemented. The three ECUs use experimental microcontroller unit (MCU) that can be reprogrammed repeatedly. The update management functions of IVN, is implemented and deployed in telematics, and all three ECUs implement the update performing function. Therefore, all three ECUs can be used as target ECUs of ASOTAU.
- **Physical vehicle chassis:** The vehicle bench integrates a physical vehicle chassis, which is an experimental electric vehicle chassis and can be driven to move. The chassis not only provides the real driving operation such as acceleration and deceleration, braking and steering, but also provides the ECUs and CAN bus of chassis domain, which is very helpful to the research and experiment for automotive cybersecurity. The CAN bus of chassis domain is depicted as CAN bus2 in Fig. 3. If an abnormal CAN command is injected into CAN bus2, the impact can be analyzed not only from the aspect of communication protocol, but also from the aspect of vehicle mechanical response and behavior.

The vehicle bench also provides a number of in-vehicle cybersecurity experiment tools, which are deployed in the tool server beside vehicle bench. The tool server is connected to CAN bus1 and CAN bus2, and can provide many operations on CAN bus, such as sniffing, injection, etc. In addition, vehicle bench also provides an ODB-II interface to facilitate other tools to access IVN.

4.3 4G Network Simulator

4G network simulator is used to simulate 4G cellular wireless communication network, which can provide 4G wireless communication network and secure access service for our vehicle bench. The user data used for secure access is stored in USIM card. In addition, in order to avoid mutual interference between the 4G wireless signal generated by simulator and the commercial 4G signal, a wireless signal shielding box is used to shield the simulated 4G signal in this box.

5 Use Example

This section will give some typical use examples of our testbed, and describe the simulation method for resources requirements in each examples, which shows the simulation capability of our testbed. The following is mainly from the attack experiment and testbed application two aspects to give 6 use examples.

5.1 Attack Experiment

This section presents attack experiments on three typical scenarios, that are OEM Cloud, ECU and CAN bus, to demonstrate the comprehensive simulation capabilities of our testbed and the conveniences for experimenter to use simulated experimental environment.

Attack OEM Cloud Experiment:

Experiment Description: An ASOTAU server is preset a vulnerability, and this experiment will exploit the vulnerability.
Experiment Setup: Experimental environment is depicted in Fig. 4 (a). An ASOTAU server and an attack computer are required and they located in the same local network. The experimental environment can be simulated using cyber range only. The simulation methods for all resource requirements are depicted in Table 1.
Execution and Result: Experimenter logs in the attack VM and use tools to perform attack operations on the simulated target server. If there is additional demand for tools, experimenter could install tools manually local or from Internet. The experimental results are shown in Fig. 4 (b).

This experiment applies the system simulation and network simulation functions of PCCR, which can quickly and easily complete simulation of software resources requirements. Experimenter only needs to log in the attack VM to perform experiments. PCCR

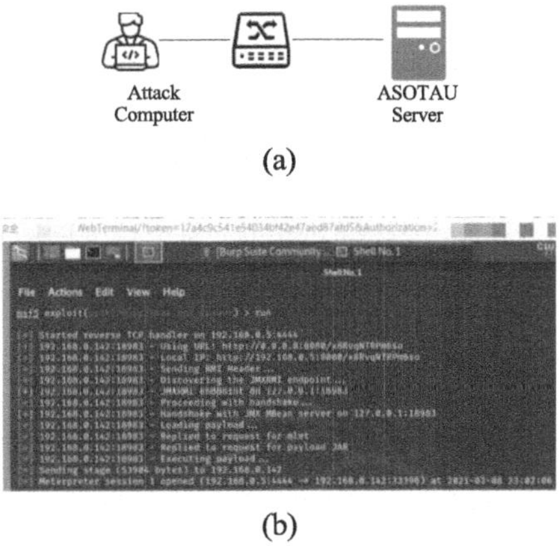

Fig. 4. Diagram of attack on OEM could experiment. (a) Network topology (b) Picture presentation of attack result

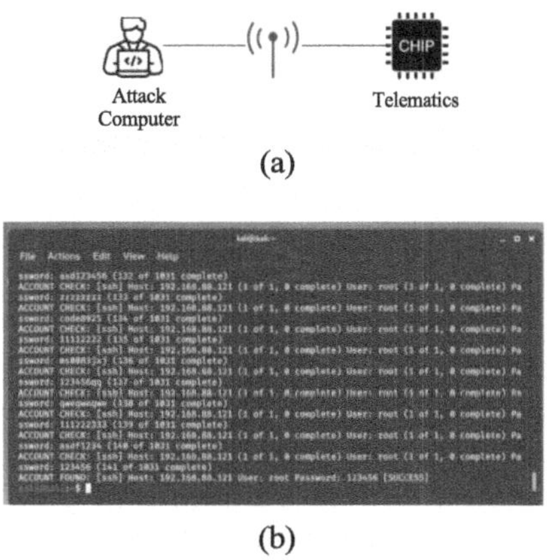

Fig. 5. Diagram of attack on telematics ECU experiment. (a) Network topology (b) Picture presentation of attack result

Table 1. Simulation of attack OEM Cloud experiment

Testbed Component	Resource Requirement	Simulation Method
Cyber Range	An ASOTAU server	Create a virtual machine (VM) using specified ASOTAU server image.
Cyber Range	An attack computer	Create an attack computer VM using an image preinstalled with some attack tools.
Cyber Range	A switch	Create a layer 2 switch and configure the experimental environment network.

Table 2. Simulation of attacks telematics ECU experiment

Testbed Component	Resource Requirement	Simulation Method
Cyber Range	An attack computer	Create an attack computer VM using an image preinstalled with some attack tools.
Vehicle Bench	A telematics	Load simulation program into telematics MCU.
4G Network Simulator	4G communication network	Start 4G network simulator and confirm USIM card data is ready. Plug USIM card into telematics. Configure connection between attack VM and 4G network simulator in cyber range.

can also provide simulation functions for more complex ASOTAU systems and network topology, so could support attack experiment for more complex OEM clouds.

Attack Telematics ECU Experiment

Experiment Description: Telematics, as the communication entrance of a vehicle, generally has access identity authentication. This experiment performs brute force on the access identity of telematics.
Experiment Setup: An attack computer is planned at external of vehicle, which connects the target telematics through wireless communication network. Experimental environment resource requirements include an attack computer, 4G communication network and telematics ECU, as depicted in Fig. 5 (a). The simulation methods for all resource requirements are depicted in Table 2.
Execution and Result: Experimenter logs in the attack VM and use tools to perform attack operations on the simulated target telematics. The experimental results are shown in Fig. 5 (b).

This experiment uses all three-layer simulation framework of our testbed, which can quickly complete the simulation of experimental environment with only loading and configuration. Experimenter could log in attack VM in laboratory to perform experiment remotely. PCCR can also simulate multiple attack computers and conduct multi-person parallel attack experiments.

Reverse CAN Protocol Experiment

Experiment Description: ECUs use CAN bus to communicate with each other, and maybe some instructions that could to control vehicle are transmitted in it, so reverse CAN protocol is a major attack method to crack the instructions. Firstly, needs to generate some traffic of operation vehicle, such as operating steering wheel, then analyze the traffic and replay analysis result to CAN bus for verifying.
Experiment Setup: This experiment needed to generate CAN traffic of operation vehicle, so simulation of the experimental environment used vehicle chassis and tools server. The simulation method for all resource requirements is depicted in Table 3.
Execution and Result: Experimenter performed the following operations: 1) Operate steering wheel to control right steering and generate CAN traffic of this operation; 2) Save traffic collected by sniffing tool, and then analyze the traffic; 3) Inject analysis instructions into CAN bus with CAN inject tool, and at the same time observe the vehicle wheel to verify the result. Using tool to inject CAN instructions as shown in Fig. 6.

This experiment uses physical chassis of vehicle bench, which not only easily generate various CAN traffic of operation vehicle chassis, but also can deliver traffic to the chassis to observe real physical response. The tools server has deployed a lot of tools for monitoring, injection and analysis of in-vehicle CAN communication, which is helpful for IVN security experiments. In addition, the chassis also provides an OBD-II interface to support traditional in-vehicle security testing approaches or access other tools into CAN bus.

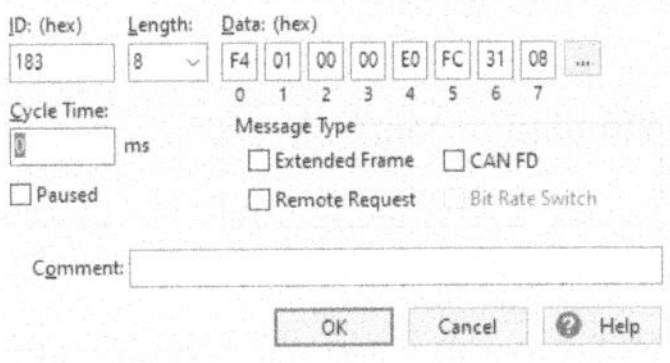

Fig. 6. Picture of CAN injection attack operation

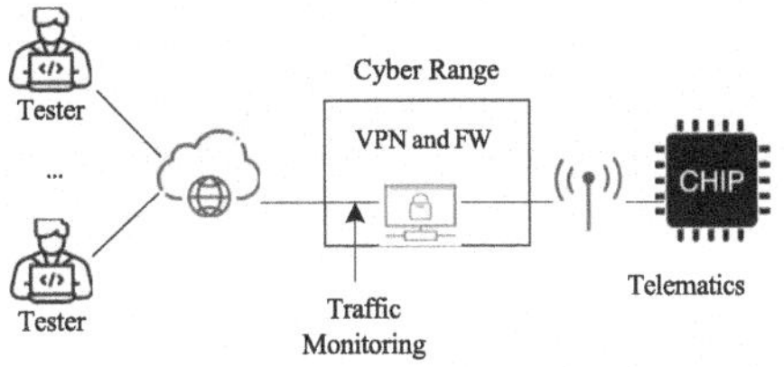

Fig. 7. Network topology diagram of Bug Bounty Program

Table 3. Simulation of reverse CAN protocol experiment

Testbed Component	Resource Requirement	Simulation Method
Vehicle Bench	Some real CAN traffic of operation vehicle	Start the vehicle chassis and perform vehicle driving operations.
Vehicle Bench	CAN sniff tool	In tools server, configure CAN sniff tool to connect to CAN bus2.
Vehicle Bench	CAN inject tool	In tools server, configure CAN inject tool to connect to CAN bus2.

5.2 Application Scenario

In this section, three typical testbed application scenarios are given on the basis of above attack experiments, including in-vehicle defense technology verification, ASO-TAU security testing and bug bounty program for vehicle cybersecurity, demonstrating our testbed applying widely.

In-Vehicle Defense Technology Verification

Experiment Description: Verification of the newly developed defense algorithm for attack detection in IVN. First, defense algorithm should to be deployed in IVN, and then simulate the specified attack to verify the effectiveness of the defense algorithm.
Experiment Setup: The defense algorithm is used to detect IP packets at the entrance of vehicle. Therefore, the defense algorithm is added to the telematics simulation program. This experiment could refer to attack experiment 2 simulating remote attack to telematics, so simulation methods for all resource requirements can refer to Table 2. The difference is that telematics program need to add the detection algorithms.
Execution and Result: Refer to attack experiment 2, experimenter performs certain specified attack operations in attack VM and checks logs to verify the algorithm.

Where there are attacks, there should be defenses. The previous sections describes a variety of attack experiment scenarios, and each attack scenario could support the corresponding defense technology research and verification.

Table 4. Simulation of ASOTAU security testing

Testbed Component	Resource Requirement	Simulation Method
Cyber Range	An ASOTAU server	Create a VM using a ASOTAU server system image.
4G Network Simulator	4G communication network	Refer to simulation method in Table 2.
Vehicle Bench	Gateway as the target ECU	Load simulation program into gateway MCU.

ASOTAU Security Testing

ASOTAU security is an important part of automotive cybersecurity, and the attack experiments and security technology verification described in previous sections are also applicable to ASOTAU. Here is an example of testing its own security mechanisms of ASOTAU.

Experiment Description: After tampering update package, upload it to ASOTAU server, then perform ASOTAU to test whether the integrity protection of the update package takes effect.
Experiment Setup: This experimental environment needs to simulate ASOTAU process, and simulation methods for all resource requirements are depicted in Table 4.
Execution and Result: Experimenter uploads the tampered update package to ASOTAU server-side system, configures an update task, update process is triggered, then observes whether to display update performing.

Our testbed simulates the CTT three-layer network functions and ASOTAU whole process, so we can provide several ASOTAU scenarios in multiple configurations for security testing and research.

Bug Bounty Program for Vehicle Cybersecurity

Experiment Description: A Bug Bounty Program is an activity in which rewards are offered for the security testing of a software system or production. In this example, testing target is our telematics ECU, and tester could connect it from Internet.
Experiment Setup: This experimental environment is depicted in Fig. 7, that could refer to attack experiment 2 simulating telematics ECU and 4G communication network. In cyber range, the simulated VPN server is used for tester to securely access from Internet and firewall is used to protect the experimental environment against attack, and they are deployed in the same VM. Ensure transparency to testing execution, the testing traffic is monitored. Simulation methods for all resource requirements are depicted in Table 5.
Execution and Result: Experimenters use the provided VPN file to remotely access testing environment and perform security testing on the target. Cyber range monitoring platform performs security analysis for testing traffic, records and monitors the whole testing process.

This activity uses the three-layer network simulation functions of our testbed, as well as remote secure access network simulation and security monitoring function of cyber range, and the testing environment can be quickly completed through configuration. Testers can access testing environment using VPN client, and perform testing using their own computer easily. We have already conducted bug bounty program for vehicle on our testbed and the target is some physical vehicles, that proves the feasibility of this application scenario [18].

Table 5. Simulation of Bug Bounty Program for vehicle cybersecurity

Testbed Component	Resource Requirement	Simulation Method
Cyber Range	Some VPN servers and firewall	Create some VMs using an image contains VPN server and firewall. Configure firewall protection functions and create VPN client files.
Cyber Range	Traffic monitoring function	Configure cyber range to monitor the traffic of entrance port of VPN.
Vehicle Bench	Telematics	Refer to simulation method in Table 2.
4G Network Simulator	4G communication network	Refer to simulation method in Table 2.

6 Conclusion

With the rapid growth of automotive software scale and requirements to communication with external of vehicle, new cybersecurity risks will be introduced, which will have potential damage to people's lives and property, even public transportation. Simulation is a major and effective approach for cybersecurity testing and research, however, most of automotive cybersecurity testbeds are for IVN, which cannot meet many cybersecurity research and testing of modern vehicles having complex communication scenario with outside. In this paper, we first proposed an automotive cybersecurity testbed applying a cyber range. Firstly, simulation model of our testbed is described, which is the CTT three-layer network architecture that can simulate a Cloud-Vehicle collaborative ASO-TAU whole process. Then the implementation framework of our testbed is presented. PCCR is used to simulate cloud network and other software experimental environment, that greatly promotes the simulation flexibility. Our testbed integrates an experimental electric vehicle chassis and ECUs that support ASOTAU, which can simulate a variety of in-vehicle cybersecurity experiment requirements. Finally, 6 typical use examples are given, simulation method and usage of each experimental environment are described, that show the comprehensive simulation ability and the richness of application scenarios of our testbed, and the convenience for experimenter to use simulated experimental environment.

Acknowledgements. This work was supported by the Major Key Project of PCL (Grant No. PCL2022A03–3), National Key research and Development Plan (Grant No. 2019QY1406), NSFC (No. 61972106).

References

1. Jo, H.J., Choi, W.: A Survey of Attacks on Controller Area Networks and Corresponding Countermeasures. IEEE Trans. Intell. Transp. Syst. **23**(7), 6123–6141 (2022)
2. Woo, S.: Hyo Jin Jo, Dong Hoon Lee, : A Practical Wireless Attack on the Connected Car and Security Protocol for In-Vehicle CAN. IEEE Trans. Intell. Transp. Syst. **16**, 993–1006 (2015)
3. Mahmoud Hashem Eiza and Qiang Ni: Driving with Sharks: Rethinking Connected Vehicles with Vehicle Cybersecurity. IEEE Veh. Technol. Mag. **12**, 45–51 (2017)
4. Miller, C., Valasek, C.: Remote exploitation of an unaltered passenger vehicle. Black Hat USA 2015 (2015)
5. S.Nie , L. Liu , Y. Du. : Free-Fall: Hacking Tesla From Wireless to Can Bus. In : Black Hat USA 2017 (2017)
6. Halder, S., Ghosal, A., Conti, M.: Secure over-the-air software updates in connected vehicles: A survey. Comput. Netw. **178**, 107343 (2020)
7. Toyama, T., Yoshida, T., Oguma, H., Matsumoto, T.: PASTA: portable automotive security testbed with adaptability. blackhat Europe 2018, London (2018)
8. Zheng, X., Pan, L., Chen, H., Pietro, R.D., Batten, L.: A Testbed for Security Analysis of Modern Vehicle Systems. In: IEEE Trustcom/BigDataSE/ICESS. pp.1090–1095 Sydney (2017)
9. Borazjani, P., Everett, C., McCoy, D.: OCTANE: an extensible open source car security testbed. In: Proceedings of the Embedded Security in Cars Conference (2014)
10. Čeleda, P., Vykopal, J., Švábenský, V., Slavíček, K.: KYPO4INDUSTRY: A Testbed for Teaching Cybersecurity of Industrial Control Systems. In: Proceedings of the 51st ACM Technical Symposium on Computer Science Education, pp.1026–1032. USA (2020)
11. Wang, C., Fang, L., Dai, Y.: A Simulation Environment for SCADA Security Analysis and Assessment. : In: 2010 International Conference on Measuring Technology and Mechatronics Automation, pp.342–347. Changsha, China, (2010)
12. He, K., Wang, C,, Han, Y., Fang, X.: Research on cyber security Technology and Test Method Of OTA for Intelligent Connected Vehicle. In: 2020 International Conference on Big Data, Artificial Intelligence and Internet of Things Engineering. pp.194–198. Fuzhou, China (2020)
13. Explanation of Firmware Over-The-Air, https://www.autosar.org/fileadmin/ standards/R21–11/CP/AUTOSAR_EXP_FirmwareOverTheAir.pdf. Last accessed 4 July 2023
14. Kuppusamy, T.K., Brown, A., Awwad, S., et al.: Uptane: Securing software updates for automobiles. In:14th ESCAR Europe. Munich, Germany (2016)
15. Dorri, A., Steger, M., Kanhere, S.S., Jurdak, R.: BlockChain: A distributed solution to automotive security and privacy. IEEE Commun. Mag. **55**, 119–125 (2017)
16. Herberth, R., Körper, S., Stiesch, T., et al.: Automated Scheduling for Optimal Parallelization to Reduce the Duration of Vehicle Software Updates. IEEE Trans. Veh. Technol. **68**, 2921–2933 (2019)
17. PengCheng Cyber Range homepage, https://pclcr.pcl.ac.cn/#/home. Last accessed 25 July 2023
18. The first intelligent Vehicle network security Competition of "Vehicle Protection Action" was successfully concluded, https://www.pcl.ac.cn/html/943/ 2021–01–15/content-3708.html. Last accessed 24 July 2023

A Fast Algorithm for Finding a Safe Steady State Point for Asynchronous RM Real-Time Systems

Guide Yang[1], Jiang Yu[1,2(✉)], Keke Tang[1], and Zhihong Tian[1]

[1] Cyberspace Institute of Advanced Technology (CIAT), Guangzhou University, Guangzhou 510006, China
jiangyu@gzhu.edu.cn
[2] Pazhou Lab, Guangzhou 510005, China

Abstract. Rate Monotonic (RM) model are widely used in real-time systems. The schedulability test is a significant problem in the design of asynchronous RM real-time systems. The test interval given by the existing researches is too large, so it needs to be optimized to reduce the test interval for improving the test efficiency, and thus it is very important to find a proper starting test point. In this paper, we propose a fast algorithm which finds a safe steady state point quickly than existing ones. Simulation results show that our algorithm outperforms existing ones.

Keywords: Real-time system · simulation · schedulability · rate monotonic · asynchronous · steady state point

1 Introduction

Real-time systems are widely used in automotive electronic systems, aircraft control systems, aerospace mission systems, digital petroleum production control systems, chemical and nuclear plant control systems [1–6], which take safety as the primary goal and real-time system technology is required.

To ensure the real-time performance of the security-first Real-time systems, the system's behavior must be predictable, which may be accomplished by schedulability analysis [1]. Specifically, it involves simulating various processes accessing system resources and using various schedulability detection algorithms to determine whether the system can schedule all simulated processes to meet the constraints of Real-time systems.

Previous researchers suggested many solutions for the aforementioned problem under various models, with Liu and Layland [7] proposed *the earliest deadline first algorithm (EDF)* under *the classical preemptive real-time scheduling model* (shortened as the *classic model* below). Because *EDF* is not easy to realize, now the rate monotonic *(RM)* have more widely used in real-time systems [8].

Choqiet-Geniet *et al.* [9] proposed a minimum schedulable interval for realtime systems of periodic tasks with offsets. As long as the task in the smallest schedulable interval is detected whether it is schedulable, the task set can be determined whether it

Z. Gu et al. (Eds.): NSE 2023, CCIS 2063, pp. 215–227, 2024.
https://doi.org/10.1007/978-981-97-4519-7_15

is schedulable. Park *et al.*[10] presented a new schedulability test for *RM* based on the response time, which is very close to the exact test and spend less time.

Jiang *et al.* [11] studied the schedulability of real-time tasks in the Priority–based Functional Reactive Programming (*P-FRP*) model under fixed priority scheduling. In [12], the authors conducted in-depth analysis and research on the problem of smallest schedulable interval, and gave the best research results for *P–FRP* model. Goossens [13] gave an upper bound on the minimum scheduling test interval for asynchronous periodic real-time tasks with the arbitrary deadline on identical multiprocessor platforms. Paolo Pazzaglia et al. [14] analyzed the simple and general methods of fixed priority schedulability optimization problems.

In the *classic model*, high-priority tasks can preempt low-priority tasks and preempted task can resume execution from the point of interruption. The *classical model* outperforms non-preemptive scheduling because it allows high-priority tasks to preempt lower-priority tasks, enabling better allocation of the resources needed in the computation. *RM* is a priority assignment that is optimal for a fixed (task) priority scheduling policy. The model we used in the paper is *RM* model, where the smaller the period, the higher the priority, and is ordered from smallest to largest period T.

In this paper, we investigate the problem of finding the minimum scheduling test interval for asynchronous periodic real-time tasks with *RM*, fixed priority, and implicit deadline (i.e., deadline equals to period). The results of previous studies on this problem show that the test schedule interval given in many cases is actually large [13]. The test schedule interval $[0, maxrT + LCM]$ is given in the classic model (The meaning of notations is seen in Sect. 2). Therefore we need to further find an earlier schedulability starting point, which we call safe steady state point *ssp*, and thus reducing the test interval from $[0, maxrT + LCM]$ to $[ssp, ssp + LCM]$. We provide a fast algorithm for finding such a safe steady state point and compare the algorithm with the traditional simulation algorithm, and the results show that our fast algorithm for finding a safe steady state point outperforms the existing algorithms.

The rest of the paper is organized as follows. For clarity, Sect. 2 introduces some basic concepts and notations that are used in this paper and system model. Section 3 introduces and analyzes our fast algorithm for finding such a *safe steady state point* for asynchronous RM real-time systems. Section 4 gives experimental results and shows that the test efficiency of our algorithm is superior to existing algorithms. Finally, Sect. 5 concludes the paper.

2 System Model

2.1 Basic Concepts and Notations

For the convenience of description, we use the Burns standard notation [15]. In this paper, we consider a hard real-time uniprocessor system in the RM model, with periodic n tasks, denoted by $\Gamma_n = \{\tau_1, \tau_2, \ldots, \tau_n\}$, under fixed priority scheduling. Each task consists of an infinite sequence of jobs $\tau_{i,j}$ $(1 \leq i \leq n, j = 1, 2, \ldots)$, where index j is the release number of τ_i and all the job have same fixed priority number when it's in the same tasks. For tasks τ_i, we set a unique priority for each task, which is defined as P_i $(1 \leq P_i \leq n)$, 1 stands for the highest priority, and in contrast, n is the lowest priority.

Each periodic task τ_i is characterized by a maximum computation time C_i, a relative deadline D_i, which is the time that the task can be executed after it arrives, a tasks period T_i, the utilization of tasks τ_i is U_i ($U_i = C_i/T_i$). Moreover, We define Φ_i is the initial release offset of the task relative to the initial time 0, or the conversion of the initial release offset. Therefore, a task τ_i can be denoted as $\tau_i = (\Phi_i, T_i, C_i, D_i)$. We define $r = \max_{1 \leq i \leq n} \Phi_i$, $U = \Sigma_{1 \leq i \leq n} U_i$, $maxrT = \max_{1 \leq i \leq n} (\Phi_i + T_i)$, $LCM = lcm(T_1, T_2, \ldots T_n)$, where *lcm* is the least common multiple of all task periods. The release time of the jth job with priority i is $r_{i,j} = \Phi_i + (j-1) \times T_i$. The *idle state* is the state where all tasks before time *t* have finished scheduling and new tasks can be added at time *t*. *sp* denotes a *steady state point*, and *ssp* denotes a *safe steady state point* which is a *steady state point* and all jobs released before the point *ssp* are completed.

According to the deadline, real-time task sets are generally divided into three categories: implicit deadline, constrained deadline, and arbitrary deadline [4, 5]. For all i, $1 \leq i \leq n$. The implicit deadline is $Di = Ti$. The constrained deadline is $D_i \leq T_i$ and arbitrary deadline indicates that D_i has no relationship with T_i. Arbitrary deadline is the most difficult situation. In this paper, we consider cases of implicit deadline.

For a task set Γ_n, a task τ_k ($1 \leq k \leq n$) is referred to as schedulable in a time interval if there is no deadline miss for all jobs of τ_k released in the same interval, and a task τ_k is referred to as schedulable if there is no deadline miss for all jobs of τ_k. A task set Γ_n is referred to as schedulable in a time interval if all tasks in the Γ_n is schedulable in the same interval, and a task set Γ_n is referred to as schedulable if all tasks in the Γ_n is schedulable. A feasibility interval of a Γ_n is a time interval $[t, t + LCM)$ with a length of *LCM* such that if the Γ_n is schedulable in the same interval, then the Γ_n will also be schedulable (in the interval $[0, Z)$: $Z \rightarrow +\infty$). Each task set in each group may be unique, as they are generated independently.

For RM hard real-time uniprocessor system, it has two characteristics: time limitation and high reliability. If a task set satisfies $U \leq 1$, it is considered that the task set is schedulable. This paper is based on the premise that all simulated task sets are scheduled, so the non-schedulable task sets are not within the scope of this paper.

2.2 System Model

In the RM model, a task has higher priority when it's period is smaller. There are two situations when a job $\tau_{i,j}$ was released. If a job of a task with lower priority than i is being executed, it will be immediately preempted, save the current running state and enter the waiting queue, and the job $\tau_{i,j}$ will start execution. After waiting for the high priority task job finished, the preempted job continues to execute the job from the interruption point. If a job of a task with higher priority than i is being executed, the job $\tau_{i,j}$ will enter the waiting queue and run again after the job with higher priority than i finishes.

The task sets are scheduled in an asynchronous manner, where asynchronous means that all tasks do not arrive at the same time. A task set starts asynchronously does not mean that the whole task set is asynchronous because it may indeed be a synchronous task set. For example, when all task periods are co-prime, they must have a common release, regardless of any imposed offsets.

To simplify the model, in this paper we do not consider the preempting time consumed by high priority preempting of low priority. The computation time C_i can be replaced by the processing time, and the processing events include the computation time as well as the time spent in switching tasks. We assume that tasks cannot be blocked from execution by another task except contending for a processor, and we do not consider the blocking time for organizing task execution due to other tasks accessing mutually exclusive shared resources [4]. The impact of these operations on the process is another issue that is out of the scope of this paper.

To better understand the RM model, we will give an example. As shown in Fig. 1, There are three tasks $\tau_1 = (\Phi_1 = 4, T_1 = 6, C_1 = 2, D_1 = T_1 = 6)$, $\tau_2 = (\Phi_2 = 0, T_2 = 9, C_2 = 2, D_2 = T_2 = 9)$, $\tau_3 = (\Phi_3 = 2, T_3 = 10, C_3 = 3, D_3 = T_3 = 10)$, $P_1 = 1, P_2 = 2, P_3 = 3$ and $maxrT = \Phi_3 + T_3 = 11$.

Because task τ_1 priority is higher than τ_3, then τ_3 is preempted by τ_1 at time point 4. Meanwhile, τ_3 still has 1 unit time to run and save the interruption point. At time point 6, τ_1 operation is completed and τ_3 starts execution from the interruption point. τ_3 finishes at time point 7.

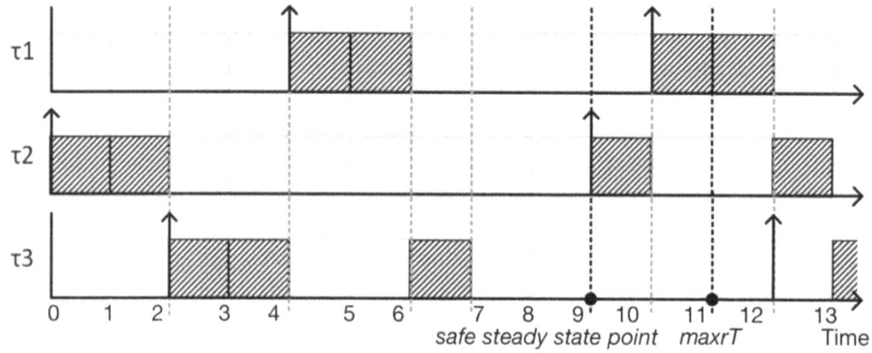

Fig. 1. Illustration of "carry in" request at time point $maxrT = 11$ and a safe steady state point at time point 9 before the time point 11.

Theoretically, we can start the schedulability test from time point *maxrT* [9]. However, it can be seen that there are still the 2nd job $\tau_{1,2}$ of task τ_1 running at time point 11 (*maxrT*), as shown in Fig. 1, i.e., some computation time of that job "carry in" the time interval behind the time point *maxrT*. Therefore, it is not feasible to start test at this point. We need to find an earlier schedulable starting point, i.e., a *safe steady state point* as shown in Fig. 1, the time point 9. It is known that the schedulability test interval of the whole task set will be shortened to $[ssp, ssp + LCM]$ instead of $[0, maxrT + LCM]$, i.e., $[9, 9 + 18 = 27]$ with a length 18 instead of $[0, 11 + 18 = 29]$ with a length 29 for the task set.

Notice that in this scenario, the time points 7 and 8 are also *safe steady state points.* However, we need not wasting time to find them because the test can start from the time point 9. Therefore, what the *safe steady state point* we need to find is the nearest time point before the point *maxrT*.

3 Our Method

In this section, we propose a fast algorithm for finding a *safe steady state point* before the time point *maxrT*. The algorithm is inspired by the important idea in [9].

According to Sect. 2, it is infeasible to start test from *maxrT* due to "carry in" jobs, the traditional simulation algorithm chooses to simulate on [0, *maxrT* + *LCM*]. Actually, the interval of *maxrT* + *LCM* is very large, we need to optimize it. Because a schedulability test needs to test the schedulability of task sets at least with *LCM* length, we choose to optimize the [0, *maxrT*] part, i.e., optimizing and shortening the test time interval [0, *maxrT*] by finding a *safe steady state point ssp*. Traditional simulation algorithms need to run each job. Our algorithm can quickly find a time point *ssp* by calculating the *preleases* and *prequests* and backward searching.

Preleases record how many jobs will be added at this time. According to the offset and period of the task, the running time of the task is accumulated at each time when the task is released. Pseudocode of calculating preleases is shown in Fig. 2 line 13. As shown in Fig. 1, before the time point *maxrT*, for task τ_1, it is released at time points 4 and 10, so $preleases[10] = preleases[4] = C_1 = 2$. For task τ_2, it is released at time points 0 and 9, so $preleases[9] = preleases[0] = C_2 = 2$. For task τ_3, it is released at time point 2, so $preleases[2] = C_3 = 3$.

Prequests record how long it still needs to execute before it enters the *idle state*, at each point in time interval [0, *maxrT*]. In other words, *prequests*[*t*] indicates how long it will take to finish running at time *t*. There are two situations when we calculate the *prequests*. If a task execution is completed before time *t* ($prequests[t - 1] = 0$), then $prequests[t] = preleases[t]$. If there are still tasks running before time *t* ($prequests[t - 1] > 0$), then $prequests[t] = prequests[t - 1] - 1 + preleases[t]$. Pseudocode of calculting prequests is shown in Fig. 2 line 22. As shown in Fig. 1 at time point 4, τ_3 is still running and τ_1 is released, so $prequests[4] = prequests[3] - 1 + preleases[4] = 2 - 1 + 2 = 3$, i.e., it still needs 3 time slots to complete released jobs.

Time *t* can be used as the schedulability test starting point only when all tasks released before this time have been executed. As shown in Fig. 2 line 7, $prequests[t] \leq preleases[t]$ indicates that all jobs have finished execution before time *t* ($prequests[t - 1] = 0$) or all jobs before time *t* are completed at time *t* ($prequests[t - 1] = 1$). Notice that in Fig. 1 the time point 2 maybe also a steady state point. However, it is no need to search because once a safe steady state point ssp is found, then the schedulability test can be conducted at [*ssp*, *ssp* + *LCM*].

Theorem 1. *Consider a schedulable task set Γ_n under fixed priority scheduling, then, for each k, $1 \leq k \leq n$, a feasibility interval of Γ_k is [ssp, ssp + LCM), where ssp is a safe steady state point of Γ_k.*

Proof. For any given *k* and Γ_k, if a schedulable system Γ_k runs till time *Z*, where *Z* is a sufficiently large positive integer ($Z \to +\infty$), we obtain a number $j = int((Z - ssp)/LCM)$ where *int*(*x*) is the integer part of *x*. Therefore, the time interval [0, *Z*) can be divided into some time intervals and the union of these consecutive time intervals equal to the time interval [0, *Z*), i.e., $[0, Z) = [0, ssp) \cup [ssp, ssp + j \times LCM) \cup [ssp + j \times LCM, Z)$, where $[ssp, ssp + j \times LCM) = [ssp, ssp + LCM) \cup \ldots \cup [ssp + (j - 1) \times LCM, ssp + j \times LCM)$.

Algorithm 1 Fast finding a safe steady state point

```
Require: an n taskset τ where τi = (φi, Ti, Ci, Di = Ti), 1 ≤ i ≤ n
Ensure: a safe steady state point ssp
begin
Initialize preleases[0 . . . maxrT] to be 0
Release(taskset, maxrT)
Request(taskset, maxrT)
ssp ← 0
for t = maxrT to 1 do
    if prequests[t] <= preleases[t] then
        break
    end if
end for
ssp ← t
end
function Release(taskset, maxrT)
    /* set task release time and computation time */
    for k = 0 to n − 1 do
        t ← Φk+1
        while t <= maxrT do
            preleases[t] ← preleases[t] + Ck+1
            t ← t + Tk+1
        end while
    end for
end function Release

function Request(taskset, maxrT)
    /* simulating task execution */
    prequests[0] ← preleases[0]
    for t = 0 to maxrT do
        if prequests[t − 1] == 0 then
            prequests[t] ← preleases[t]
        else
            prequests[t] ← prequests[t − 1] − 1 + preleases[t]
        end if
    end for
end function Request
```

Fig. 2. Fast finding a safe steady state point algorithm.

By Lemmas 3.1 and 3.2 from [11], both the release and the effective execution patterns of Γ_k in time intervals $[ssp, ssp + LCM)$, $[ssp + LCM, ssp + 2LCM)$,..., $[ssp + (j-1) \times LCM, ssp + j \times LCM)$ will be the same. Thus, schedules of Γ_n in these j time intervals will be the same. Due to the periodicity of tasks, schedules in time intervals $[ssp + j \times LCM, Z)$ and $[ssp + (j-1) \times LCM, ssp + (j-1) \times LCM + (Z - (ssp + j \times LCM)))$ will also be the same.

Therefore, if task set Γ_n is schedulable in time interval $[ssp, ssp + LCM)$, it will also be schedulable in intervals $[ssp + LCM, ssp + 2LCM), \ldots, [ssp + (j - 1) \times LCM, ssp + j \times LCM)$, and $[ssp + j \times LCM, Z)$. Since all tasks before *sp* are finished, i.e., scheduling is complete, the task set Γ_n is still schedulable on $[0, ssp)$. Hence, the schedulability of Γ_n in the time interval $[ssp, ssp + LCM)$ will determine the schedulability of Γ_n in the interval $[0, Z)$.

In other words, the above theory proves that we can start the schedulability test from any a *safe steady state point ssp*.

In order to find an earlier schedulable starting point, i.e., a *safe steady state point*, we start from the theoretical point *maxrT* and search backward to point 0. Once the schedulable starting condition ($prequests[t] \leq preleases[t]$) is met, this point is then a *safe steady state point*. We have theoretically analyzed the validity of the algorithm for finding a *steady state point* above, and we will illustrate the efficiency of the algorithm with experimental results in the next section. Notice that this algorithm is not applicable to asynchronous *P-FRP* real time systems.

4 Experimental Results

In this section, we present the performance of traditional array simulation algorithm, the newest linked list simulation algorithm [12], and our fast algorithm for finding a safe steady state point in a task set using three kinds typical periods generation methods: expRandRSL, harmonicRand, and magnRand. Experimental results show the efficiency on the running time. In addition, considering results in [12] we notice that the linked list algorithm has high efficiency for the *PFRP* model compared to the *classic* model, especially for simulation in a short length.

For particular tasks, such as those with harmonic period length, there are exact polynomial time algorithms for testing schedulability [16]. In the following, we will generate different task sets with periods using the expRandRSL, magnRand, and harmonicRand methods, and compare the running efficiency of the newest linked list simulation algorithm, the traditional array simulation algorithm and our fast algorithm for finding a safe steady state point.

Using ExpRandRSL to generate periods stands for that periods are exponential distribution, which can generate a task set with *n* tasks and periods within $[min\_p, max\_p]$, where min_p is a given minimum periods, and max_p is a given maximum periods, The task periods are divided into $int(k)$ layers, where $k = \log(max\_p/min\_p)$. When $int(k)$ is 0, then randomly generating n periods within $[min\_p, max\_p]$. When $int(k)$ is not 0, for two adjacent layers of period, the interval for generating periods of the latter layer is *e* times that of the previous layer. When the error between *k* and $int(k)$ is within 0.1, set $k_1 = 0$ and $k_n = int(k)$; otherwise, set $k_1 = 1$ and $k_n = int(k) + 1$. Then, when $k_n \geq n$, for *i* from 1 to $n - 1$, randomly generating first $n - 1$ periods within $int([e^{(i-1)}, e^i]min\_p)$, and then randomly generating a period within the interval $([e^{(n-1)}, int(max\_p/min\_p)]min\_p)$. When $k_n < n$, we first generate one task period to be *max_period* for a task in the task set, then generating rest $n - 1$ period in the following steps: (1) for *i* from 1 to *k*, randomly generating $int((n - 1)/(k + k_1))$ periods within the interval $int([e^{(i-1)}, e^i] \times min\_p)$; when k_1 is 1, then randomly generating another period within $[e^{int(k)}, int(max\_p/min\_p)] \times min\_p$. At this step, $int((n-1)/(k+k_1)) \times k$ periods are generated. (2) randomly generating rest $(n-1) - int((n-1)/(k+k_1))(k+k_1)$ periods within the interval $int([e^{(i-1)}, e^i]min\_p)$.

Using MagnRand to generate periods stands for that periods are of magnitude characteristics. We use parameter *mag* to indicate the number of types in a task sets. For example, if $mag = 1$ there is only one magnitude periods and if $mag = 2$ there will be two magnitudes periods in a task set. In the scenario of $mag > 1$, the difference between the two adjacent magnitude period is an order of magnitude based on 10. We use parameter *mn* to indicate the starting order of magnitude of the period. For example, if $mn = 1$, the starting order of magnitude of the period is 10^1 and if $mn = 2$, the starting order of magnitude of the period is 10^2. If $mag = 3$ and $mn = 1$, then there will be three magnitudes periods in a task set and $int(n/mag)$ periods will be randomly selected from intervals $[10^1, 10^2 - 1]$, $[10^2, 10^3 - 1]$, and $[10^3, 10^4 - 1]$, respectively, and the rest n mod *mag* periods will be randomly selected in sequence from the first different $mag - 1$ intervals.

Using HarmonicRand to generate periods stands for that periods are harmonic numbers, i.e., for each pair of periods, one period divides the other. In this paper, periods for this type of task sets are randomly selected from the set {16, 32, 64, 128, 256, 512, 1024}, and the minimum and maximum periods is set to 16 and 1024 respectively.

For the offset Φ_i of a task τ_i in an n-task set $(1 \le i \le n)$, it is randomly selected from $[0, T_i]$ where T_i is the generated period of the task τ_i. At this step, randomly set some offsets to be 0, i.e., assume some tasks are initially released at time point 0.

In order to make the total task set utilization be a given value, the *UUniFast* algorithm [17] is used to generate computation time C of a task in the task set, and the total utilization of each task set in this experiment is set to 0.69. The generated task sets are all schedulable, because we are comparing the efficiency of the traditional test and the fast algorithm for finding a safe steady state point, not the judgment of whether a task set is schedulable or not, which needs still by testing in an interval of the operation length of *LCM*.

We aim to optimize the simulation time within $[0, maxrT]$ to the time spent for finding a safe steady state point, so this simulation is performed on $[0, maxrT]$. Otherwise, if finding a *safe steady state point* takes a lot of time, it would be better to test it directly on $[0, maxrT]$. Therefore, we need to compare the time spent for finding a *safe steady state point* and the traditional simulation on $[0, maxrT]$ in order to better illustrate the efficiency of the fast algorithm for finding a *safe steady state point*. The following are the detailed setup for this experiment.

Experiment setup: Each task set group contains 10,000 task sets. The utilization rate of the task sets are 0.69, and the tasks are asynchronous. RM scheduling is employed, the smaller the period, the higher the priority. The tasks in the task set are sorted from small to large according to the period. Then let the computation time C_i to be $T_i \times U_i$. For the expRandRSL task sets, tasks with periods within [5, 40] are generated and 8 groups task sets are generated with the number of task per task set increasing from 3 to 10. For the magnRand task sets, we set $mag = 1$, $mn = 1$ and thus tasks with periods within [10, 99] are generated, so there is only one order of magnitude in this scenario. Eight groups task sets are generated with the number of task per task set increasing from 3 to

10. For the harmonicRand task sets, tasks with periods of {16, 32, 64, 128, 256, 512, 1024} are generated and 12 groups task sets are generated with the number of task per task set increasing from 3 to 10, and then 16, 32, 64, 128. For all task sets, each of task sets in each group is likely to be unique since they are independently generated. The experiment is conducted on a PC with CPU i7-8750h 3.9GHz and 24GB memory.

In order to evaluate performance, we compare the running efficiency between different algorithms by calculating the average running time of 10,000 task sets for each task group, i.e., number of tasks per task set from 3 to 128.

The efficiency comparison of the newest linked list simulation algorithm (LinkedList), traditional array simulation algorithm (TradiArray) and our fast algorithm (Fast) for finding a *safe steady state point* is shown in Fig. 3, Fig. 4 and Fig. 5 for expRandRSL, magnRand, and harmonicRand task sets respectively. The corresponding average running time in second is shown in Tables 1, 2 and 3, respectively.

As shown in Fig. 3, Fig. 4, and Fig. 5, it can be seen that for finding a safe steady state point the performance of the fast algorithm is significantly better than the two existing simulation algorithms. With the increase of the number of tasks per task set, *the Fast* is more excellent, and the range ahead of the LinkedList and TradiArray is expanding. In addition, the TradiArray is a few better than the LinkedList, which is likely due to that within a length of *maxrT*, operations on an array are faster than that on a linked list because of memory allocation of list nodes.

Table 1. Performance comparison of algorithms.

Number of tasks per task set	Running time (ms) of expRandRSL task sets		
	Linkedlist	TradiArray	Fast
3	0.03347974	0.03305616	0.02529676
4	0.04260497	0.04136442	0.03122187
5	0.05871601	0.05615304	0.04110601
6	0.06800413	0.0653411	0.04708508
7	0.08304944	0.07885231	0.05575849
8	0.09170673	0.08712705	0.06166125
9	0.10576804	0.09938558	0.06958884
10	0.11633746	0.10881598	0.07664881

Notice that in the harmonicRand scenario in Fig. 5, the difference between the Fast and the other two algorithms is enlarging while the number of tasks per task set is larger than 16. This does not mean that the difference will enlarge in an exponential way. In fact it is due to the exponential increase of the number of tasks.

In the harmonicRand scenario, it can be seen in Table 3 that for the two groups where the number of tasks per task set is 3 and 4, the TradiArray algorithm lags slightly behind the LinkedList algorithm. We give the following two reasons. First, although the LinkedList algorithm needs memory allocation for list nodes, the periods in this scenario

Table 2. Performance comparison of algorithms.

Number of tasks per task set	Running time (ms) of expRandRSL task sets		
	Linkedlist	TradiArray	Fast
3	0.02936106	0.02853822	0.02300449
4	0.04017175	0.03913886	0.03063998
5	0.05021831	0.04822825	0.03698942
6	0.06109527	0.05917712	0.04411891
7	0.07256408	0.06949141	0.05147003
8	0.08478052	0.08115385	0.05913839
9	0.09785358	0.0937385	0.06807538
10	0.10864055	0.10314018	0.07398998

Table 3. Performance comparison of algorithms.

Number of tasks per task set	Running time (ms) of expRandRSL task sets		
	Linkedlist	TradiArray	Fast
3	0.17717182	0.19011312	0.11941586
4	0.22258752	0.22787788	0.14223901
5	0.27316551	0.26875449	0.16841356
6	0.32516627	0.31755765	0.19592497
7	0.38170614	0.36590151	0.2252568
8	0.44567131	0.42704856	0.26131086
9	0.50641923	0.48046268	0.29359343
10	0.58857036	0.54594259	0.33666898
16	0.92659778	0.84073236	0.50512604
32	1.6902544	1.49335187	0.89510265
64	2.13971136	1.87208117	1.1314363
128	2.61848987	2.26565053	1.39783789

are harmonic and thus less new linked list nodes are needed, resulting in less running time spent. Second, the LinkedList algorithm needs only to record the start and end time points, while the TradiArray algorithm needs to simulate at every time point one by one within [0, $maxrT$] and thus will take much time. When the number of tasks increases, the LinkedList algorithm needs more list nodes and thus it's running time increases.

To sum up, compared with the newest LinkedList and TradiArray algorithms, the Fast algorithm for finding a safe steady state point is not only effective but also high efficient in running time.

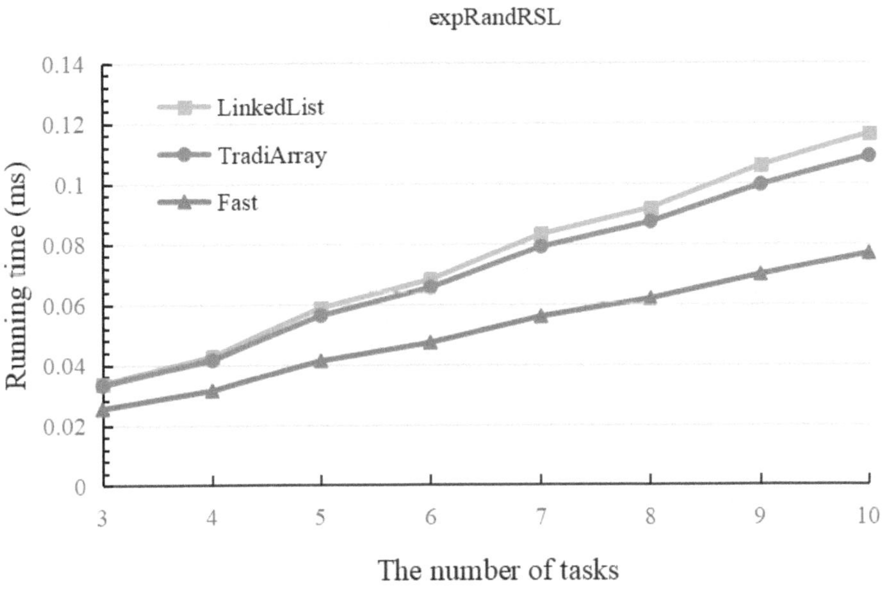

Fig. 3. Average time of each algorithm in the expRandRSL scenario.

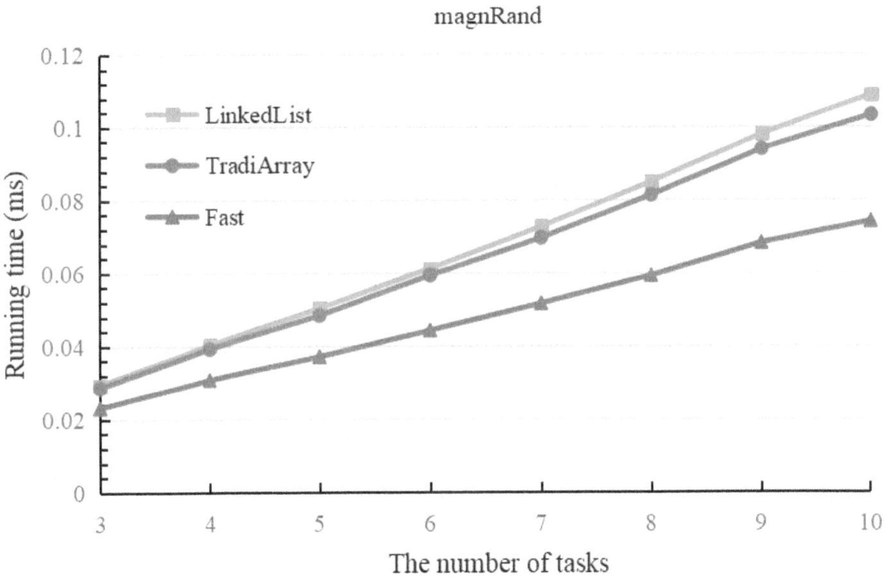

Fig. 4. Average time of each algorithm in the magnRand scenario.

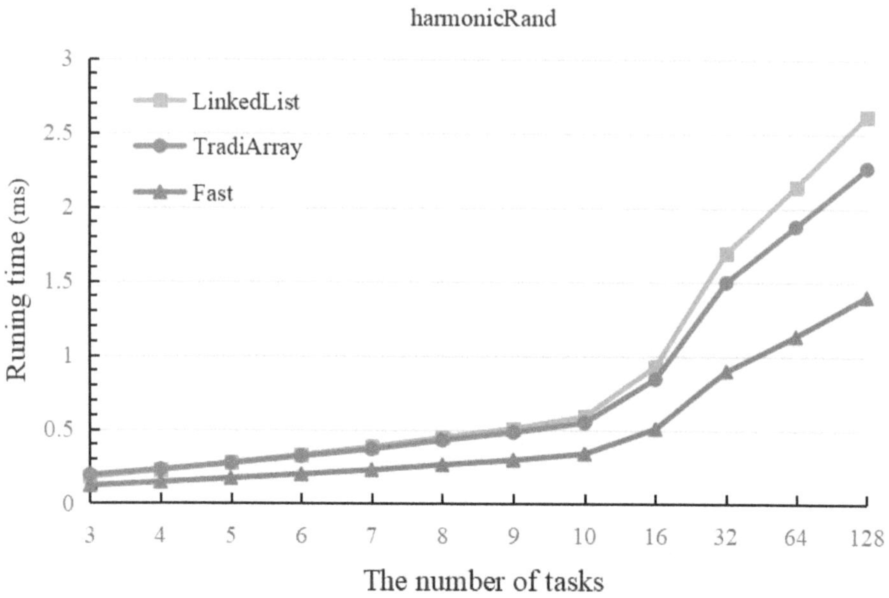

Fig. 5. Average time of each algorithm in the HarmonicRand scenario.

5 Conclusion

In this paper, we propose a fast algorithm for finding a *safe steady state point ssp* for asynchronous periodic RM real-time systems with fixed priority and implicit deadlines. We conduct experiments with three kinds of typical periods: expRandRSL, magnRand, and harmonicRand, and the results show that the fast algorithm for finding a *safe steady state point* takes much less time than the newest linked list algorithm and the traditional array algorithm. Starting from the *safe steady state point ssp*, the follow-up schedulability tests can be conducted within an optimal test interval $[ssp, ssp + LCM]$ with a length LCM instead of within the whole interval $[0, maxrT + LCM]$ with a length $maxrT + LCM$. Next, we will extend our method to the asynchronous RM real-time systems with constrained deadlines.

Acknowledgements. This work is supported in part by Guangdong Student's Platform for Innovation Training Program of China (No. S202011078044) and Cultivation Project of Pazhoulab (No. PZL2022KF0013), and the corresponding author is Yu Jiang, jiangyu@gzhu.edu.cn.

References

1. Audsley, N.C., Burns, A., Davis, R.I.: Fixed Priority scheduling: A historical perspective. Real-Time Systems **8**(2), 173–198 (1995). https://doi.org/10.1007/BF01094342
2. Sha, L., Abdelzaher, T., Arzen, K.E.: Real time scheduling theory: A historical perspective. Real-Time Systems **28**(2), 101–155 (2004). https://doi.org/10.1023/B:TIME.0000045315.61234.1e

3. Buttazzo, G.C.: Rate monotonic vs. EDF: Judgment day. Real-Time Systems, **29**(1), 5–26. (2005). https://doi.org/10.1023/B:TIME.0000048932.30002.d9
4. Davis, R.I., Burns, A.: A survey of hard real-time scheduling for multi–processor systems. ACM Comp. Surveys **43**(4), 1–44 (2011). https://doi.org/10.1145/1978802.1978814
5. Davis, R.: A review of fixed priority and EDF scheduling for hard real-time uniprocessor systems. ACM Comp. Surveys **11**(1), 8–19 (2014). https://doi.org/10.1145/2597457.2597458
6. Tian, Z.H., Shi, W., Wang, Y.H., Zhu, C.S., Du, X.J., Su, S., Sun, Y.B., Guizani, N.: Real-time lateral movement detection based on evidence reasoning network for edge computing environment. In IEEE Trans. Ind. Informatics, pp. 4285–4294. IEEE **15**(7) (2019)
7. Liu, C.L., Layland, L.W.: Scheduling algorithms for multi-programming in a hard-real-time environment. Journal of ACM **20**(1), 46–61 (1973). https://doi.org/10.1145/321738.321743
8. Bini, E., Buttazzo, G.C., Buttazzo, G.M.: Rate monotonic analysis: the hyperbolic bound. IEEE Trans. Comput. **52**(7), 933–942 (2003). https://doi.org/10.1109/TC.2003.1214341
9. Choqiet-Geniet, A., Grolleau, E.: Minimal schedulability interval for real-time systems of periodic tasks with offsets. Theoretical Computer Sci **310**(1), 117–134 (2004). https://doi.org/10.1016/S0304-3975(03)00362-1
10. Park, M., Park, H.: An efficient test method for rate monotonic schedulability. IEEE Trans. on Computers **63**(5), 1309–1315 (2014). https://doi.org/10.1109/TC.2012.296
11. Jiang, Y., Cheng, A.M.K., Zou, X.L.: Schedulability analysis for real-time p-frp tasks under fixed priority scheduling. In IEEE RTCSA 2015, pp. 31–40. IEEE (2015)
12. Qin, Y., Zou, X., Cheng, A.M.K., Jiang, Y.: Finding a steady state point for fixed priority independent periodic real-time tasks with arbitrary given release offsets. In: 2017 IEEE 20th International Symposium on Real-Time Distributed Computing (IEEE ISORC 2017), pp. 203–206. IEEE (2017)
13. Goossens, J., Grolleau, E., Cucu-Grosjean, L.: Periodicity of real-time schedules for dependent periodic tasks on identical multiprocessor platforms. Real-Time Systems. **52**(6), 808–832 (2016). https://doi.org/10.1007/s11241-016-9256-1
14. Pazzaglia, P., Biondi, A., Di Natale, M.: Simple and general methods for fixed–priority schedulability in optimization problems. In: 2019 Design, Automation & Test in Europe Conference & Exhibition, pp. 1543–1548. IEEE (2019)
15. Davis, R.I., Burns, A.: Burns standard notation for real-time scheduling. In: Real-Time Systems: The past, the present, and the future. Audsley, N., Baruah, S.K. (eds.), pp. 38–41. Citeseer (2013)
16. Bonifaci, V., Marchetti-Spaccamela, A., Megow, N., Wiese, A.: Polynomial-time exact schedulability tests for harmonic real-time tasks. In 2013 IEEE 34th Real–Time Systems Symposium, pp. 236–245. IEEE (2013)
17. Bini, E., Buttazzo, G.C.: Measuring the performance of schedulability tests. Real-Time Systems **30**(1–2), 129–154 (2005). https://doi.org/10.1007/s11241-005-0507-9

RiskLens: A Novel Way to Quantify the Risk for Big Data Platform Enhanced by Machine Learning

Haomou Zhan[1], Jiawei Yang[1], Zhenyang Guo[1,2], Jin Cao[1,2](✉), Xingwen Zhao[1,2], Wei You[1,2], and Hui Li[1,2]

[1] School of Cyber Engineering, Xidian University, Shaanxi 710126, China
{hmzhan,jwyang_zz,zyguo}@stu.xidian.edu.cn, {jcao, wyou}@xidian.edu.cn, lihui@mail.xidian.edu.cn

[2] The State Key Laboratory of Integrated Service Network, Shaanxi 710126, China

Abstract. Big data, characterized by its large volume, diverse formats, and rapid processing speed, possesses remarkable value. However, it is also highly susceptible to security incidents such as data theft, tampering, and destruction during the transmission, storage, and processing stages. Existing methods for assessing security risks in information systems, constrained by human resources and computational methodologies, are not directly applicable to big data platforms. Therefore, researching a quantitative risk assessment scheme tailored to big data platforms, identifying potential risks, and quantitatively analyzing potential losses is a crucial task. To this end, this paper proposes an automated theory for quantitatively assessing risks in big data platforms, integrating traditional quantitative risk assessment methods with the unique features of big data. To validate this theory, a simulation experiment platform was constructed, and the experimental results demonstrate that the proposed scheme can automatically and efficiently quantify risks related to both assets and data processing procedures in big data platforms.

Keywords: Big data platform · Quantitative risk assessment · Machine learning

1 Introduction

In recent years, the rapid development of information technology has led to the widespread adoption of big data in all sectors of society. Big data technologies have demonstrated great potential for every scene, like commodity recommendation and decision analysis. Despite the benefits of big data technology, security concerns related to its use have become increasingly prevalent. Recently, some major Internet companies such as Google[1] and Facebook[2], all of which have caused a large number of user data leakage problems due to their reasons. These incidents are a stark reminder of the criticality of implementing effective security measures to mitigate the risks associated with big data platforms.

Haomou Zhan and Jiawei Yang have equal contribution.

[1] https://firewalltimes.com/google-data-breach-timeline/.

[2] https://firewalltimes.com/facebook-data-breach-timeline/.

Z. Gu et al. (Eds.): NSE 2023, CCIS 2063, pp. 228–242, 2024.
https://doi.org/10.1007/978-981-97-4519-7_16

Due to the openness of Internet technology, many companies rely on different open-source programs to build different types of big data platforms with different functions. Wu et al. [11] employ the Delphi method and develop a comprehensive security evaluation index system for ensuring enterprise big data security. Zhu et al. [17] introduce information entropy to calculate the weight of each risk index and incorporated the fuzzy comprehensive evaluation method to assess and predict the privacy risk of social networks quantitatively. Also, the security requirements for user data vary from country to country. For example, Europe has implemented the General Data Protection Regulation and America has enacted Sarbanes–Oxley Acts, and China has implemented the Cybersecurity Law of the PRC and the Personal Information Protection Law of the PRC. It can be seen that the existing big data security standard system urgently needs to be improved. A general quantitative risk assessment mechanism for big data platforms needs to be proposed.

This paper aims to develop risk assessment theories, technologies, and tools specifically adapted to the big data environment to address this issue. It analyzes potential threats to big data platforms from various perspectives and comprehensively discusses the risk identification, quantification, and assessment processes. We summarise the main contributions of this paper shown as follows:

1. Based on the existing risk assessment theory and combined with the characteristics of big data, we propose a big data quantitative risk assessment theory. Thus, an evaluation framework for big data platform assets and data is constructed. Enterprises can effectively improve their risk identification capabilities for their big data platforms by using this framework.
2. Based on the proposed theoretical system, an evaluation index system is constructed for the assets of the big data platform and the data of the big data platform. These index systems provide a specific and practical process for the risk assessment of big data platforms.
3. To verify the correctness of our theory, we build a test platform using the common way of building a big data platform and used the scheme proposed in this paper to carry out the risk assessment of assets and data. The experimental results show that benefiting from the powerful computing power of computers and the excellent learning ability of machine learning algorithms. Our proposed method is more objective and has significantly lower labor and time cost, a better ability to process a large volume of data continuously, and more importantly, our method is highly adaptable for different scenarios and allows for iterative updates and refinements.

The organization of this paper is shown as follows: We delve into related work in the field of risk quantification in Sect. 2. The risk quantitative assessment scheme and risk quantitative index system for big data platforms are explained in Sect. 3. Subsequently, a comparative evaluation of the effectiveness of the proposed scheme is in Sect. 4. The paper culminates with conclusions and future plans in Sect. 5.

2 Related Work

Risk assessment is a critical process for securing a system and involves identifying program vulnerabilities, potential threats, and losses resulting from the above. In the field of information system risk assessment, the current research directions can be divided into three categories: **risk assessment index system**, **risk assessment model**, and **vulnerability scoring system**.

Regarding the **risk assessment index system**, existing studies focus on selecting risk indicators and constructing a risk index system by analyzing the threats posed to the target information system. Peng et al. [10] utilize the Delphi method to identify risk indicators for big data transmission processes and then use the Analytic Hierarchy Process (AHP) to calculate the weight of each indicator. Zhao et al. [16] use a fuzzy evaluation method to calculate the probability and impact of risk events and then use an entropy weight coefficient method to assess the contribution of each risk factor to the overall risk assessment. Lu et al. [7] conduct a quantitative risk study on industrial control systems and improve the analytic hierarchy process by introducing fuzzy AHP to address the problem of judgment matrix consistency. Quantitative risk index systems intuitively demonstrate the potential risks associated with the target system and enable a more comprehensive analysis of risk factors. However, many existing approaches rely heavily on subjective data sources such as questionnaires or expert evaluations, which may introduce some level of subjectivity into the quantification process, thereby affecting accuracy and credibility.

In terms of **risk assessment model**, existing research has focused on extracting risk factors from information systems, establishing models, and converting risks into variables in the model for analysis to achieve quantitative risk. Zhang et al. [15] propose a quantitative risk assessment approach for host system security based on vulnerability scanning. They construct a vulnerability correlation graph of the host system for a quantitative risk assessment. Xie et al. [13] propose an attack tree model that analyzes the threat vectors of each leaf in the attack tree and calculate the threat vector of the entire attack path to calculate the risk value. Zhang et al. [14] propose a fuzzy radial basis function neural network model to quantify network security risk factors and output risk levels. Nan et al. [3] propose a security risk analysis model that constructs a Bayesian network to deduce the probability and severity of security events and use an ant colony optimization algorithm to calculate the risk propagation path. Li et al. [6] propose an Analytic Hierarchy Process-Genetic Algorithm Back Propagation model, which determines the structure of the BP neural network according to the given risk index system structure and optimizes the parameters of the neural network with a genetic algorithm. Dacier et al. [1] establish an IT system vulnerability privilege graph model and transform it into a Markov chain to quantitatively assess risks under different attack scenarios. Patel et al. [9] propose an enhanced fault tree method to quantify the vulnerability and threats' impact on information systems. While modeling the target system can provide real and objective risk data, the establishment and analysis of quantitative models will become complicated when the scale of the system becomes large.

The existing **vulnerability scoring system** establishes predefined risk indicators and evaluation levels for vulnerabilities and uses these models to quantify vulnerability threats. Currently, widely used vulnerability scoring models in scientific research and

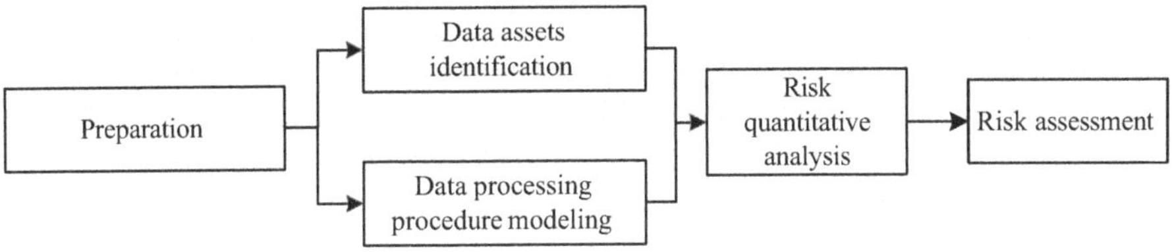

Fig. 1. Risk assessment process of the big data platform.

production include the Common Vulnerability Scoring System [4], Threat Assessment and Remediation Analysis [12], and Common Weakness Scoring System [8]. These models are used to assess and quantify the severity of specific vulnerabilities within an information system.

In summary, current risk assessment research primarily focuses on selecting risk indicators and constructing a risk index system by analyzing threats faced by the target information system. Existing methods for quantifying risk heavily rely on subjective data sources, thereby introducing potential biases and affecting the accuracy and reliability of the results. Moreover, modeling the target system presents substantial challenges. Therefore, it is essential to develop an automated, universal, big data platform-specific quantitative risk assessment scheme for big data platforms.

3 System Design

3.1 Workflow

The proposed quantitative risk assessment process for big data platforms is outlined in Fig. 1 and can be broken down into the following steps:

1. Preparation: First, the enterprise needs to analyze its big data platform to clarify the target and scope of the assessment.
2. Risk identification: This is a key step in the risk assessment process, including the following aspects:
   (a) Data assets identification: Identify key data assets within the big data platform and determine their value, source, storage location, access rights, sensitivity, correlation, and other attributes to conduct risk analysis on them.
   (b) Data processing procedures modeling: The data processing procedures of the big data platform is divided into six stages: data acquisition, transmission, storage, calculation, exchange, and destruction. Establish data processing models for different stages and associate them with equipment, technology, personnel, and other factors.
3. Risk quantitative analysis: At this stage, the system comprehensively considers the possibility, impact, and tolerance of risks, constructs a risk indicators system, and calculate the weights of risk indicators with machine learning algorithms.

4. Risk assessment: Once the risk index system has been established and the weights of risk indexes have been determined, it can be effectively utilized to systematically evaluate the risks associated with the big data platform under evaluation. Subsequently, the overall risk value of the platform, along with its specific risk level, can be calculated by applying weights of risk indexes.

3.2 Preparation

In the initial stage, the enterprise conducts a comprehensive analysis of the big data platform and clarifies the goals and scope of the evaluation. The assessment objectives include understanding potential risks associated with big data platforms, whether appropriate risk management strategies have been implemented, and ensuring compliance with relevant laws and regulations. The evaluation scope encompasses big data assets within the enterprise, data processing procedures, data services provided by external vendors, and data shared with third parties. Additionally, the enterprise needs to select methods and tools for quantitative risk assessment to identify, analyze, and evaluate risks in subsequent stages.

3.3 Data Assets Identification

User profiling, as a typical service within big data platforms, consists of four stages: data acquisition, data storage and calculation, model training and prediction, and user profile visualization [2]. In this section, the user profiling service is elaborated as an example, and the involvement of personnel within the aforementioned service is also taken into consideration.

In the data acquisition phase, the responsibility lies in the real-time capturing, aggregating, and streaming of vast amounts of log data. Major big data assets utilized in this process include data acquisition servers and tools such as Flume and Kafka. In the data storage and processing phase, data storage servers and related storage tools, such as HDFS, manage and process big data. Databases equipped with management systems like MySQL or MongoDB serve as vital assets, offering persistent storage, querying capabilities, and comprehensive management of accumulated data. Docker is employed for packaging and deploying applications, and it facilitates rapid computation and analysis of big data at this stage. Computational analysis servers contribute by providing the essential computing resources needed to perform operations related to model training, prediction, and other analytical tasks. Additional assets like load-balancing servers, operating systems, personal computers, switches, routers, and various interfaces also play a role in the efficiency and stability of user profiling services.

In the data calculation phase, key big data assets for model training and prediction encompass Spark.MLlib, computational analysis servers, data storage servers, and database servers. These assets collaborate to support model training, prediction, and related computational tasks. For user profile visualization, big data assets such as web servers, web service tools like Nginx, database servers, and data visualization tools like MySQL, MongoDB, and Tableau facilitate data presentation and interaction. Ensuring the security of user profiling services involves a collection of big data assets, such as

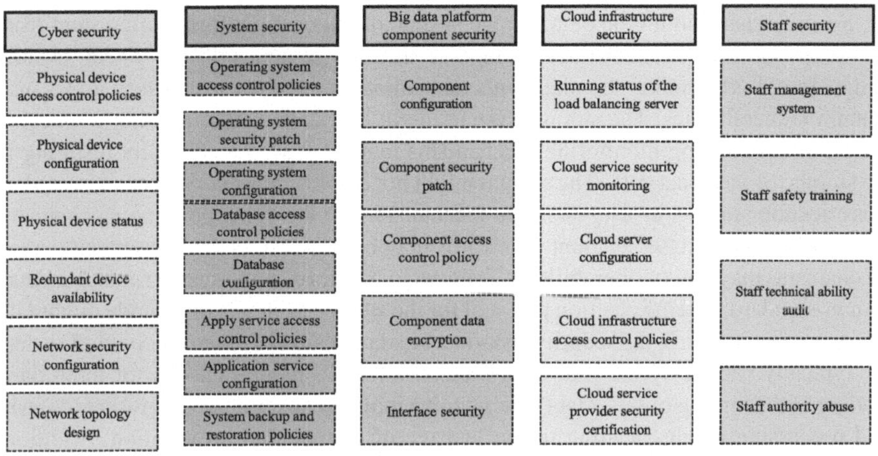

Fig. 2. Risk quantitative evaluation index system of User portrait business assets.

web application firewalls, bastion hosts, firewalls, user authentication, and access control systems, and log system platform configurations. These assets collectively assure the security of the services. In terms of personnel, operation and maintenance staff, business technology experts, security management staff, and outsourced service personnel are also crucial big data assets. Their responsibilities and skills play a significant role in successful operations.

In summary, on the basis of relevant standard documents for big data platform asset risk assessment, such as GB/T 37550–2019, this paper identifies the assets related to user profile business in big data platforms mentioned above. Building on this foundation, as shown in Fig. 2, five primary indicators and 28 secondary indicators are constructed to establish a quantitative assessment index system for the asset risks of e-commerce big data platforms.

3.4 Data Processing Procedures Modeling

As described in the Sect. 3.3, data processing procedures in a big data platform typically encompass stages such as data acquisition, data storage, and data computation. Additionally, considerations must be made for phases including data transmission, data sharing, and data destruction.

From the perspective of data calculation workflow, vulnerabilities may exist in the data acquisition process of big data platforms. Collection devices, with limited processing capability, can only conduct elementary preprocessing of the raw data, leading to the infiltration of meticulously crafted malicious data by attackers into the big data platform.

The data transmission process of the big data platform is vulnerable to security risks. Big data platforms may use less secure transport and authentication protocols for non-critical services. Additionally, big data platforms may selectively encrypt or verify the integrity of transmitted data to reduce overhead and optimize performance. The complex and changeable transmission environment may interfere with the normal operation of the transmission link.

Security risks within the data storage system of big data platforms often stem from the following aspects: mechanical failures in the storage nodes of the system, leading to data loss. Furthermore, the platform's authentication mechanism for new nodes may contain vulnerabilities. The storage system might lack data quality monitoring mechanisms, or the existing monitoring mechanisms may be flawed, potentially resulting in the storage of incorrect data. The system might not employ data encryption storage, data classification storage, or user isolation mechanisms, or the corresponding mechanisms might be defective. User account passwords may be exposed, and client authentication mechanisms may have vulnerabilities. Training for platform authorization managers may be non-standard, and the election protocol for the storage system's main node might fail.

As shown in Fig. 3, the security risks in a big data platform's data calculation system are typically caused by the following factors: a lack of quality checks for input data, leading to the mixing of false data with genuine input, or an absence of scrutiny for tasks and processing results, leading to the leakage of confidential information. Additionally, passwords for regular users and administrators may be disclosed due to improper management, and client identity authentication mechanisms may contain vulnerabilities. Owing to a lack of formal training, management staff may make mistakes or execute incorrect operations. After extended periods of intense computing, reading, writing, and data transmission, the failure rates of computation nodes, local disks, memory, and transmission links may increase.

Security risks in the data application and sharing process may arise from factors such as a lack of secure transmission or authentication protocols and insufficient privacy and integrity protection for public data. Additionally, vulnerabilities in the platform's audit mechanisms, such as only recording the type, time, and content of operations without identifying the operator, could lead to security issues within the big data platform during data application and sharing processes [5].

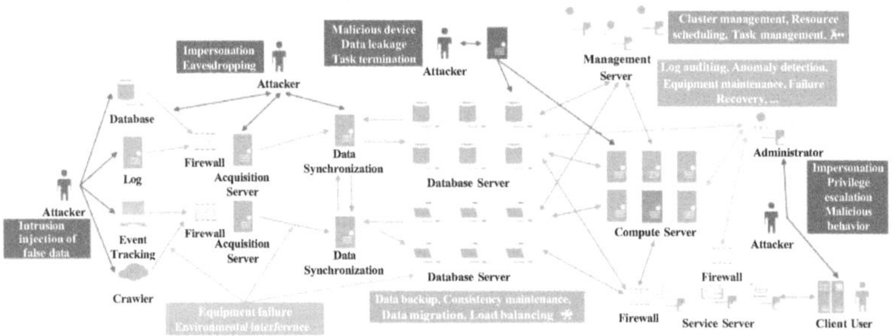

Fig. 3. Risks in data processing procedures.

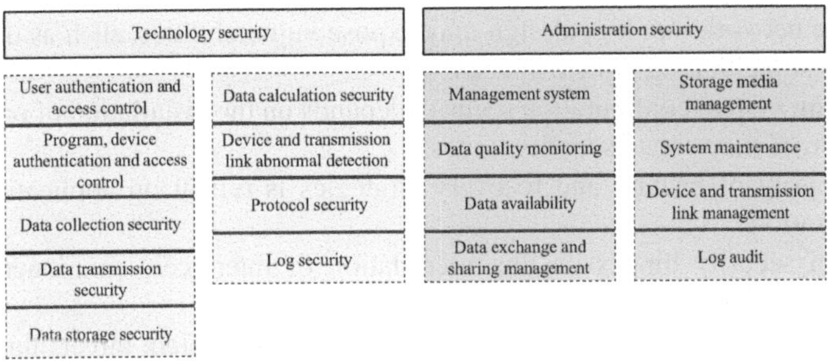

Fig. 4. Risk quantitative evaluation index system of data processing procedures of user portrait business.

During the data destruction process, false data destruction or inadequate management of the personnel handling the destruction may result in the leakage of confidential information.

Based on the threat modeling analysis results above, the security of the big data platform is divided into two secondary risk indicators: technical security and management security, as illustrated in Fig. 4. Technical security includes nine tertiary indicators, while management security encompasses eight tertiary indicators.

3.5 Risk Quantitative Analysis

For the risk assessment of big data platforms, an expert scoring system is typically utilized, and its data format is shown in Table 1. To study the weight calculation of the risk indicators for big data assets and processing procedures, this paper investigates the correlation between the big data platform risk indicators and their relevance to business scenarios. Subsequently, a series of logical relationships are extracted to analyze the expert review data.

Table 1. Format of expert review data.

Index1	Index 2	…	Index *n*	Total Risk grade
Medium	*High*	…	*Medium*	***Medium***

Upon analysis, we have discovered that the evaluations within the expert system generally exhibit the following relationships:

1. Faulty equipment access control policies may enable unauthorized access, risking device configuration.
2. Improper device configuration can degrade performance or render devices unusable.

3. Insecure network topology design may expose vulnerabilities, such as unprotected entrances, facilitating system attacks.
4. Operating system configuration security depends on the installation of patches, and lack thereof may create security issues.
5. The security of backup and recovery strategies is reliant on application service configuration.
6. Platform security hinges on the interrelation of interfaces, component security patches, and configurations.
7. Inadequate data encryption risks interception and decryption, threatening platform interface security.
8. Cloud server configuration influences load-balancing server operations and cloud service security monitoring.
9. A cloud provider's security certification reflects its security management capability, and lower levels may lead to infrastructure issues.
10. Staff technical ability audits correlate with security training, and audit issues may signify training deficiencies.
11. Inadequate personnel management systems can risk improper employee rights management and potential abuse of power.

Subsequently, we attempted to determine the weights of various indicators by analyzing the correlations within simulated expert review data. The credibility of risk values and the ultimate quantitative risk assessment conclusions directly depend on the reasonableness of indicator weights. Although linear regression and Lasso regression are popular machine learning methods for learning weights, they are not suitable for this problem. Linear regression assumes a linear relationship between data, but in risk assessment, the relationships between various indicators are often nonlinear. While Lasso regression can perform feature selection, it primarily uses the L1 regularization term and performs best when there is significant collinearity between features, but strong collinearity may not exist between risk indicators. Therefore, we employed the random forest algorithm for computation. Random forests can evaluate feature importance, effectively handle a large number of features to prevent overfitting, and manage nonlinear relationships to determine the weights of each indicator.

3.6 Risk Assessment

To calculate the risk value of the big data platform, two matrices are first defined: X is the matrix after quantifying each risk index, and V is the calculated weight matrix. The X matrix contains the quantitative risk values of each risk index, which were obtained from the preprocessed data set. The V matrix contains the weights of each risk indicator, which were learned from the random forest model. By multiplying X and V matrices, the risk value of each indicator can be obtained. The detailed calculation process will be presented as follows.

Let $X = [x_1, x_2,..., x_n]$, where x_i represents the quantized value of the i th risk indicator. Similarly, let $V = [v_1, v_2,..., v_n]$, where v_i represents the weight of the i th risk indicator. The quantified risk value of each risk indicator can be obtained by matrix multiplication, that is, $R = X \times V = [r_1, r_2,..., r_n]$, where $r_i = x_i \times v_i$ represents the quantified risk value of the i th risk indicator. Finally, the quantitative risk values of all risk indicators are added together, and the sum obtained is the final total risk value, as shown in Eq. 1.

$$R_{total} = \sum_{i=1}^{n} r_i = \sum_{i=1}^{n} x_i \times v_i \tag{1}$$

4 Experiment

4.1 Setup

To evaluate the method proposed in this paper, we utilized a simulated cluster built with Hadoop and Spark, consisting of three CentOS7 nodes. We configured the network information of these nodes and established non-encrypted logins between them. To obtain experimental data, we built a user portrait function for e-commerce big data in the above cluster and made it run automatically to collect data.

Table 2. The relationships between the overall risk level and the proportion of risk items.

Overall risk level	meaning
high	***very high*** *risks* $\geq$ 10% or ***high*** *risks* $\geq$ 30%
medium	***medium*** *risks* $\geq$ 30%
low	otherwise

4.2 Weights Calculation by Random Forest

In the data preprocessing stage, the paper quantified the risk level from ***very low*** to ***very high*** into numerical values of 10, 30, 50, 70, and 90, preserving the relative magnitude relationship between risk levels and efficiently encoding them. According to *GB/T 31509–2015*, the overall risk level is determined by the percentage of risk levels for each indicator. As shown in Table 2, a **high overall risk level** signifies that the proportion of ***very high*** risks is not less than 10%, or the proportion of ***high*** risks is not less than 30%. A **medium overall risk level** means that the proportion of ***medium*** risks is not less than 30%. In other cases, the overall risk rating is **low overall risk level**. Please note that this research established two separate experiments. The first was designed to calculate the weights of risk indicators associated with big data platform assets, while the second was to determine the weights of risk indicators related to the data processing procedures. All experiments were performed on Intel i58265U@1.6 GHz.

In the first experiment, the implementation of the random forest algorithm in this paper used the sklearn library and a grid search method was used to find the best number of forests. The value of n_estimators was chosen as 210 for the number of trees, using accuracy on the test set and time cost as the primary evaluation metrics. Finally, the random forest algorithm was used to determine the weights of each index, and the top 10 indicators with the highest weight are shown in Table 3. The weights of these indicators increase with the number of risk indicators associated with them. This is intuitive because an increase in the number of correlated items means that the metric has direct or indirect effects on more other metrics, and therefore its importance increases accordingly.

In the second experiment, this paper collected and preprocesses the data generated or output during the running of the simulation program. Then the data were analyzed to obtain the scores of various risk quantification indicators and integrate the total loss of platform assets, risk level, and risk quantification indicator scores within the same period into one risk data. Multiple risk data were combined into a risk data set, which was then used for the weight calculation of subsequent risk quantification indicators.

The pre-processed big data platform risk data set was used to train decision trees, random forests, and gradient-boosting decision trees. The effects of the three machine learning models were comprehensively compared, and the weight of the big data platform risk quantization index was extracted from the model with the best training effect. After

Table 3. Weight of risk indexes of big data platform assets.

big data platform assets[1]	Risk index	$\Re_1$	$\Re_2$	$\Re_3$	$\Re_4$	$\Re_5$	$\Re_6$
	Index weight	0.0350	0.0359	0.0345	0.0349	0.0364	0.0369
	Risk index	$\Re_7$	$\Re_8$	$\Re_9$	$\Re_{10}$	$\Re_{11}$	$\Re_{12}$
	Index weight	0.0386	0.0365	0.0365	0.0358	0.0352	0.0353
	Risk index	$\Re_{13}$	$\Re_{14}$	$\Re_{15}$	$\Re_{16}$	$\Re_{17}$	$\Re_{18}$
	Index weight	0.0372	0.0356	0.0348	0.0357	0.0353	0.0357
	Risk index	$\Re_{19}$	$\Re_{20}$	$\Re_{21}$	$\Re_{22}$	$\Re_{23}$	$\Re_{24}$
	Index weight	0.0361	0.0355	0.0351	0.0350	0.0356	0.0358
	Risk index	$\Re_{25}$	$\Re_{26}$	$\Re_{27}$	$\Re_{28}$		
	Index weight	0.0359	0.0352	0.0346	0.0355		

Note[1]: $\Re_1$ presents Personnel privilege abuse, $\Re_2$ presents Personnel technical competency audits, $\Re_3$ presents Personnel security training, $\Re_4$ presents Personnel management system, $\Re_5$ presents Cloud service provider security authentication, $\Re_6$ presents Cloud infrastructure access control policies, $\Re_7$ presents Cloud Server Configuration, $\Re_8$ presents Cloud Service Security Monitoring, $\Re_9$ presents Load Balancing Server Operation, $\Re_{10}$ presents Platform Interface Security, $\Re_{11}$ presents Platform Component Data Encryption, $\Re_{12}$ presents Platform Component Access Control Policy, $\Re_{13}$ presents Platform Component Security Patches, $\Re_{14}$ presents Platform Component Configuration, $\Re_{15}$ presents System Backup And Recovery Policies, $\Re_{16}$ presents Application Service Configuration, $\Re_{17}$ presents Application Service Access Control Policies, $\Re_{18}$ presents Database Configuration, $\Re_{19}$ presents Database Access Control Policy, $\Re_{20}$ presents Operating System Configuration, $\Re_{21}$ presents Operating System Security Patches, $\Re_{22}$ presents Operating System Access Control Policy, $\Re_{23}$ presents Network Topology Design, $\Re_{24}$ presents Network Security Configuration, $\Re_{25}$ presents Spare Device Availability, $\Re_{26}$ presents Physical Device Operational Status, $\Re_{27}$ presents Physical Device Configuration, $\Re_{28}$ presents Physical Device Access Control Policy.

a comprehensive comparison of the learning effects of several models, the weight of the risk quantification index of the big data platform was extracted from the random forest model, which is shown in Table 4. The accuracy of the three models on the training set and the validation data set and the training time were obtained as shown in Table 5. The method proposed in this paper combines empirical judgment with data-driven models and utilizes the optimization ability of machine learning algorithms to produce a more objective risk assessment index system. The experimental results show that the system proposed in this paper can more comprehensively and accurately describe the current risks faced by the big data platform.

Table 4. Weight of risk indexes of data processing procedures.

technical security[1]	Risk index	$\mathfrak{R}_1$	$\mathfrak{R}_2$	$\mathfrak{R}_3$	$\mathfrak{R}_4$	$\mathfrak{R}_5$	$\mathfrak{R}_6$
	Index weight	0.0143	0.0627	0.0068	0.0087	0.0010	0.0176
	Risk index	$\mathfrak{R}_7$	$\mathfrak{R}_8$	$\mathfrak{R}_9$			
	Index weight	0.0012	0.5173	0.01726			
administration security[2]	Risk index	$\mathfrak{T}_1$	$\mathfrak{T}_2$	$\mathfrak{T}_3$	$\mathfrak{T}_4$	$\mathfrak{T}_5$	$\mathfrak{T}_6$
	Index weight	0.1021	0.0055	0.0032	0.0036	0.0002	0.0088
	Risk index	$\mathfrak{T}_7$	$\mathfrak{T}_8$				
	Index weight	0.0085	0.0695				

Note[1]: $\mathfrak{R}_1$ presents User authentication and access control, $\mathfrak{R}_2$ presents Program, device authentication and access control, $\mathfrak{R}_3$ presents Data collection security, $\mathfrak{R}_4$ presents Data transmission security, $\mathfrak{R}_5$ presents Data storage security, $\mathfrak{R}_6$ presents Data calculation security, $\mathfrak{R}_7$ presents Device and transmission link abnormal detection, $\mathfrak{R}_8$ presents Protocol security, $\mathfrak{R}_9$ presents Log security.
Note[2]: $\mathfrak{T}_1$ presents Management system, $\mathfrak{T}_2$ presents Data quality monitoring, $\mathfrak{T}_3$ presents Data availability, $\mathfrak{T}_4$ presents Data exchange and sharing management, $\mathfrak{T}_5$ presents Storage media management, $\mathfrak{R}_6$ presents System maintenance $\mathfrak{T}_7$ presents Device and transmission link management, $\mathfrak{T}_8$ presents Log audit.

Table 5. Training results of three machine learning algorithms.

Model	Accuracy of train set	Accuracy of test set	Training time
Decision tree	82.33%	82.07%	2.75 s
Random forest	82.53%	82.13%	288.31 s
Gradient boosted decision tree	83.46%	81.73%	5293.33 s

Table 6. Distribution and the threshold value of total risk level of the big data platform.

Big data platform assets	Risk level	**high**	**medium**	**low**
	Ratio	4.6%	40.8%	54.6%
	Threshold value[1]	43.4740	35.5236	0
Data processing procedures	Risk level	**high**	**medium**	**low**
	Ratio	9.8%	11.5%	78.7%
	Threshold value[2]	39.1054	32.3718	0

Note[1]: If the risk value of big data platform assets is lower than 35.5236, it is classified as low risk. When the risk value is not less than 35.5236 and less than 43.4740, it is medium risk. When the risk value is not less than 43.4740, it is high risk.
Note[2]: If the risk value of data processing procedures is lower than 32.3718, it is classified as **low** risk. When the risk value is not less than 32.3718 and less than 39.1054, it is **medium** risk. When the risk value is not less than 39.1054, it is **high** risk.

Table 7. Comparative analysis of different approaches.

Approach	$\mathfrak{B}_1$	$\mathfrak{B}_2$	$\mathfrak{B}_3$	$\mathfrak{B}_4$	$\mathfrak{B}_5$
The AHP	∘	∘	∘	∘	∘
The Delphi method	∘	∘	∘	∘	∘
The fuzzy judgment method	∘	∘	∘	∘	∘
Ours	•	•	•	•	•

Note[1]: $\mathfrak{B}_1$ *presents objectivity,* $\mathfrak{B}_2$ *presents low-cost,* $\mathfrak{B}_3$ *presents processing with large amount of data,* $\mathfrak{B}_4$ *presents applicable to a variety of scenarios,* $\mathfrak{B}_5$ *presents easy to update and iterate.*
Note[2]: • *presents TRUE,* ∘ *presents FALSE*

4.3 Risk Values Calculation

By integrating the asset quantification index system in the big data platform and the risks in the data processing procedures, 40,000 instances of simulated expert review data of platform assets and data processing procedures risks were generated. The simulated data is then analyzed, mapping the risk values derived from Eq. 1 to specific risk levels: low, medium, and high. This process yields the overall risk level distribution for the big data platform's assets and data processing procedures. The specific mapping rules and the total risk level distribution are presented in Table 6.

4.4 Comparative Analysis

The machine learning approach proposed in this study offers significant advantages in risk assessment. It ensures objectivity through a data-driven approach and reduces operational costs through automated processing and analysis, a process that does not require in-depth analyses and interpretations by experts, which greatly saves human resources. In addition, machine learning models are capable of handling large-scale data and are highly adaptable to meet the requirements of quantitative risk assessment for different big data scenarios. Importantly, as the data is expanded and updated, the model can be continuously trained and optimized to adapt to new situations, which contrasts with hierarchical, Delphi, and fuzzy judgment methods that require re-analysis and computation. Therefore, the machine learning approach proposed in this study provides a more efficient, flexible, and sustainable solution for risk assessment compared to traditional methods as shown in Table 7.

5 Conclusion

This paper studies the risks faced by big data platforms, analyzes the possible loopholes or threats in platform assets and data processing procedures, and proposes a quantitative assessment theory and index system for big data platform risks. The purpose is to provide reasonable and scientific reference indicators for the quantitative assessment of big data risks. To verify the correctness of the proposed scheme, a simulated big data platform was built to verify the scheme. Its experience shows that this scheme has sufficient advantages in objectivity, cost, generality, and more compared with the existing schemes.

Acknowledgements. This work is supported by the National Key R&D Program of China (No. 2022YFB3103401), and the National Natural Science Foundation of China (No. 62172317).

References

1. Dacier, M., Deswarte, Y., Kaâniche, M.: Models and tools for quantitative assessment of operational security. Springer (1996)
2. Eke, C.I., Norman, A.A., Shuib, L., Nweke, H.F.: A survey of user profiling: State-of-the-art, challenges and solutions. IEEE Access **7**, 144907–144924 (2019). https://doi.org/10.1109/ACCESS.2019.2944243
3. Feng, N., Wang, H.J., Li, M.: A security risk analysis model for information systems: Causal relationships of risk factors and vulnerability propagation analysis. Inf. Sci. **256**, 57–73 (2014). https://doi.org/10.1016/j.ins.2013.02.036
4. FIRST.org: Common Vulnerability Scoring System v3.1: Specification Document (2019), https://www.first.org/cvss/v3-1/, accessed: Aug 6, 2023
5. Fu, X., Gao, Y., Luo, B., Du, X., Guizani, M.: Security threats to hadoop: Data leakage attacks and investigation. IEEE Network **31**(2), 67–71 (2017). https://doi.org/10.1109/MNET.2017.1500095NM
6. Li, Z., Xu, W., Shi, H., Zhang, Y., Yan, Y.: Security and privacy risk assessment of energy big data in cloud environment. Computational Intelligence and Neuro–science **2021**, 1–11 (2021). https://doi.org/10.1155/2021/2398460
7. Lu, H., Chen, D., Peng, Y., Wang, H.: Quantitative research on risk assessment for information security of industrial control system. Automation Instrumentation **35**(10), 21–25 (2014)
8. MITRE Corporation: Common Weakness Scoring System (CWSS) v1.0.1 (2023), https://cwe.mitre.org/cwss/cwss_v1.0.1.html, accessed: Aug 6, 2023
9. Patel, S.C., Graham, J.H., Ralston, P.A.: Quantitatively assessing the vulnerability of critical information systems: A new method for evaluating security enhancements. Int. J. Inf. Manage. **28**(6), 483–491 (2008). https://doi.org/10.1016/j.ijinfomgt.2008.01.009

10. Peng, Y., Song, S., Fang, D.: Research on risk identification and evaluation indicators for big data dissemination process. Science and Technology Management Research **38**(10), 78–82 (2018)
11. Wu, J., Zhang, Y., Xiong, S.: Construction of enterprise big data security evaluation index system based on shel model. Science and Technology Management Research **41**(17), 144–151 (2021)
12. Wynn, J., Whitmore, G., Upton, L., Spriggs, D., McKinnon, R., McInnes, R., Graubart, L., Clausen, J.: Threat assessment & remediation analysis (tara) methodology description version 1.0. Bedford, MA (2011)
13. Xie, F., Lu, T., Guo, X., Liu, J., Peng, Y., Gao, Y.: Security analysis on cyber–physical system using attack tree. In: 2013 Ninth International Conference on Intelligent Information Hiding and Multimedia Signal Processing. pp. 429–432 (2013). https://doi.org/10.1109/IIH-MSP.2013.113
14. Zhang, J.: Network security risk dynamic evaluation method research. Computer Simulation **33**(10), 356–360 (2016)
15. Zhang, Y., Fang, B., Yun, X.: A quantitative risk assessment approach for host system security. Comput. Eng. **31**(14), 147–149 (2005)
16. Zhao, D., Zhang, Y., Ma, J.: Fuzzy risk assessment of entorpy-weight coefficient method applied in network security. Comput. Eng. **30**(18), 21–23 (2004)
17. Zhu, G., Feng, M., Chen, Y., Yang, J.: Research on fuzzy evaluation of privacy risk for social network in big data environment. Information science **34**(9), 94–98 (2016). https://doi.org/10.13833/j.cnki.is.2016.09.019

5G-NSSSP: A Simulation Platform for Enhanced 5G Network Slice Security Mechanism

Shujuan Gao, Jianjiang Xu(✉), Jiayi Wang, Yulong Fu, Jin Cao, and Hui Li

Xidian University, Xian, Shanxi, China
xujj@stu.xidian.edu.cn

Abstract. With the gradual commercialization of 5G network slicing, slicing security has become the focus of research. In this paper, we propose a 5G network slicing security simulation platform (5G-NSSSP), which implements special authentication and authorization functions in the 5G network slicing access process. We also propose two improvement solutions for the security vulnerabilities in the existing 3GPP network slicing authentication process: the introduction of a new slicing shared key generation step and a service subscription key. These solutions have been experimentally verified on 5G-NSSSP. The results show that they circumvent the access of illegal slicing requests to some extent and improve the security of the network slicing authentication process.

Keywords: 5G network slice · NSSAA · Simulation Platform

1 Introduction

With the commercial deployment of 5G networks [1], network slicing technology has demonstrated its unique advantages in complex and diverse service scenarios. Through virtualization technology, network slicing [2] can achieve the coexistence of multiple logical networks [3] on limited physical resources, providing users with a rich variety of network service options. However, as the application scenarios of network slicing increase, ensuring the secure and efficient operation of network slicing becomes crucial.

In existing research, most simulation techniques related to 5G network slicing mainly focus on a single scenario or specific problems, lacking a comprehensive and systematic simulation platform. This makes it difficult to evaluate the performance and security of network slicing in complex scenarios. Moreover, considering the multiple authentication processes and details involved in 5G network slicing, an accurate and comprehensive simulation platform is an indispensable tool in research.

Based on the aforementioned background, this study aims to build a comprehensive 5G network slicing simulation platform, with a focus on researching the initial registration of User Equipment (UE), primary authentication, and specific network slicing authentication processes in 5G networks [4]. This platform will provide a powerful tool and foundation for further research and experimentation.

Z. Gu et al. (Eds.): NSE 2023, CCIS 2063, pp. 243–255, 2024.
https://doi.org/10.1007/978-981-97-4519-7_17

The main contributions of this article are as follows:

- We have built a simulation platform(5G-NSSSP) for 5G network slicing authentication based on the 3GPP standard specifications [5]. It integrates custom security modules to make the simulation environment more realistic and targeted.
- We have designed a shared slicing key Kslice generation process. This key is saved by the UE and the 5G core network element NSSAAF and is used to detect whether the UE passes the primary authentication process [6] during the NSSAA process.
- We have optimized the existing NSSAA process specified by 3GPP. We introduce a service subscription key K_{af} shared between the UE and a thirdparty server, enhancing the resistance to attacks during the bi-directional authentication process between the UE and the third-party server.
- We have conducted security evaluation experiments on the optimized NSSAA process and conducted an in-depth analysis of the experimental results.

2 Background

Network slicing as a virtual logical network connecting users and application servers, the user's legal access to network slicing is directly related to the safe use of subsequent slicing services.The freeze on R17 [7] standard announced in March 2022 newly defined Network Slicing Admission Control Function (NSACF) for monitoring and controlling the number of UE accesses to each network slice. We believe that the key to slice access management is the authentication and management of the various types of slice identifiers associated with UEs. 3GPP has proposed relevant concepts and processes to standardize slice access authentication.

2.1 Network Slice Selection Assistance Information——NSSAI

NSSAI is composed of several S-NSSAIs to identify a class of network slices. NSSAI can be categorized into different classes for different usage objects [8], e.g., UE has Requested NSSAI, Subscribed NSSAI and Allowed NSSAI. This section focuses on the NSSAI related to slice authentication.

- **Subscribed NSSAI:** A collection of S-NSSAIs determined in different network environments based on subscription information agreed between the user and the operator, stored in the 5G core network element Unified Data Management (UDM).
- **Requested NSSAI:** The set of S-NSSAIs of the network slices that the user expects to access that are sent to the network when initiating the initial registration operation.
- **Allowed NSSAI:** The set of S-NSSAIs of network slices that the network allows users to access.
- **Pending NSSAI:** The set of S-NSSAIs for network slices waiting for or in the process of NSSAA.
- **Rejected NSSAI:** The set of S-NSSAIs of network slices for which the network denies user access.

2.2 Network Slice-Specific Authentication and Authorization——NSSAA

The essence of network slices is a virtual logical network running on shared physical resources, so terminal access to network slices also has the security requirements of authentication and authentication. 3GPP proposes that after the terminal completes the master authentication of the registration process, the 5G core network needs to perform specific authentication and authentication of the network slices requested by the terminal [9], i.e., the NSSAA process. The NSSAA process is an authentication process designed based on the EAP framework. The third-party authentication server (AAA-S) interacts with the network slice user through the 5G core network element NSSAAF and AMF for authentication to determine whether the user is authorized to use the current network slice. The AMF or AAA-S can trigger the NSSAA process, and the network slice that passes the NSSAA process can be configured as the user's Allowed NSSAI. It is noteworthy that not all requested network slices require authentication forensics, which is related to the endpoint's slice subscription information.

3 Platform Design

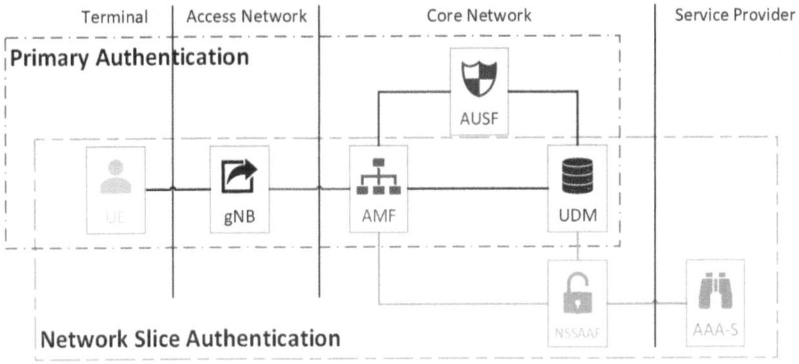

Fig. 1. Platform Architecture Diagram.

In the era of rapid 5G technological evolution, we have successfully crafted a cutting-edge 5G-NSSSP shown as Fig. 1. Meticulously written in C Programming Language, this platform taps into the potent capabilities of the C language for system-level programming. Going a step further, we utilized the Redis database to ensure stable data storage and through the epoll architecture, accomplished high concurrent processing capabilities. Notably, all our entities are implemented at the network layer. For instance, the "air interface transmission" of base stations is simulated via network layer broadcasting. Communication between entities is facilitated through TCP, guaranteeing stable and reliable information exchange. Additionally, we've designed a comprehensive suite of logging and monitoring tools to aid researchers in deep analysis and debugging. Adherence to the 3GPP and related standard guidelines remains our steadfast principle, ensuring a high alignment between our simulation environment and real-world applications. In the following sections, we will delve into the platform's core components and pivotal functionalities.

3.1 Platform Components

This section will provide a detailed description of all entities within the platform, including their functions, interactions, and the roles they play within the system,as shown in the Fig. 2:

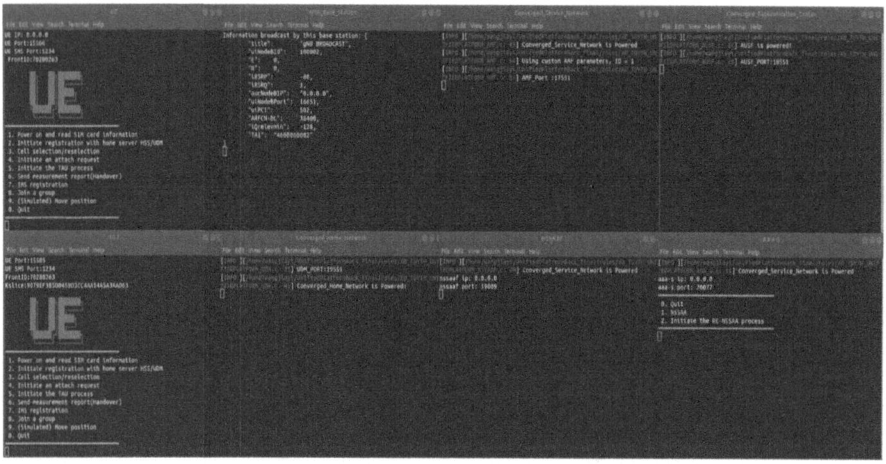

Fig. 2. All Platform Components

- **User equipment** (UE): We simulate the core functions of the UE, implementing the UE's SIM card reading, network registration attachment, cell selection and base station residency, 5G access authentication, 5G handover, and slice authentication process in accordance with the 3GPP standard specifications.
- **gNodeB**(gNB): The design of the gNB omits the details of the physical layer, instead opting to simulate its key behaviors. All essential parameters, such as frequency and power, are encoded into a JSON file. This means that, unlike RF signal broadcasts in the real world, our gNB broadcasts these parameters at the network layer, allowing the UE to perceive and interact with it. This approach not only simplifies the implementation but also offers a flexible means for researchers to modify and test various parameter configurations easily.
- **Mobility Management Module** (AMF): Within the simulation environment, we integrated the Security Anchor Function (SEAF) into the AMF, which is responsible for handling the UE's authentication tasks. Additionally, the AMF implements the UE handover functionality based on the XN interface, ensuring the process's consistency and accuracy.

- **Authentication Management Module** (AUSF): The Authentication Service Function (AUSF) is used to receive the request from AMF to authenticate UE, by requesting the key from UDM and then forwarding the key issued by UDM to AMF for authentication processing.
- **Entity Identity Management Module** (UDM): In the simulation system, the unified data management function(UDM) is responsible for managing user identities, subscription data, and authentication data. Additionally, it oversees the registration management of user service elements (such as the current AMF and SMF serving the UE). For instance, when a user switches to a different AMF, the UDM initiates a deregistration message to the previous AMF, instructing it to delete relevant user information.
- **Network Slice Specific Authentication and Authorization Module** (NSSAAF): In our 5G-NSSSP, the NSSAAF collaborates with the AAA Server (AAA-S) to deliver dedicated authentication and authorization services for network slices. To ensure communication security, it also manages the storage of shared keys between the UE and the slice, as well as between the UE and the AAA-S.
- **AAA server** (AAA-S): The AAA server in the simulation environment acts as a virtual application server tailored for end-to-end services. Its primary function is to handle the authentication and authorization of network slices. Crucially, to ensure secure communication, it implements the same end-to-end key-sharing mechanism as in real-world scenarios, safeguarding communication security between the terminal and the network.

3.2 Platform Functions

Based on the 3GPP standard specifications, our 5G-NSSSP boasts a range of core features:

- **USIM Reading**: The platform can simulate how the UE reads the USIM card, laying the foundation for subsequent network operations.
- **Registration Attachment**: Supports the UE in completing the registration attachment process, enabling initial connectivity to the network.
- **Cell Selection & Base Station Residency**: Ensures the UE can appropriately select cells and remain in base stations based on network conditions and policies.
- **5G AKA Access Authentication Protocol**: The platform incorporates the 5G AKA access authentication protocol, safeguarding the UE's security during the access process [10].
- **5G EAP Authentication Framework**: To comprehensively support various authentication strategies, we have implemented the 5G EAP authentication framework on the platform, especially the 5G EAP AKA' protocol.
- **Network Slice Authentication**: This function ensures the UE receives professional and secure authentication processing when requesting a specific network slice [11].
- **Slice Management & Configuration**: This function allows users to define, configure, and manage different network slices, including the lifecycle management of slices, such as creation, modification, activation, dormancy, and deletion.

4 Enhancing Slice Selection Security

While the 3GPP NSSAA process is steadfast, it remains susceptible to manin-the-middle assaults and impersonation risks. To rectify these susceptibilities and augment the security of the mutual authentication process between the UE and third-party servers, we introduce a refined NSSAA procedure [12]. This methodology incorporates a shared service subscription key, Kaf, between the UE and the third-party servers, thereby strengthening the security architecture of the network-slicing environment. In the subsequent sections, we expound on the enhanced NSSAA procedure, underscoring the superior anti-attack capabilities it offers.

4.1 Network Slice Shared Key Generation Process

In the current 3GPP network slicing environment, the UE needs to validate whether the S-NSSAI restricted by NSSAA is in its requested NSSAI after undergoing primary authentication. In this process, it is vulnerable to man-in-the- middle attacks by unauthorized users. For example, an external attacker, not authenticated primarily, may impersonate an AMF to initiate a slice authentication request to NSSAAF, potentially resulting in the attacker illicitly gaining network slice access authorization. To solve this issue, we propose adding a process to generate a slice shared key, K_{Slice}, between the primary authentication and the NSSAA procedures.

K_{Slice} is used to verify the legality of the S-NSSAI authentication request initiated by the AMF to NSSAAF. The specific process is shown in Fig. 3:

- **Step1-Step3: AMF < - > UDM/UDR:**

The AMF, after primary authentication, requests the UE slice context from the UDM. It then decides whether to proceed with the NSSAA based on the user's slice subscription information, classifying the NSSAI into Pending and Allowed categories.

- **Step4: AMF- > UDM:**

The AMF sends the S-NSSAI(s) constrained by NSSAA in the Requested NSSAI to the UDM, requesting the UDM to generate the K_{Slice} shared with the UE for it.

- **Step5-Step6: UDM- > NSSAAF**

The UDM generates a shared symmetric key K_{Slice} between the UE and the slice for each of the S-NSSAI(s) sent by the AMF, and eventually generates List_K_{slice}.

$$K_{slice}i = \mathrm{KDF}(K_{\mathrm{AUSF}}, SUPI, S - \mathrm{NSSAI}_i) \tag{1}$$

$$\mathrm{List\_}K_{slice} = \{\{S - \mathrm{NSSAI}_1, K_{\mathrm{Slice}}i\}, \ldots, \{S - \mathrm{NSSAI}_m, K_{\mathrm{Slice}}m\}\} \tag{2}$$

Then, the SUPI and List_K_{slice} collections are sent to the NSSAAF. The NSSAAF receives and stores them.

- **Step7: NSSAAF- > AMF:**

The NSSAAF confirms the readiness of the K_{slice} by sending a feedback message to the AMF.

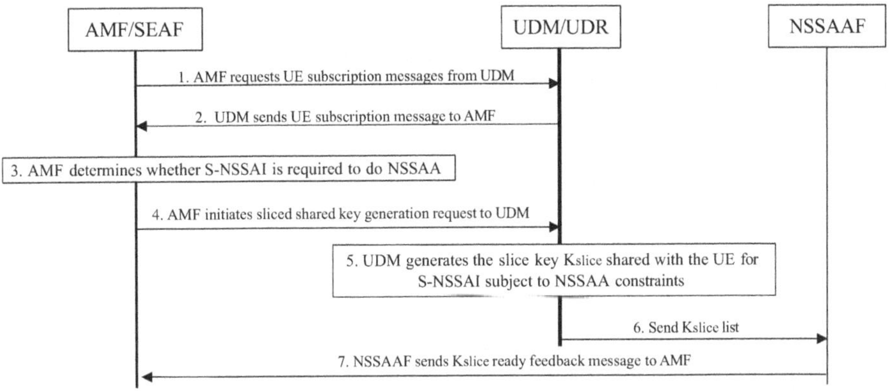

Fig. 3. Network slice shared key generation process.

4.2 Enhanced Security in the NSSAA Process

Based on the 3GPP NSSAA process, we have added a service subscription key K_{af} shared between UE and third-party servers to enhance the anti-attack capability of the mutual authentication process between UE and third-party servers [13]. K_{af} is generated during the service subscription phase between UE and third-party service providers and will not be discussed in this article. The specific process of the improved NSSAA is as follows in Fig. 4:

- **Step1-Step2: AMF < - > UE**

The AMF requests the UE's identity information based on the S-NSSAI. Upon receiving the request, the UE generates a random number Rand1, and returns it along with the identity information GPSI and S-NSSAI to the AMF.

- **Step3: AMF- > NSSAAF**

After receiving the UE's identity information, the AMF initiates a slice authentication request to NSSAAF.

- **Step4: NSSAAF- > AAA-s**

The NSSAF retrieves the corresponding K_{Slice} based on the UE's SUPI and S-NSSAI. It then forwards the slice authentication request to the AAA-S.

- **Step5-Step6: AAA-S- > UE**

The AAA-S retrieves the shared cipher K_{af} with the UE based on the GPSI. Next, AAA-S generates a random number RAND2, calculates

$$H1 = MAC(GPSI, Rand_1, Rand_2, S - NSSAI, K_{Slice}, K_{af}), \tag{3}$$

and then returns Rand2 and H1 to the UE.

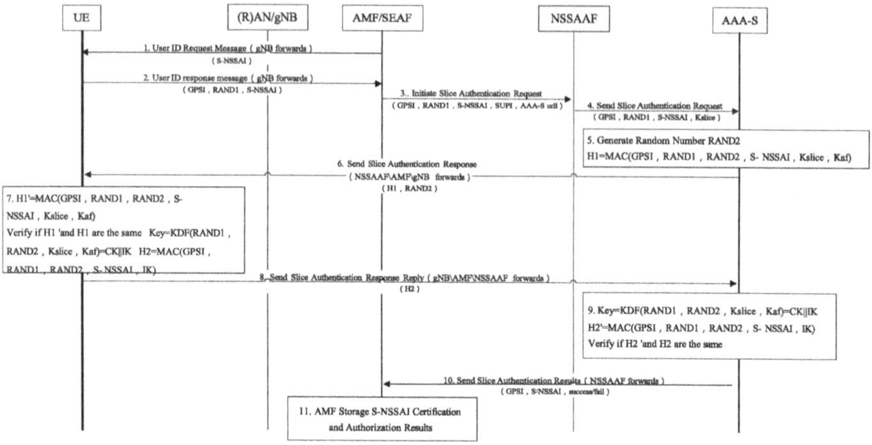

Fig. 4. Improved NSSAA process.

- **Step7-Step8: UE- > AAA-s**

Upon receipt, the UE initially calculates H1' based on the received Rand2 and locally shared key K_{af}. If H1 equals H1', it then computes the Key, using the first 128 bits as the encryption key (CK) and the last 128 bits as the integrity protection key (IK). Finally, it computes H2 and sends it to the AAA-S.

$$H1' = \mathrm{MAC}(GPSI, Rand_1, Rand_2, S - NSSAI, K_{Slice}, K_{af}) \tag{4}$$

$$\mathrm{Key} = \mathrm{KDF}(Rand_1, Rand_2, K_{Slice}, K_{af}) \tag{5}$$

$$H2 = \mathrm{MAC}(GPSI, Rand_1, Rand_2, S - NSSAI, IK) \tag{6}$$

- **Step9-Step10: AAA-S- > AMF**

Using the same algorithm, the AAA-S calculates the values of Key, CK, IK, and H2'. If H2' equals H2, the AAA-S successfully completes the authentication and authorization of the UE. If not, it denies the UE access to the slice. Then AAA-S sends the slice authentication result to AMF.

$$H2' = \mathrm{MAC}(GPSI, Rand_1, Rand_2, S - NSSAI, IK) \tag{7}$$

- **AMF:**

AMF stores the authentication and authorization results of S-NSSAI and classifies them. The S-NSSAI that has passed the authentication and authorization is classified as "Allowed NSSAI," the S-NSSAI that has not passed the authentication and authorization is classified as "Rejected NSSAI," and the "pending NSSAI" is modified accordingly.

Then, the authentication and authorization process of the next S-NSSAI is triggered, and the process continues until all S-NSSAIs subject to NSSAA constraints have completed the authentication and authorization process.

5 Simulation Experiments

In this section, we leverage our constructed 5G-NSSSP to conduct experimental simulations of service slice key distribution and authentication methods assisted by the 5G core network. These simulations serve to demonstrate the effectiveness and practicality of the proposed method. We provide detailed explanations from the perspectives of experimental environment setup, experimental procedures, and analysis of experimental results, thus validating the performance of our proposed solutions in practical applications.

5.1 Experimental Environment

The experiment was conducted on a personal computer with a main frequency of 2.80 GHz and 32 GB of memory, using an Ubuntu 18.04 Linux system running on a virtual machine to establish the simulation environment. By using it to run 5G-NSSSP, the slices can be authenticated.

5G-NSSSP built in the simulation environment used in this experiment can simulate real network slicing security events of users. In the simulated environment, there is no need to worry about economic losses caused by network damage caused by attacks.

5.2 Experimental Process

Based on the virtual network simulation tool and 5G-NSSSP, this paper designs the following three network slicing security experiments:

- **The authentication process of an invalid third-party authentication server**

First, start 5G-NSSSP and then instantiate UE, gNB base stations, AMF, UDM, AUSF, NSSAAF, AAA-S(third-party authentication server, not forged), and an unauthorized forged AAA-S server.

Secondly, in the process of NSSAA, after receiving the slicing authentication request, AAA-S searches the shared secret K_{af} between UE and AAA-S of third-party according to GPSI, then generates a random number RAND2, and then according to equation (3), generates the message authentication code H1 and sends it to UE.

However, The forged, unauthorized AAA-S forges an integrity message authentication code H1 and sends it to UE, as shown in the Fig. 5a:

Finally, after receiving the integrity message authentication code H1 from AAAS, UE uses its own key material, calculates H1' according to equation (4), and compares H1' with the received H1. UE's identification results of illegally forged AAA-S are shown in Fig. 5b:

- **The authentication process of unauthorized user**

First, start our 5G-NSSSP and then instantiate UE(normal), GNB base station, AMF, UDM, AUSF, NSSAAF, AAA-S(third-party authentication server, not forged), and an unauthorized UE. Secondly, in the process of NSSAA, after receiving the integrity message authentication code H1 from AAA-S, UE uses its own key material to calculate H1' according to equation (4) and compares H1' with the received H1. Then, UE derives the

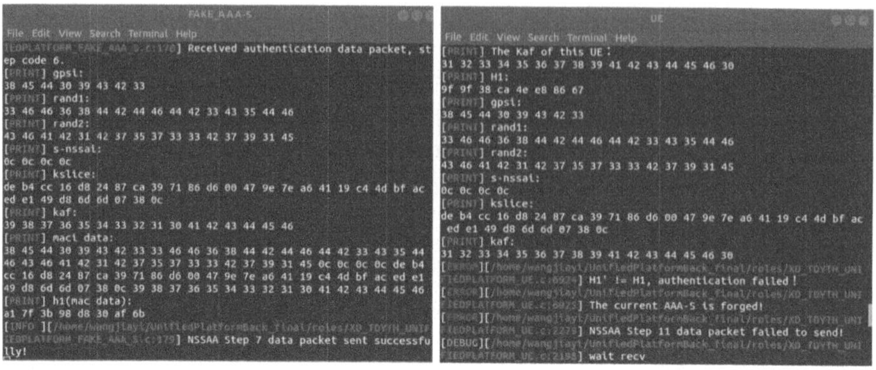

(a) Invalid AAA-S authentication parameters (b) Authentication result of unauthorized AAA-S

Fig. 5. The authentication process of an invalid third-party authentication server

session Key Key between UE and AAA-S based on equation (5) . Next, UE generates a message authentication code H2 according to equation (6), then sends it to AAA-S.

However, after receiving the integrity message authentication code H1 from AAA-S, the unauthorized UE does not verify the accuracy of H1, disguises its own H1'=H1, and sends a fake H2 to AAA-S, as shown in the Fig. 6a:
Finally, after receiving the integrity message authentication code H2 from UE, AAA-S uses its own key material to calculates H2' based on equation (7) to compare with the received H2. The result of AAA-S identification of unauthorized UE is shown in Fig. 6b:

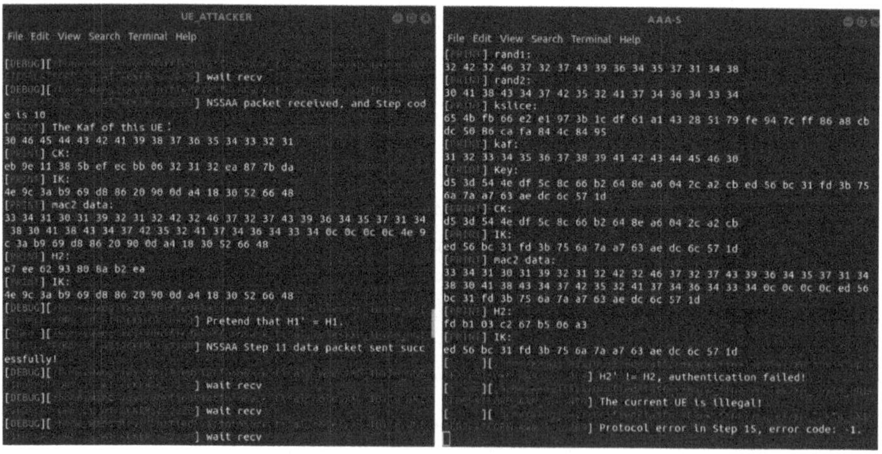

(a) Invalid UE authentication parameters (b) Authentication result of unauthorized UE

Fig. 6. The authentication process of unauthorized user

- **The authentication process of unauthorized user access slicing**

A legitimate user, without permission to subscribe to this slice, forges a slice sharing key, namely K_{slice}, to access this slice, as shown in Fig. 7a:

If UE wants to access the slice and attempts to pass the authentication illegally, after receiving the integrity message authentication code H1 from AAA-S, it disguises its H1'=H1 and sends a forged H2 to AAA-S due to the lack of its own key material. Finally, after receiving the integrity message authentication code H2 from UE, AAA-S uses its own key material to calculates H2' based on equation (7) to compare with the received H2. The authentication result of AAA-S on UE unauthorized access slicing is shown in Fig. 7b:

5.3 Analysis of Experimental Results

- **Analysis of unauthorized third-party authentication server verification results**

During the NSSAA process, the illegally forged AAA-S cannot generate the correct integrity message authentication code H1 due to the lack of K_{af}. It can only generate an incorrect H1 by forging K_{af} or directly forging H1 in an attempt to pass authentication as a legitimate UE. When the UE receives the integrity message authentication code H1 from AAA-S, it calculate H1' based on equation (4), and then compares H1' with the received H1. Since the illegal AAA-S has an incorrect H1 value, it fails to authenticate with a legitimate UE, and only a legitimate AAA-S can pass the authentication. At the same time, the UE completes the one-way authentication of AAA-S. The results prove that our 5G-NSSSP can effectively identify unauthorized third-party authentication servers in network slicing security authentication.

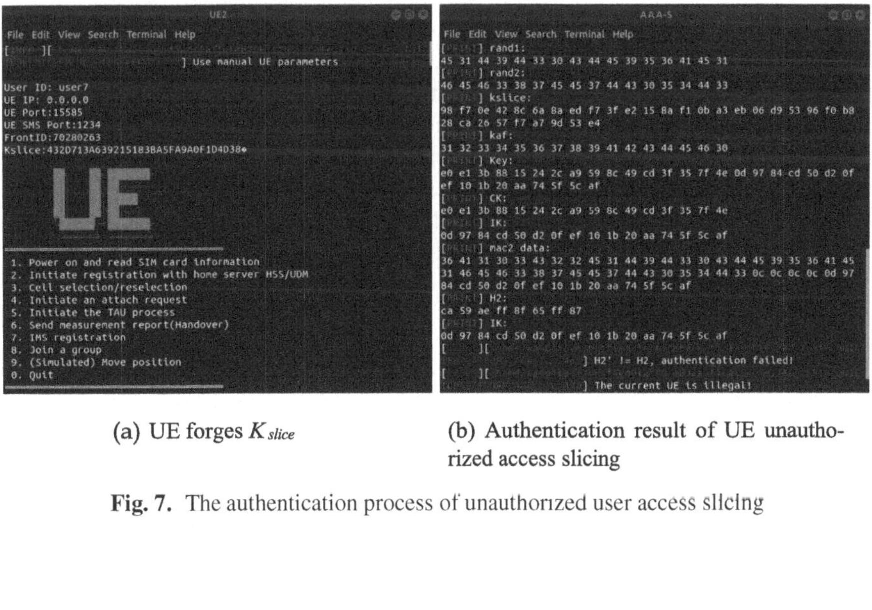

(a) UE forges K_{slice}

(b) Authentication result of UE unauthorized access slicing

Fig. 7. The authentication process of unauthorized user access slicing

- **Analysis of unauthorized user authentication results**

During the NSSAA process, the unauthorized UE cannot generate the correct integrity message authentication code H2 due to the lack of K_{af}. It can only generate an incorrect H2 by forging K_{af} or directly forging H2 in an attempt to authenticate with the AAA-S. When the AAA-S receives the integrity message authentication code H2 from the UE, it calculates H2' based on equation (7) to compare with the received H2. Since the unauthorized user has an incorrect H2 value, it fails to authenticate with the AAA-S, and only a legitimate UE can pass the authentication. At the same time, the AAA-S completes the one-way authentication of the UE. The results prove that our 5G-NSSSP can effectively identify unauthorized users in network slicing security authentication.

- **Analysis of unauthorized user access to slice behavior identification results**

During the NSSAA process, when the UE illegally accesses the slice for authentication, it cannot generate the correct integrity message authentication code H2 due to the lack of K_{slice}. It can only generate an incorrect H2 by forging K_{slice} or directly forging H2 in an attempt to authenticate with the AAA-S. When the AAA-S receives the integrity message authentication code H2 from the UE, it calculates H2' based on equation (7) to compare with the received H2. Since the H2 value of the user's unauthorized slice access behavior is incorrect, it fails to authenticate with the AAA-S, and only a UE with a legitimate slice access behavior can pass the authentication. At the same time, the AAAS completes the one-way authentication of the UE. The results prove that our 5G-NSSSP can effectively determine the legitimacy of user access to slices in network slicing security authentication.

5.4 Summary of This Chapter

In this chapter, multiple virtualized network simulation tools and our 5G-NSSSP were utilized to design three experiments on network slicing security. The experimental results were analyzed to demonstrate the effectiveness and practicality of the proposed methods for service slice key distribution and authentication under the assistance of the 5G core network. After network slicing generation, it is necessary to establish an authentication and authorization mechanism specific to each network slice. This ensures that only authenticated and authorized users can access and utilize specific slice resources while also guaranteeing that users can connect to legitimate slice networks.

6 Conclusion

Aiming at the problems of single research scenario of the existing 5G network slicing simulation technology and insufficient security of the slice authentication process, this paper improves the existing NSSAA process, which effectively defends against Man-in-the-MiddleAttack and Replay Attacks, etc. in the slice authentication process. Based on the improved NSSAA process, this paper builds a 5G-NSSSP that includes initial user registration, primary authentication, and enhanced NSSAA process, and proves through experiments that this platform can provide a powerful security authentication tool for the research of 5G network slicing. In the future, we hope to diversify the experimental scenarios of this authentication platform to improve the security capability of the authentication process, reduce the operation overhead of the platform, and improve the competitiveness of the platform's actual deployment.

Acknowledgements. This work was supported by the National Key R&D Program of China (No. 2022YFB2902205) and the Natural Science Basis Research Plan in Shaanxi Province of China (No. 2021JM137) and the Fundamental Research Funds for the Central Universities (No. YJSJ23007).

References

1. Subedi, P., et al.: Network slicing: A next generation 5G perspective. In: Journal of EURASIP Journal on Wireless Communications and Networking, no. 1, p. 102 (2021). https://doi.org/10.1186/s13638-021-01983-7
2. 3GPP TS 23.501: System Architecture for the 5G System
3. 3GPP TS 33.501: Security architecture and procedures for 5G system
4. 3GPP TS 23.502: Procedures for the 5G System
5. Ni, J., Lin, X., Shen, X.S.: Efficient and secure service-oriented authentication supporting network slicing for 5G-Enabled IoT. IEEE J. Sel. Areas Commun. **36**(3), 644–657 (2018). https://doi.org/10.1109/JSAC.2018.2815418
6. Zhang, Y., Wu, A., Chen, Z.: Flexible and anonymous network slicing selection for C-RAN enabled 5G service authentication. Comput. Commun. **166**, 165–173 (2021). https://doi.org/10.1016/j.comcom.2020.12.014
7. Fan, C.-I., Shih, Y.-T., Huang, J.-J., Chiu, W.-R.: Cross-Network-Slice Authentication Scheme for the 5th Generation Mobile Communication System. In: IEEE Transactions on Network and Service Management, vol. 18, no. 1, pp. 701–712 (2021). https://doi.org/10.1109/TNSM.2021.3052208
8. Fan, C.-I., Zheng, K.-Y., Shih, Y.-T., Zhuang, E.-S., Huang, J.-J.: An efficient secure handover scheme supporting cross-network slicing for multi-operator environments. In: 2021 Thirteenth International Conference on Mobile Computing and Ubiquitous Network (ICMU). Tokyo, Japan pp. 1–7 (2021). https://doi.org/10.23919/ICMU50196.2021.9638843
9. Chen, Z., Ao, J., Luo, W.: A dual-factor access authentication scheme for IoT terminal in 5G environments with network slice selection. J. Info. Secu. Appl. **68** (2022). https://doi.org/10.1016/j.jisa.2022.103247
10. Yadav, A.K., Wijethilaka, S., Braeken, A., Misra, M., Liyanage, M.: An enhanced cross-network-slice authentication protocol for 5G. In: IEEE Transactions on Sustainable Computing. https://doi.org/10.1109/TSUSC.2023.3283615
11. Ren, Z., et al.: Network slicing in 6G: an authentication framework for unattended terminals. IEEE Network **37**(1), 78–86 (2023). https://doi.org/10.1109/MNET.112.2100738
12. 3GPP TR 21.917: Summary of Rel-17 Work Items
13. Wong, S., Han, B., Schotten, H.D.: 5G Network Slice Isolation. Network https://doi.org/10.3390/network2010011

A Certificate Transparency-Based Certificate Monitoring Scheme for the Power Grid

Liang Yin[1], Hongjie Zhang[2], Jianing Peng[2], Zhiyan Ning[3], Zhijun Zhang[3], Kaiyuan Zhang[4], and Yueyu Zhang[4(✉)]

[1] Electric Power Research Institute of State Grid Ningxia Electric Power Co., Ltd., Yinchuan, Ningxia Hui Autonomous Region, People's Republic of China

[2] State Grid Ningxia Electric Power Co., Ltd., Yinchuan, Ningxia Hui Autonomous Region, People's Republic of China

[3] NARI Group Corporation (State Grid Electric Power Research Institute), Beijing Kedong Electric Power Control System Co., Ltd., Beijing, China

[4] Xidian University, Xi'an, Shannxi Province, People's Republic of China

kyzhang_31@stu.xidian.edu.cn, yyzhang@xidian.edu.cn

Abstract. With the exponential evolution and development of the Internet, the security of infrastructure and network information has become increasingly important. As a critical component of infrastructure, the power grid plays a crucial role in ensuring the supply of electricity. However, there are many challenges and issues in the issuance and use of certificates in the context of power grid scenarios, lacking a fair and tamper-proof certificate monitoring scheme to ensure the security of the certificates. This paper proposes a certificate monitor scheme for power equipment by using certificate transparency logs, Ethereum's distributed database technology, and machine learning detection algorithms to achieve transparent management and monitoring device certificates. By uploading certificate transparency logs to the Ethereum blockchain, the proposal ensures the transparency of certificate operations by certificate authorities, preventing malicious tampering and inappropriate behavior. Furthermore, the certificate detection based on machine learning algorithm provides real-time monitoring of certificate validity and integrity, and further detecting potential certificate issues and security threats. The analysis shows that this scheme enhances the trustworthiness and security of certificate monitor for power grid, and provides assurance for the stable operation of the power grid.

Keywords: Certificate Monitor · Certificate Transparency · Ethereum

1 Introduction

1.1 Background

In today's digital and internet-driven social environment, power grid, as a critical part of the infrastructure, bear the essential task of ensuring a reliable power supply [1]. Not only do they control the normal operation of the power grid, but they also play a

Z. Gu et al. (Eds.): NSE 2023, CCIS 2063, pp. 256–270, 2024.
https://doi.org/10.1007/978-981-97-4519-7_18

crucial role in responding to emergencies and ensuring the safety and stability of the power supply. Within the power grid, certificate auditing and monitoring play a pivotal protective role. Certificates, issued by Certificate Authorities (CAs), serve as a means of ensuring the trustworthiness of power equipment [2]. This is especially vital in power grid, as effective digital certificates can prevent network attackers from masquerading as legitimate nodes or power equipment, thereby safeguarding power monitoring system devices, data, and control instructions from being tampered with or intercepted. However, potential security issues may also lie within the Certificate Authorities [3]. For example, if a CA falls under the control of an attacker, or if CA administrators conduct operations that are not compliant with standards, attackers could forge certificates and launch attacks on power grid. Therefore, the auditing and monitoring of certificates, including checking their legality and validity, are significant in preventing certificate misuse and protecting the safety of power grid [4].

1.2 Challenges

In 2011, an attacker managed to compromise the Public Key Infrastructure (PKI), gaining control over the DigiNotar certificate authority [5], and subsequently issuing certificates for several fraudulent websites under the name of high-profile companies such as Google.com. This allowed the attacker to impersonate any google.com website and deceive victims. Additionally, current research continues to delve into potential cyberattacks on PKI or Certificate Authorities (CAs). Tian et al. [6] proposed a DNS cache poisoning attack, demonstrating that even a weak attacker could use this method to deceive CAs into issuing fraudulent certificates for domains that the attacker did not legitimately own. In 2021, Zhang et al. [7] discovered significant shortcomings in the implementation of hidden root CAs and certificates, such as pervasive issues with weak keys and signature algorithms. This revealed a vast and dynamic ecosystem of hidden root CAs, which could lead to various security issues. In the same year, Markus Brandt et al. [8] conducted a simulation using a dataset of 230 popular internet domains. The results indicated that even if the certificate issuing authority is secure, network adversaries could still issue counterfeit certificates, introducing a weak link into the PKI ecosystem.

We reveals serious vulnerabilities and security risks in the Public Key Infrastructure (PKI) and Certificate Authorities (CA) in the context of power grid, which could potentially threaten the stability and safety of the entire power supply system. Therefore, finding more reliable PKI and CA security solutions for power grid has become an urgent challenge today.

2 Related Works

2.1 Certificate Transparency

Certificate Transparency (CT) is a technology invented by Google [9]. The purpose of CT is to protect users from erroneously or maliciously issued certificates by ensuring that all certificates are recorded in append-only logs that can be publicly audited. CT logs are public, available for audit and monitoring by the public: anyone can monitor suspicious certificates or improper behavior of Cas [10]. The CT framework, showed in Fig. 1, includes a Certificate Transparency Log (CTL), a publicly auditable database that can only append certificates. In the context of power monitoring systems, Certificate Transparency Logs can help system administrators, device managers, and researchers track and audit the issuance and use of certificates, detect potential security threats such as unauthorized certificate issuance, and certificate misuse [11].

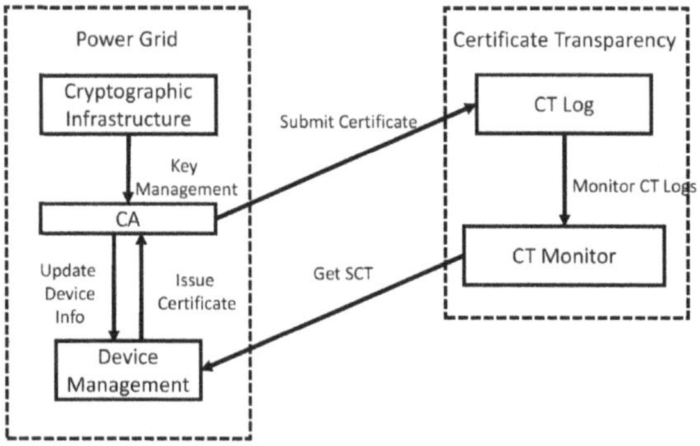

Fig. 1. The Certificate Transparency framework for the power grid.

The implementation of Certificate Transparency Logs primarily relies on a data structure called Merkle Tree. This is a tree structure in which the hash value of each non-leaf node is calculated from the hash values of its child nodes [12]. The advantage of this data structure is that only a small amount of information (Merkle proof) is needed to verify whether an element is in the tree. When a Certificate Authority (CA) issues a new certificate, it submits the certificate to one or more CTLs. After the Certificate Transparency Log server receives a new certificate, it checks the certificate for legality and completeness. Then, the server generates a new Merkle Tree node, stores the certificate information in it, and calculates a new root hash value. Subsequently, the CTL server creates a Signed Certificate Timestamp (SCT) and returns it to the CA. The SCT is a proof that the certificate has been added to the log, containing certificate information, a timestamp, and the server's signature.

For any queries to the CTL, the server will provide a Merkle proof, which proves whether a particular certificate is in the log. Users or auditors can use the Merkle proof and the known root hash value to verify the existence and integrity of the certificate.

2.2 Ethereum and Smart Contracts

Blockchain is a decentralized distributed technology used for recording transactions [13]. It offers many benefits to the blockchain such as decentralization, durability, anonymity, and auditability [14]. Therefore, blockchain technology has developed rapidly in recent years, and its application scenarios have expanded from the initial issuance and trading of virtual currencies to fields such as finance, medicine, the Internet of Things, and software engineering [15].

Ethereum is a smart contract platform [16] based on a Turing-complete language and is also the earliest Turing-complete smart contract. The Ethereum system provides the Ethereum Virtual Machine (EVM) [17], and contract code runs within the EVM. Ethereum users write smart contract code in a specific language and compile it into EVM bytecode to run.

2.3 Related Research

As new technologies appear, some extensive research expands into the protection strategies and technical solutions for PKI and CAs. Yu et al. [18] proposed a Distributed Transparent Key Infrastructure (DTKI) in 2016, addressing CA security issues and eliminating dependence on trusted parties by mitigating the monopolistic control of service providers. They also formalized the public log data structure and conducted a formal analysis of the security guarantees provided by DTKI. Liand his team [19] designed and implemented a new method to discover suspicious SSL man-in-the-middle attack hosts in the wild in 2020, which found 322,831 counterfeit certificates vulnerable to these attacks. Talamo et al. [20] proposed a contract algorithm by using blockchain technology. Their consensus algorithm retained all the functionality surrounding X.509 certificates and demonstrated a mechanism to achieve fast consensus. This enables consensus to be reached even if not all participating PKI members are involved in the transaction, and it doesn't require high-level trust protocols between PKIs.

However, traditional power grid scenarios and current Internet scenarios have slight differences. Firstly, power grid scenarios do not use the traditional PKI key generation mode. They have special encryption algorithms and dedicated cryptographic infrastructure. Secondly, multi-level CAs in power grid are all located in strictly physically isolated local area network environments. Their security can only be ensured by system employees, without any other thirdparty institutions or organizations for supervision and management. Lastly, due to the confidentiality and data privacy of power grid, multiple manual operations by administrators are required, which differ from traditional Internet certificate management [21].

Therefore, to address the current certificate security issues in power grid, we propose a certificate monitoring scheme based on certificate transparency and blockchain.

3 Proposed Approach

According to the features and challenges of the power grid, it is vulnerable to the following certificate related threats.

Malicious CA. The current PKI system allows CAs to issue digital certificates, but these CAs may have potential malicious behavior. For instance, a malicious CA might issue unauthorized certificates, forge certificates, or discover that the CA's private key has been stolen but fail to notify the users.

Unauthorized Certificates. In certain situations, a CA may mistakenly issue unauthorized certificates that do not belong to the device. Attackers can use unauthorized certificates to spoof administrators and gain access to critical power grid information.

Single Point of Failure. Traditional centralized CT server is centralized, and if this server fails, the entire system becomes paralyzed.

Data Tampering. Data is typically stored in the central CT server, which could be attacked by hackers or internally manipulated.

In the context of power grid, certificate transparency technology can effectively record and monitor certificates of multi-level CAs themselves, as well as those issued for power grid equipment entities, ensuring system security. However, unlike the standard internet environment, due to the strict physical security control of the power grid's local area network, and the complexity of the network and the magnitude of equipment certificates which cannot be compared with the internet environment, the probability of the CA being compromised is lower than in the internet environment. Power grid However, the power grid is a critical national infrastructure, often considered a high-value target by hackers, so from an external perspective, it still faces considerable security risks. In addition, compared to the internet environment, the power grid network needs to consider the possibility of deliberate damage or penetration by internal personnel.

And against the above threats, the use of blockchain technology can be a good solution to this problem. Due to its consensus mechanisms and hash function mechanism, blockchain can significantly protect the integrity and timeliness of data. If the certificate transparency log and the certificate are published on the blockchain simultaneously, it can not only ensure the security and supervision of the CA but also protect and monitor the certificate transparency log server. This proposal provides a method for certificate auditing and monitoring in the context of power grid, based on technologies such as Certificate Transparency and blockchain. The proposal mainly involves the following parts:

CA Server. The CA, using the cipher machine independently managed in the power grid, signs the power monitoring equipment and other entities, and generates certificates.

CT Log Server. Collects and validates the device entity certificate chain issued by the Certificate Authority. If validation is successful, the Certificate Transparency log saves the certificate information, generates a hash, and stores it in its own data structure. Administrators can query related certificates through the interface.

Ethereum Server. Ethereum deploys a smart contract of the Certificate Transparency log, which stores the Certificate Transparency log information in the Ethereum server, achieving data decentralization and ensuring the integrity of the Certificate Transparency log.

Automated Detection Method for Administrators. To conveniently validate and manage the integrity and non-tampering of the certificate log, this automated detection method uses the Certificate Transparency log as a data set. It extracts key information of the device and certificate chain and uses machine learning algorithms to generate a model that can automatically detect fake certificates. The entire process of the scheme is shown in Fig. 2, which we can divide into the following five steps.

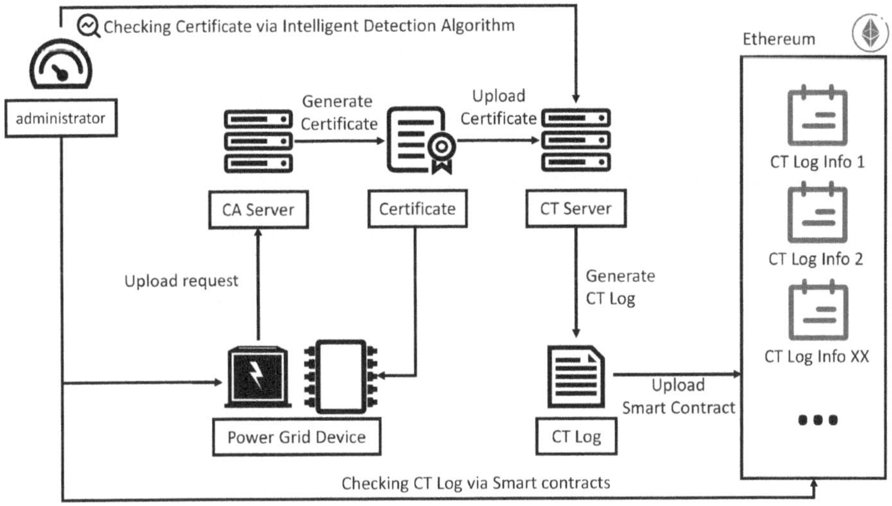

Fig. 2. The workflow of the scheme.

The administrator of a certain device first needs to apply to the CA server of the power grid to provide a digital certificate for its power device A. The application for the certificate contains information such as the device ID, MAC address etc.

After layers of checks and verification by the CA server administrator, a digital certificate is returned to the device administrator. The administrator can now deploy the certificate to device A in its system.

At the same time, this digital certificate is also uploaded to the Certificate Transparency log server in the power grid. The log server signs the information of its certificate A to generate a certificate log A containing two hash values (leaf input and extra data).

The Certificate Transparency Monitor will proactively detect whether there is any incorrect or counterfeit information in log A and generate a report.

The log server will also use this log to call a smart contract and upload it to Ethereum. All users and administrators in this system can query the content and status of the certificate log by consuming Ethereum fuel through the smart contract.

We analyze the attack surface of device certificates to demonstrate the feasibility of our proposal. Without the certificate monitoring provided by our scheme, both external attackers and internal personnel can bypass the Certificate Authority (CA) and forge certificates, threatening the security of power equipment without the CA administrator's knowledge. This proposal ensures certificate security using Certificate Transparency (CT) technology and Ethereum. To achieve the effect of forging certificates, an attacker would need to simultaneously alter both the hash signature database of the CT used in our solution and the hash certificate information stored on Ethereum. However, to modify the certificate information is nearly infeasible.

4 Implementation

4.1 Certificate Transparency Server

The CT server is deployed on a local area network and opens RPC ports 6962 and 6963 according to the RFC 6962 [10], and also opens the long process port 6965 for viewing. Its domain name is set to following the standard, and the following interfaces are opened:

- "add-chain": This interface is used to add new certificates to the CT log and returns related SCT and CT log ID information.
- "get-sth": This interface retrieves the latest signed header information of the Merkle Tree data structure.
- "get-entries": This interface retrieves detailed information from the first to the n-th CT log.

4.2 Etheum Server

An Ethereum private chain accessible only within the local network is deployed on the server, mainly using the open-source framework Geth for implementation. It includes the Ethereum client and Ethereum network, which contain a command parsing module, EVM, and P2P module, as well as numerous Ethereum nodes using P2P communication. The mining difficulty and gasLimit settings are quite friendly, and the alloc parameter also allocates some Ether to various accounts. This proposal focuses more on data integrity and privacy when using Ethereum, and its performance should be optimized as much as possible. After testing, an empty block will be mined every 5–10 s.

The Solidity language is used for smart contract development and use in this proposal. The smart contracts mainly include the creation of a string list and operations on the elements of this list. They require the addition of Certificate Transparency log strings to the list, reading certain log content, reading all log content (returning the list). The pseudo code of the smart contract can be found in Table 1.

Table 1. Pseudo code of the smart contract

Pseudo code of the smart contract

```
// Create a public list of strings
List<String> list = new List<String>()

// Define a function to add a string to the list
Function add(String value):
  list.append(value)
End Function

// Define a function to read a string from the list by its index
Function read(Integer index):
  if index >= list.length or index < 0:
    throw Error("Index is out of bounds.")
  else:
    return list[index]
  end if
End Function

// Define a function to read all strings from the list
Function readAll():
  return list
End Function
```

4.3 Machine Learning Based Automated Detection Method

This proposal utilizes machine learning algorithms to automate the detection of irregular behaviors in the logs generated by Certificate Transparency. As the Certificate Transparency log contains the complete certificate chain information, i.e., all signature information from the root CA to the device entity certificate, it carries more authority and research significance than individual certificate information.

The machine learning detection algorithm is deployed in the certificate monitoring module in the Certificate Transparency server. It can query, edit, and delete the Certificate Transparency logs uploaded to Ethereum by invoking the smart contract functions of the Ethereum smart contract server. Afterward, it checks whether there are inaccuracies in the timestamps of the Certificate Transparency logs. The key fields in its certificate content can be found in Table 2.

Table 2. The key fields in its certificate content

Certificate Authority	Device Entity	Device Certificate
CA's ID CA Validity Period	Device ID MAC Firmware Version Power Consumption	Certificate ID Certificate Index Certificate Validity Period

In the feature extraction stage, we manually selected and extracted important fields and parameters from the Certificate Transparency logs based on experience to serve as features for the algorithm.

We extract these key fields as the main training features for machine learning. For Certificate Authorities (CAs), the validity periods of different levels of CAs are our primary focus. Due to the strict management of CAs in the power grid, the likelihood of issuing certificates across levels is relatively low, and the methods for generating and encrypting keys are not our focus.

For device entities, firmware version, and power consumption are the key points of concern. Some older devices have not updated their firmware for a long time or have been operating for a long time without maintenance, which are areas that attackers often focus on. The power consumption are important because due to the confidentiality and security of the power grid's devices, this information is typically inaccessible to attackers and can only be estimated based on the attacker's experience or randomly selected. Therefore, if a device's power consumption is abnormal, there is a high likelihood that its certificate is fraudulent.

For device certificates, similar to the issuing institutions, the validity period they issued is more likely to cause problems. For instance, an attacker might assign a 99-year or up to 2099-year usage period to their counterfeit certificates for their own purposes.

In the feature engineering stage, we employed a range of methods to calculate or transform the original fields. This allows each valid Certificate Transparency log to be converted into a feature vector, thereby enabling more effective training and testing of our machine learning model.

For the CA validity periods and certificate validity periods, we first extracted the valid start and end dates from the certificates and calculated the number of valid days. Subsequently, we normalized these days so that the feature value falls between 0 and 1.

For device power consumption, we categorized it into high, medium, and low types based on the type of the device. Specifically, we utilized the Z-score algorithm for standardization, where x is the original power consumption data, μ and σ are the mean and standard deviation of all device power consumption data, respectively. The standardized power consumption data is used as one of the feature inputs for the machine learning model.

$$z = \frac{x - \mu}{\sigma} \tag{1}$$

For the CA IDs, device entity IDs, device MAC addresses, firmware versions, as well as device certificate IDs and indices, we employed label encoding. Specifically, we used the

LabelEncoder class from the Scikit-learn library, which maps each non-numeric label to a unique integer.

In the model training phase, we selected some of the most commonly used machine learning models in the field of anomaly detection to train and compare performance. For each machine learning algorithm, we fine-tuned various parameters such as learning rate and optimizer to maximize the model's performance. Details on performance evaluation will be covered in Sect. 5.

5 Performance Evaluation

5.1 Validation

The cryptographic infrastructure, CA, and certificates in the power grid are different from those on the Internet, for ease of understanding, this section uses a traditional Internet X.509 certificate to verify this system.

First, we use a traditional X509 certificate, the content of which after base64 encryption is shown in Fig. 3. Then, we place the certificate content of the CA into the certificate file to form a certificate chain file and call the add-chain interface of the certificate transparency server to upload it to the server. The server will audit the entire certificate chain and generate a certificate transparency log for certificates that pass the audit. Afterwards, we can obtain its log information by calling the get-entries interface, as shown in Fig. 4. By decoding the content in Fig. 4, we can obtain the certificate information stored in the certificate transparency log, as shown in Fig. 5.

If attackers want to upload a certificate signed by an unauthenticated CA/expired CA, or an expired certificate, it will be rejected by the certificate transparency server. Once a certificate is uploaded to the certificate transparency server, it will be directly uploaded to the blockchain using the CA administrator's Ethereum account. All Ethereum users can view the certificate transparency logs in it. We can view and edit the certificate transparency logs already stored on Ethereum, by using the functions encapsulated in the smart contract.


```
-----BEGIN CERTIFICATE-----
MIIDGzCCAgMCFB7SRl7yTC5PJqtF1z+7hpu+lKXcMA0GCSqGSIb3DQEBCwUAMEox
CzAJBgNVBAYTAjAxMRowGAYDVQQIDBFmMzpiMTo2Nzo1MzowZjozMTEQMA4GA1UE
BwwHMjAyMy0wMTENMAsGA1UECgwENTBkVzAeFw0yMzA2MjkwNjI2MzVaFw0yNDA2
MjgwNjI2MzVaMEoxCzAJBgNVBAYTAjAxMRowGAYDVQQIDBFmMzpiMTo2Nzo1Mzow
ZjozMTEQMA4GA1UEBwwHMjAyMy0wMTENMAsGA1UECgwENTBkVzCCASIwDQYJKoZI
hvcNAQEBBQADggEPADCCAQoCggEBANZZYUnvQ77PFlCIFzGW8A8VyTn5eoEZ1LiX
sWXz9AfczhwMRxPAVQmvCvhNzm5il+7KSUaDqrdPnOlQaUrkJmuajcK/kl0/jt5t
SFtn0KW8BJa2FoELb/GfRP2X3hI+M9ePqGdJVGuymVkayupuQedIQs96TysYLFQ1
/J10Bm/hbJDVgVKvUVkP59bPmG7AGxAYV4SFkCx12aj7WeFf+AbSIEiVr5zqiYtD
JBmthIOtItm+95nElb90Lkkg5sh9IlSFm8EOpjImkcjp6Y0IRBLmArqKK0pH+VK0
/Ezwfx5MqIdUpOqNayDaZX2I9F3oSkfbY/JkD7sHHIj+KsjRxbMCAwEAATANBgkq
hkiG9w0BAQsFAAOCAQEAXA3w71YwMI8rOJrIcpWxmqnPHr3SibIGBzInbzJz+Fgy
jW+kKZ0uRouuzRJ+aQGlM+dovBGfY8tL+YPfFtJLcWkSUujGXeFejIosv4sHr1RC
3EP1HuWFTG+Mhmfv46Xi7BTYmECg78YgHNcRX6vzpnGlOLiQmNxjc8UwKjd3x4hd
uLpLIBV8P+JgqQ1Srh20mqBP9E4+4vFN65bvplEiKunDZui+I7jMgHrmbGyyo3cl
OAcQ4OkJ4+vXod/aHClDi2Yjx32c+omo/dmi24ARpXZYx+8ORg/tNHv5pPXdcGbF
B7Rhtw5PVkHIOKZH4zcC/ZwpSVjm+BdxHHO2jQ5wpw==
-----END CERTIFICATE-----
```


Fig. 3. The example of certificate.

{
"leaf_input" :
"AAAAAAGJBbRmuQABsPuJV4Eoofjsca5SMxbU7wJWoJiC6gtHzHrIsgyV+SUAAhkwggIVoAMCAQICBQDerb7vMAoGCCqGSM49BAMCMHIxCzAJBgNVBAYTAkdCMQ8wDQYDVQQIEwZMb25kb24xDz
ANBgNVBAcTBkxvbmRvbjEPMA0GA1UEChMGR29vZ2xlMQwwCgYDVQQLEwNFbmcxIjAgBgNVBAMTGUZha2VJbnRlcm1lZGlhdGVBdXRob3JpdHkwHhcNMTgwMzI5MTcwNTExWhcNMjUwNTExMT
cwNTExWjBWMQswCQYDVQQGEwJHQjEPMA0GA1UECAwGTG9uZG9uMQ8wDQYDVQQKDAZHb29nbGUxDDAKBgNVBAsMA0VuZzEXMBUGA1UEAwwObGVhZjAxLmNzci5wZW0wWTATBgcqhkjO
PQIBBggqhkjOPQMBBwNCAATrN05SRZxG1ai4xe1YuTAppnCKaaAmXJ4vbrhrI2yE4UY6mDaCRKWKF4tBgjL0LeAIW34HOFL8R1YoJ5vtYluso4GzMIGwMA8GA1UdDwEB/wQFAwMH+YAwgYsGA1UdD
gSBgwSBgGk2r+g3vMY3MIkq7grmOO04+QYQBgrFrjNAw6EKXjv9zUu4K22292z4PJI7zE9TxUD6nmmjfq3OI/0zEodJkAZqnHJFtWrdxUzNyKl9BeO+ADNUYzViQ4NpI5vvnuHuuOYwcZLKRHGdWLgZmE
rWgJUT1pzXC2qqGgItTN71M1OvMA8GA1UdIwQIMAaABAUGBwgAAA== "
"extra_data":
"AAMYMIIDFDCCArmgAwIBAgIFAN6tvu8wCgYIKoZIzj0EAwIwfzELMAkGA1UEBhMCR0IxDzANBgNVBAgTBkxvbmRvbjEPMA0GA1UEBxMGTG9uZG9uMQ8wDQYDVQQKEwZHb29nbGUxDDAKBgNVB
AsTA0VuZzEvMC0GA1UEAxMmRmFrZUludGVybWVkaWF0ZUF1dGhvcml0eVByZWNlcnRJc3N1ZXIwHhcNMTgwMzI5MTcwNTExWhcNMjUwNTExMTcwNTExWjBWMQswCQYDVQQGEwJHQjEPMA
0GA1UECAwGTG9uZG9uMQ8wDQYDVQQKDAZHb29nbGUxDDAKBgNVBAsMA0VuZzEXMBUGA1UEAwwObGVhZjAxLmNzci5wZW0wWTATBgcqhkjOPQIBBggqhkjOPQMBBwNCAATrN05SRZxG1ai4
xe1YuTAppnCKaaAmXJ4vbrhrI2yE4UY6mDaCRKWKF4tBgjL0LeAIW34HOFL8R1YoJ5vtYluso4IBSTCCAUUwDwYDVR0PAQH/BAUDAwf5gDCBiwYDVR0OBIGDBIGAaTav6De8xjcyWSruCuY47Tj5BhAGC
sWuM0DDoQpeO/3NS7grbbb3bPg8kjvMT1PFQPqeaaN+rc4j/TMSh0mQBmqcckW1at3FTM3IqX0F474AM1RjNWJDg2mXm++e4e645jBxkspEcZ1YuBmYStaAlRPWnNcLaqoaAi1M3vUzU68wgY4GA1
UdIwSBhjCBg4CBgBTn3XQVDsSveQi/uotdbbSR1ZwBTxHe1kBtb0pTRcseLMflHfYFl5h0MRH+4v8+AGW7cAHe/eWPHCPBOGNTQQ04kJ9XukH2KdYlqBuQLAeVqTJR+vbM62TDJrdsn/axXEiw4ExwNccj
V6Ow49I8RBC8RJUTe6Z+UW9iLJkcAp83MBMGCisGAQQB1nkCBAMBAf8EAgUAMAoGCCqGSM49BAMCA0kAMEYCIQDFgzFagfKVcsyc84dMWBlRomG5RcZSeBj8awC7GIQAMglhAMBoGTFo83arm
B8tPvfqJkYrDjeG3rdHodxtAMHgHdTGAAdDAALaMIIC1jCCAnygAwIBAgIEQkJCQjAKBggqhkjOPQQDAjByMQswCQYDVQQGEwJHQjEPMA0GA1UECBMGTG9uZG9uMQ8wDQYDVQQHEwZMb25kb24
xDzANBgNVBAoTBkdvb2dsZTEMMAoGA1UECxMDRW5nMSIwIAYDVQQDExlGYWtlSW50ZXJtZWRpYXRlQXV0aG9yaXR5MB4XDTE4MDMyOTE3MDUxMVoXDTI4MDIwNTE3MDUxMVowfzELMAkG
A1UEBhMCR0IxDzANBgNVBAgTBkxvbmRvbjEPMA0GA1UEBxMGTG9uZG9uMQ8wDQYDVQQKEwZHb29nbGUxDDAKBgNVBAsTA0VuZzEvMC0GA1UEAxMmRmFrZUludGVybWVkaWF0ZUF1dGhvc
ml0eVByZWNlcnRJc3N1ZXIwWTATBgcqhkjOPQIBBggqhkjOPQMBBwNCAAS9UsHCAu5XDE81gEs1lw1J1f90K3N/V0uTvi3t/eMWxoACHfMK6p8LACJAoaaunWgc8kyZlN4YOtaMxcn8B3Nxo4HyMIHv
MA8GA1UdDwEB/wQFAwMH/4AwKQYDVR0lBCIwIAYIKwYBBQUHAwEGCCsGAQUFBwMCBgorBgEEAdZ5AgQEMBIGA1UdEwEB/wQIMAYBAf8CAQAwgYsGA1UdDgSBgwSBgBTn3XQVDsSveQi/uot
dbbSR1ZwBTxHe1kBtb0pTRcseLMflHfYFl5h0MRH+4v8+AGW7cAHe/eWPHCPBOGNTQQ04kJ9XukH2KdYlqBuQLAeVqTJR+vbM62TDJrdsn/axXEiw4ExwNccjV6Ow49I8RBC8RJUTe6Z+UW9iLJkcAp83
MA8GA1UdIwQIMAaABAUGBwgwCgYIKoZIzj0EAwIDSAAwRQIhAKl5/JnOR2DYHDUvhh8kDLJSSDIKnUCVVpzbaBkYUa8dAIBAP9zkq+5G1AwJvcJF9QofoNx+Czuz9xBwkqaJEjGRxwACQDCCAjwwggHh
oAMCAQICBEJCQkIwCgYIKoZIzj0EAwIwcTELMAkGA1UEBhMCR0IxDzANBgNVBAgTBkxvbmRvbjEPMA0GA1UEBxMGTG9uZG9uMQ8wDQYDVQQKEwZHb29nbGUxDDAKBgNVBAsTA0VuZzEhMB8GA
1UEAxMYRmFrZUNlcnRpZmljYXRlQXV0aG9yaXR5MB4XDTE4MDMyOTE3MDUxMVoXDTI4MDIwNTE3MDUxMVowcjELMAkGA1UEBhMCR0IxDzANBgNVBAgTBkxvbmRvbjEPMA0GA1UEBxMGTG9
uZG9uMQ8wDQYDVQQKEwZHb29nbGUxDDAKBgNVBAsTA0VuZzEiMCAGA1UEAxMZRmFrZUludGVybWVkaWF0ZUF1dGhvcml0eTBZMBMGByqGSM49AgEGCCqGSM49AwEHA0IABPG/LeiMZkDjq
NFU4EJJAsvdRwiFwmdBTOv3h82NownIGMwuMFMWMqrVnAhzxnb6+jo46TQ1nFHR7hKBXZhfXV2jZjBkMA0GA1UdDgQGBAQFBgcIMA8GA1UdIwQIMAaABAECAwQwEgYDVR0TAQH/BAgwBgEB/
wIBADAPBgNVHQ8BAf8EBQMDB/+AMB0GA1UdJQQWMBQGCCsGAQUFBwMBBggrBgEFBQcDAjAKBggqhkjOPQQDAgNJADBGAiEAlb9Is8drOb2Ol2u/SybetYIIRdFEKANQqgrYtCL3oeACIQD7ZBjIwD
gSqpxe2S7KcrLGhujgi3m0K893M+xh37DG3AACIDCCAhwwggHBoAMCAQICBAQGyv4wCgYIKoZIzj0EAwIwcTELMAkGA1UEBhMCR0IxDzANBgNVBAgTBkxvbmRvbjEPMA0GA1UEBxMGTG9uZG9uM
Q8wDQYDVQQKEwZHb29nbGUxDDAKBgNVBAsTA0VuZzEhMB8GA1UEAxMYRmFrZUNlcnRpZmljYXRlQXV0aG9yaXR5MB4XDTE2MTIwNzE1MTMzNloXDTI2MTIwNTE1MTMzNlowcTELMAkGA1UEB
hMCR0IxDzANBgNVBAgTBkxvbmRvbjEPMA0GA1UEBxMGTG9uZG9uMQ8wDQYDVQQKEwZHb29nbGUxDDAKBgNVBAsTA0VuZzEhMB8GA1UEAxMYRmFrZUNlcnRpZmljYXRlQXV0aG9yaXR5MFkw
EwYHKoZIzj0CAQYIKoZIzj0DAQcDQgAE8tMH737fz8709ApbvJ4/yxz9DEbchfvB9tOyuh1R8ZhsSKgVRkVjyt/WyazPYDvHTt240haroAkkHQlmHk3roaNHMEUwDQYDVR0OBAYEBAECAwQwDwYDVR0jB
AgwBoAEAQIDBDASBgNVHRMBAf8ECDAGAQH/AgEKMA8GA1UdDwEB/wQFAwMH/4AwCgYIKoZIzj0EAwIDSQAwRgIhAKYoSTlDb4DkQ6YeO6qJXsIlYCrhOb1VQ65NXKmm76xlAiEAycUIxlmTtIZwp
WtUK1v8DIhrsCMHK8cMJ96HLZaA1VY="
}

Fig. 4. The example of CT log.

```
Issuer: CN=FakeCertificateAuthority,OU=Eng,O=Google,L=London,ST=London,C=GB
Subject: CN=FakeCertificateAuthority,OU=Eng,O=Google,L=London,ST=London,C=GB
SerialNumber: 67554046
NotBefore: 2016-12-07 15:13:36 +0000 UTC
NotAfter: 2026-12-05 15:13:36 +0000 UTC
(EXTRA int=-1) (EXTRA time.Time=0001-01-01 00:00:00 +0000 UTC)Issuer: CN=FakeCertificateAuthority,OU=Eng,O=Google,L=London,ST=London,C=GB
Subject: CN=FakeIntermediateAuthority,OU=Eng,O=Google,L=London,ST=London,C=GB
SerialNumber: 1111638594
NotBefore: 2018-03-29 17:05:11 +0000 UTC
NotAfter: 2028-02-05 17:05:11 +0000 UTC
(EXTRA int=-1) (EXTRA time.Time=0001-01-01 00:00:00 +0000 UTC)Issuer: CN=FakeIntermediateAuthority,OU=Eng,O=Google,L=London,ST=London,C=GB
Subject: CN=leaf01.csr.pem,OU=Eng,O=Google,ST=London,C=GB
SerialNumber: 3735928559
NotBefore: 2018-03-29 17:05:11 +0000 UTC
NotAfter: 2025-05-11 17:05:11 +0000 UTC
```

Fig. 5. The information of certificate in CT log.

5.2 Machine Learning Performance Analysis

We adopted four algorithms for the approach in this paper: XGBoost, RF (Random Forest), SVM (Support Vector Machine), and DP. After testing, we observed that the XGBoost algorithm performed the best, achieving an F-measure of 0.9548, significantly surpassing the other three algorithms. Detailed results are presented in Table 3. One possible reason for this superiority of the XGBoost algorithm is its capability to handle a wide range of outlier data in the dataset, thus improving the predictive accuracy of the model.

Table 3. Machine learning algorithms performance comparison

Algorithms	Accuracy	Precision	Recall	F-measure
XGBoost	0.9450	0.9474	0.9222	0.9548
RF	0.8846	0.8723	0.8555	0.8456
SVM	0.6612	0.6750	0.2924	0.4525
DP	0.5212	0.4879	0.2242	0.3386

5.3 Performance Evaluation

The Certificate Transparency server and Ethereum server are both hosted on virtual machines (VMs). The VMs provide a virtualized environment where the servers can operate independently, isolated from the host operating system and other virtual machines running on the same physical server.

The Certificate Transparency server utilizes a 4-core CPU with 8GB of RAM and runs on the Ubuntu 20.04 operating system. It primarily operates the Go language version of the Certificate Transparency project. The project uses two RPC microservice endpoints and one functional endpoint to provide feature interfaces. We primarily calculate the performance of the CT server based on the following indicators, and regarding CPU utilization, we specifically focus on the user space. Detailed performance can be found in Table 4.

Table 4. Performance of CT server

Performance Item	Value
CPU utilization	26.5%
Memory utilization	70.48%
Network Traffic Utilization	48.4%
Processing speed for 100 certificates	1.48 s

The Ethereum server is equipped with an 8-core CPU, 16GB of RAM, and runs on the Ubuntu 20.04 operating system. Ethereum is built using Solc 0.4.24, and Solidity version is 0.8.1. Our scheme is different from traditional Ethereum applications, due to the strict physical network isolation required by the power grid, uses a private Ethereum chain. Hence, this system can customize parameters related to system performance, such as the gas cost of transactions. This customization can significantly improve the efficiency of miners and reduce the gas and power consumption required for transactions, such as submitting logs, retrieving logs, and modifying logs. Consequently, this significantly improves the performance of the Ethereum system, helping to enhance the overall efficiency of our solution. We mainly measure the performance of the Ethereum server based on the following indicators, in this regard, we only calculate the gas cost and specific consumption of Ethereum currency required when adding CT logs, as the cost of queries will increase with the increase in the number of logs, making it difficult to estimate. Detailed performance can be found in Table 5.

Table 5. Performance of Ethereum server

Performance Item	Value
CPU utilization	78.3%
Memory utilization	56.3%
Network Traffic Utilization	63.1%
Gas fees of adding	0.0000705ETH

By utilizing a private Ethereum chain, the solution takes advantage of the security and decentralization provided by blockchain technology while being able to tailor performance to the specific needs of the power grid infrastructure. This approach provides a good balance between system performance and the benefits offered by distributed ledger technology.

6 Conclusion

In this paper, we address the issue of certificate monitoring within power grid systems by introducing a certificate monitoring scheme based on Certificate Transparency (CT) and Ethereum. We employ Certificate Transparency technology to implement the core functionality of certificate monitoring within the power grid environment and utilize Ethereum smart contracts to ensure the integrity and security of the certificate transparency logs. Additionally, through machine learning algorithms, we have developed an intelligent algorithmic system that assists system administrators in discerning and managing device certificates by detecting counterfeit or non-compliant ones. The complete operational procedure of the scheme is provided, validating its feasibility and security. A thorough analysis of the system's computational overhead and network overhead, among other performance metrics, has been conducted. Despite challenges related to the primary design for a single power system local area network environment, the need for enhanced security protection, and potential impacts on system performance.

Overall, this paper not only makes meaningful contributions in the field of power system security, but also offers new research perspectives and solutions in the areas of cybersecurity and blockchain applications. Despite some limitations, these issues are likely to be overcome through performance optimization, resource management, large-scale testing, and simulated attacks.

Acknowledgements. This work is supported by State Grid Corporation of ChinaScience and Technology Project Funding (No: 5108-202235042A-1-1-ZN) "Research and Application of Key Technologies for Cryptographic Infrastructure in The Dispatchingand Control Field of Power Monitoring System".

References

1. Merabti, M., Kennedy, M., Hurst, W.: Critical infrastructure protection: A 21 st century challenge. 2011 International Conference on Communications and Information Technology (ICCIT), 1–6. IEEE (2011)
2. Aas, J., Barnes, R., Case, B., et al.: Let's Encrypt: an automated certificate authority to encrypt the entire web. In: Proceedings of the 2019 ACM SIGSAC Conference on Computer and Communications Security, pp. 2473–2487 (2019)
3. Soghoian, C., Stamm, S.: Certified lies: Detecting and defeating government interception attacks against SSL (short paper). In: International Conference on Financial Cryptography and Data Security, pp. 250–259. Springer Berlin Heidelberg, Berlin, Heidelberg (2011)
4. Wang, M., Qian, C., Li, X., et al.: Collaborative validation of public-key certificates for IoT by distributed caching. IEEE/ACM Trans. Networking **29**(1), 92–105 (2020)
5. Kasten, J., Wustrow, E., Halderman, J.A.: Cage: Taming certificate authorities by inferring restricted scopes. In: Financial Cryptography and Data Security: 17th International Conference, FC 2013, Okinawa, Japan, April 1-5, 2013, Revised Selected Papers 17, pp. 329-337. Springer Berlin Heidelberg (2013)
6. Dai, T., Shulman, H., Waidner, M.: Off-Path Attacks Against PKI. In: Proceedings of the 2018 ACM SIGSAC Conference on Computer and Communications Security, pp. 2213–2215 (2018)
7. Zhang, Y., Liu, B., Lu, C., et al.: Rusted Anchors: A National Client-Side View of Hidden Root CAs in the Web PKI Ecosystem. In: Proceedings of the 2021 ACM SIGSAC Conference on Computer and Communications Security, 1373–1387 (2021)
8. Brandt, M., Shulman, H., Waidner, M.: Evaluating Resilience of Domains in PKI. In: Proceedings of the 2021 ACM SIGSAC Conference on Computer and Communications Security, pp. 2444–2446 (2021)
9. Stark, E., DeBlasio, J., O'Brien, D.: Certificate transparency in google chrome: Past, present, and future. IEEE Secur. Priv. **19**(6), 112–118 (2021)
10. Laurie, B., Langley, A., Kasper, E.: RFC 6962: Certificate transparency (2013)
11. Dowling, B., Günther, F., Herath, U., et al.: Secure logging schemes and certificate transparency. In: Computer Security–ESORICS 2016: 21st European Symposium on Research in Computer Security, Heraklion, Greece, September 26-30, 2016, Proceedings, Part II 21, pp. 140–158. Springer International Publishing (2016)
12. Szydlo, M.: Merkle tree traversal in log space and time. In: Advances in Cryptology EUROCRYPT 2004: International Conference on the Theory and Applications of Cryptographic Techniques, Interlaken, Switzerland, May 2–6, 2004. Proceedings 23, pp. 541–554. Springer Berlin Heidelberg (2004)
13. Nakamoto, S.: Bitcoin: A peer-to-peer electronic cash system. Decentralized business review (2008)
14. Zheng, Z., Xie, S., Dai, H.N., et al.: Blockchain challenges and opportunities: A survey. Int. J. Web Grid Serv. **14**(4), 352–375 (2018)
15. Nofer, M., Gomber, P., Hinz, O., et al.: Blockchain. Business Information Systems Engineering **59**, 183–187 (2017)
16. Chen, T., Li, Z., Zhu, Y., et al.: Understanding ethereum via graph analysis. ACM Transactions on Internet Technology (TOIT) **20**(2), 1–32 (2020)
17. Wood, G.: Ethereum: A secure decentralised generalised transaction ledger. Ethereum Project Yellow Paper **2014**(151), 1–32 (2014)
18. Yu, J., Cheval, V., Ryan, M.: DTKI: A new formalized PKI with verifiable trusted parties. Comput. J. **59**(11), 1695–1713 (2016)

19. Li, Z., Xiong, G., Guo, L.: Unveiling SSL/TLS MITM hosts in the wild. In: 2020 IEEE 3rd International Conference on Information Systems and Computer Aided Education (ICISCAE), pp. 141–145. IEEE (2020)
20. Talamo, M., Arcieri, F., Dimitri, A., et al.: A Blockchain based PKI validation system based on rare events management. Future Internet **12**(2), 40 (2020)
21. Guan, Z., Si, G., Zhang, X., et al.: Privacy-preserving and efficient aggregation based on blockchain for power grid communications in smart communities. IEEE Communications Magazine **56**(7), 82–88 (2018)

Crowd-Sourced Testing of Electrical Substation Based on Network Emulation Platform

Sha Tao, Haiyan Wang(✉), and Weihong Han

Department of New Networks, Peng Cheng Laboratory, Shenzhen 518055, China
{taosh,wanghy01,hanwh}@pcl.ac.cn

Abstract. With the rapid evolution of networks and the continual advancement of information technology, the power industry is progressively reliant on network infrastructures. However, this amplified dependence on networks gives rise to novel cybersecurity challenges. Notably, network attacks present a significant threat to the secure operations of such enterprises, leading to economic setbacks for businesses and potential risks to human lives. This paper aims to delve into the current landscape of network security within power industry substations, both at domestic and international levels, while also exploring pertinent research in this domain. It encompasses an in-depth analysis of the feasibility surrounding the utilization of network emulation platforms for simulating substations, and undertakes a comprehensive review of the challenges that necessitate resolution during the implementation phase of such emulation platforms. Drawing from the preceding analysis, we propose a crowd-sourced testing methodology designed specifically for substation emulation systems leveraging network emulation platforms. By conducting rigorous verification tests, our proposed approach showcases significant potential in its application for comprehensive network security evaluations in critical substations. This initiative ultimately aims to fortify the security of substation networks, thereby safeguarding critical infrastructure and minimizing potential vulnerabilities.

Keywords: the power industry · Crowd-sourced testing · substation · Network emulation platform

1 Introduction

Given the relentless advancement in network technology, the power industry is increasingly leaning on intricate information technology and network infrastructures to proficiently manage and oversee its electrical systems. However, this burgeoning reliance on networks has ushered in a new era of challenges, particularly concerning network security. An illustrative instance is observed in the smart grids and automation systems, where a myriad of devices is interconnected via networks. Any vulnerability in network security holds the potential to pave the way for unauthorized access, thereby disturbing the stability and dependability of power systems. The power system, serving as one of the

This work is supported by the Major Key Project of PCL (No. PCL2022A03).

Z. Gu et al. (Eds.): NSE 2023, CCIS 2063, pp. 271–282, 2024.
https://doi.org/10.1007/978-981-97-4519-7_19

vital national infrastructures, is intricately interwoven with the fabric of modern society's production capacities and daily existence. Consequently, it stands susceptible to potential threats endangering national security and acts of terrorism. The ramifications of an unexpected power outage could indeed be incomprehensible. In recent years, a discernible surge in cyberattacks has been witnessed within the power industry, underscoring the imperative need for conducting comprehensive network security assessments, particularly within substations. These assessments serve as critical safeguards in fortifying the resilience of the power infrastructure against potential threats and vulnerabilities. The following are major network attacks in the power industry in the past decade:

In August 2023, Canada's second-largest urban power system was attacked by ransomware, which forced the government to rebuild its IT infrastructure [1]. In October 2022, the hacker group Everest invaded the South African state owned power company ESKOM. Require the company to pay 200000 [2].

In early November 2021, the Delta Montrose Power Association (DMEA) in Colorado, USA, was hit by a cyber attack that affected all support systems, payment processing tools, billing platforms, and other tools provided by the company to customers, resulting in 90% of its internal systems being paralyzed and 25 years of historical data loss [3].

In mid-November 2021, Danish wind turbine giant Vestas Wind Systems was hit by a cyber attack that damaged some of its internal IT infrastructure, resulting in unspecified data breaches and forcing the company to shut down IT systems in multiple business departments and locations to prevent the damage from spreading [3].

In late November 2021, Australian power supplier CS Energy was attacked by ransomware. The ransomware has damaged devices on its company network [4]. In February 2021, two large Brazilian power companies, Eletrobras and Copel, announced that they had suffered ransomware attacks. Both ransomware attacks resulted in operational disruptions, forcing these companies to suspend some systems [5].

In April 2020, EDP, a Portuguese multinational energy company (natural gas and electricity), was attacked by ransomware [5].

In September 2019, the intranet of India's nuclear power plant was infected with malicious software, and the backdoor Trojan infected the management network of the nuclear power plant [5].

In July 2019, a major cybersecurity accident occurred at a nuclear power plant in Ukraine [5].

From 2019 to 2020, the Venezuelan power system experienced multiple network attacks within two years, resulting in large-scale power outages [5].

In July 2019, Johannesburg Electric Power Company in South Africa was attacked by ransomware [5].

In March 2019, hackers used known vulnerabilities in Cisco's firewall to launch a denial of service (DoS) attack against the US power company [5].

In June 2018, French company Ingerop encountered network attacks, resulting in the leakage of sensitive data at the Fessenheim nuclear power plant [5].

In March 2018, Russian hackers attacked US nuclear power plants and water supply facilities [5].

From the above-mentioned attack events, it can be observed that the major network attacks that have occurred in the power industry in recent years mainly include but are not limited to the following:

1. Distributed Denial-of-Service (DDoS) Attacks: Intended to render the power system inaccessible by inundating it with a deluge of network requests, thereby disrupting services.
2. Malware Infections and Viral Intrusions: Deliberate introduction of malicious software or viruses into the power system via network attacks, aimed at disrupting normal operations or exfiltrating sensitive information.
3. Internal Threats and Data Breaches: Occurring due to employee misconduct or malicious actions, leading to data breaches or system compromises.
4. Phishing and Social Engineering: Targeting power system personnel through deceptive tactics to obtain login credentials or sensitive information, facilitating unauthorized access.
5. Zero-Day Exploits: Leveraging undisclosed system vulnerabilities to infiltrate power systems, which are yet to be rectified by the manufacturers.

Therefore, we can see that the network attacks suffered by the substation not only include network attacks on the hardware equipment of the substation but also extend to the network itself and the management network of the office area where it is connected. Therefore, placing network attack protection solely on the substation itself cannot meet the increasingly stringent requirements of network security protection.

2 Related Work

In recent years, a series of measures have been taken to deal with network attacks on substations, including real-time network monitoring, hardened network defense, access control policy, data backup and disaster recovery, network isolation, and incident response service. The state and various organizations have introduced a series of regulations. The State Economic and Trade Commission issued the "Regulations on Security Protection of Computer Monitoring Systems and Dispatching Data Networks in Power Grids and Power Plants" [6] on May 8, 2002. On February 1, 2005, the Chairman's Office Meeting of the State Electricity Regulatory Commission passed the "Regulations on Security Protection of Electric Power Secondary Systems" [7]. The "Guidelines for Information Security Protection of Industrial Control Systems" [8] by the Ministry of Industry and Information Technology clearly states that the network security protection system of power monitoring systems must have a dynamic real-time perception function. With the help of dynamic real-time perception function, the ability to upgrade network security can be improved. According to changes in the network environment, the security protection system can be upgraded to the highest level of management. On May 1, 2018, the country implemented the GB/T 34871-2017 Intelligent Substation Relay Protection Inspection and Testing Specification [9], proposing specifications for intelligent substation relay protection detection and testing. The standard specifies the conditions, items, methods, and judgment methods for the inspection and testing of relay protection devices and related equipment (analog input merging units, intelligent terminals) that undergo

GOOSE tripping (including direct and network tripping) using analog sampling, digital direct sampling, and digital network sampling methods in intelligent substations. The "Secondary System Security Protection Plan for Substations" [10], "Implementation Guidelines for Security Level Protection of Power Information Systems" [9], "Basic Requirements for Network Security Level Protection of Information Security Technology" [11], and "Evaluation Requirements for Network Security Level Protection of Information Security Technology" [12] all provide important guidance and specifications for the security protection of substations.

The 57th Technical Committee of the International Electrotechnical Commission has formulated the IEC62351 standard. The IEC 62351 standard is very important among the current reference standards for substation safety protection both domestically and internationally. In 2009, the United States issued the "Strategy for Protecting Industrial Control Systems" [13], covering the safety of industrial control systems in 14 industries such as transportation and energy. In 2015, the National Institute of Standards and Technology of the United States released the second edition of the "Industrial Control System Security Guidelines" [14], which discussed how to adjust traditional information security protection technologies to meet the reliability and security requirements of industrial control systems. At the same time, the security system architecture, risk management, and implementation plans were updated. In May 2017, US President Trump signed the Executive Order "Enhancing Cybersecurity in Federal Government Networks and Critical Infrastructure", which stipulated cybersecurity enhancement measures in three areas: the federal government, critical infrastructure, and the state.

Scholars and engineers have developed their own testing and protection methods for the safety of substations from various entry points. Li Jin et al. proposed using a mobile computer device to meet all the operation and maintenance work in the substation for the connection of external computers or USB drives during the production and maintenance process of the substation [15]; Shi Chang et al. analyzed various aspects of network security protection in power monitoring systems and proposed the establishment of a comprehensive governance system for network security in power monitoring systems [16]; You Xiaoming and others have developed a substation internal network security monitoring device to achieve real-time monitoring of the operation status of the main equipment, network equipment, and security protection equipment of the power secondary system [17]; Chai Jiwen et al. used attack testing based on TCP/IP and IEC 61850 protocols to detect system vulnerabilities [18]; Noce et al. optimized using the GEESE method to create various network attacks to verify system vulnerabilities [19]; Igor Nai Fovino et al. built a platform that integrates many network attack tools to detect the security of SCADA systems [20]; Elbez et al. proposed a testing platform based on IEC 61850 [21]; Liu Shanmei et al. conducted safety assessments on equipment and operation and maintenance management at various levels of the substation [22]; Cao Yijia et al. established a ranking of the impact of various security threats on the information security of substation communication systems [23].

The testing and protective measures mentioned above primarily focus on specific aspects, with a heavy emphasis on substation terminal equipment and substation monitoring systems. However, these two components are only a part of the entire substation system, not the entirety of it. Recent security incidents have shown that many attacks originate from the office areas of employees and the management information systems, and even extend to the entire utility company. Currently, the information network management aspects of substation systems lack comprehensive network security testing, with most existing security testing being conducted based on isolated industrial control environments. This leads to the following issues:

1. The testing content focuses on the functions of the substation system itself, and there is insufficient testing of the network security where the substation is located;
2. The testing quality is limited to the on-site testing engineers(white hats), more experienced testing engineers should be involved;
3. The testing scope is limited to the industrial control network, which is only part of the entire network, it should be extended to the employee office area. Current test scope didn't consider the overall network security.
4. Base on the real substation system, testing engineers (white hats) can't perform destructive testing to verify the subsequent of a certain operation.

To address the various issues mentioned above, while considering factors such as the security, cost-effectiveness, repeatability, flexibility in experimental design, and rapid iteration within the entire substation system, a prerequisite for a systematic testing method of the entire substation system is to emulate it. Simulating the substation system offers several advantages; it allows testing within a controlled environment without impacting the actual production system, reducing potential risks and security vulnerabilities. It also lowers testing costs and conserves resources, enables easy replication of the same testing conditions for continuous testing and improvement, operates within a virtual environment without the need for physical equipment, provides an isolated testing environment without affecting the actual production system, and can be scaled to test system performance and scalability in large-scale deployments. In summary, simulating the substation system for testing purposes is a flexible, secure, and cost-effective approach that aids in identifying and resolving system issues while enhancing system performance and reliability.

In this paper, based on the environment and resources provided by the Pengcheng Cyber Range, we propose a novel testing method for substation systems. Our contributions can be summarized as follows:

1. we establish a small-scale network emulation platform that aligns with the power system environment.
2. We place the entire power company's network on this emulation platform to monitor and safeguard the system boundaries of the entire power system.
3. We propose the crowd-sourced testing method for substation systems based on a network emulation platform. The experimental results indicate that our method is flexible, secure, and universal.

3 Emulate Substation in Network Emulation Platform

In the "Network Visualization Editing and Automated Deployment System Based on OpenStack" [24], a network emulation system is proposed. The system's front-end is used to provide users with an interface to quickly create network topology maps. The system's back-end imports information parsed from the network topology map and creates corresponding virtual network layers and emulation network layers on the OpenStack cloud platform. It then connects the physical network layer and emulation network layer with the virtual network layer, respectively. Using this system, network emulation of the monitoring area and employee office area within the entire substation system can be done quickly, flexibly, and at a low cost. Inspired by these ideas, in this paper, based on the environment and resources provided by the Pengcheng Cyber Range, we have established a small-scale network emulation platform that aligns with the power system environment. We place the entire power company's network on this emulation platform to monitor and safeguard the system boundaries of the entire power system.

Before conducting network emulation, we first partition the entire system where the substation is located. It can be roughly divided into several parts as shown in Fig. 1: the primary system of the substation (generator, transformer, motor, switch (circuit breaker), isolation switch, etc.); the Substation secondary system (various automation systems that monitor and control the production and operation process of the power system); DMZ area of the substation (official website server of the power company, etc.); Substation employee office area (including the management information system of the substation, employee office computers, etc.).

For the DMZ area and employee office area of the substation, the respective network topology and configuration information, operating system, and installation and deployment software of each node can be sorted out. The OpenStack based network visualization editing and automation deployment system can be used for flexible deployment, emulating the configuration of physical devices and installing and deploying corresponding software in the virtual machines.

For the secondary system of the substation, it can be divided into two parts, one is the physical control equipment, the other is the control system software.

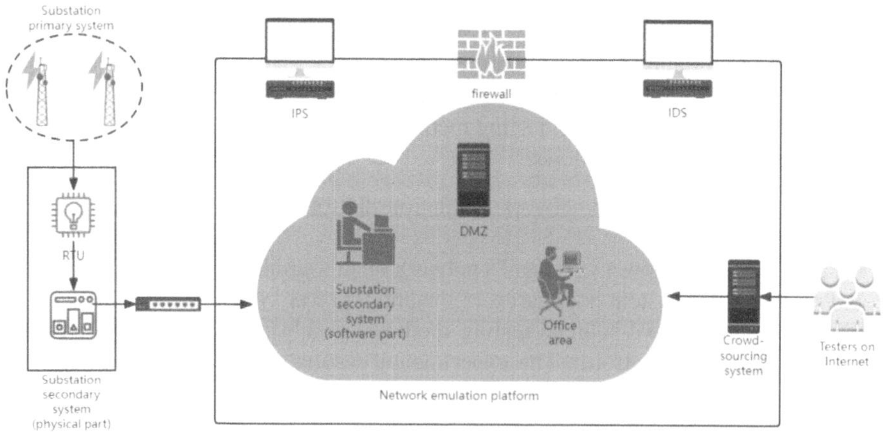

Fig. 1. Substation emulation in Network Emulation Platform.

As for the physical control equipment, currently it can't be comprehensive virtualized. In the invention – "Virtual-Real Interconnection Method, Device, Equipment, and Storage Medium for Network Range Platform" [25], a method, device, equipment, and storage medium for the virtual-real interconnection of a network range platform are disclosed. The method includes: obtaining request information from a physical machine or a virtual machine; based on the request information, respectively querying target physical machines and target virtual machines to be interconnected from the physical machine and the virtual machine; establishing a connection between the target physical machines and the target virtual machines, where each physical machine is pre-connected to any virtual machine, enabling the interconnection of the target physical machines and the target virtual machines. This approach facilitates the flexibility of physical machine-virtual machine interconnection by directly updating the connected physical machines and virtual machines based on request information. This virtual-real interconnection method can be employed to integrate non-virtualizable custom devices into a network visualization editing and automatically deployed system based on OpenStack. This allows the physical control system to interconnect with the virtualized DMZ zone and substation employee office network on OpenStack. In remote scenarios, remote virtual-real interconnection can be achieved through VPN technology. Besides the physical control system equipment, for the control system software, we can create image and install them in those virtual machines in OpenStack-based network visualization editing and automation deployment system.

For the primary system of the substation, we can substitute a simulated screen for the primary system substation, or alternatively, achieve integration with our OpenStack virtual environment through virtual-real interconnection, allowing for virtual representation of the substation's primary system.

At this point, a complete power company's entire network system including its physical equipment has been placed on the virtual platform of OpenStack, with low-cost and efficient methods, providing a foundation for our next crowd-sourced testing.

In this simulation environment, integration with a crowd-sourced testing system is established. The crowd-sourced testing system provides a means for white hats to access the simulation platform targets, specifying testing tasks, duration, and facilitating the submission of test results and reports. The crowd-sourced testing system is open to the internet, designed to attract white hats by offering rewards and through invitations to experienced individuals for participation in penetration testing. It grants control over the testing time slots and access to the simulation environment, ensuring the overall test is under control.

On this virtual platform within the power company, we can deploy firewalls at the network edge to isolate the simulation system from the internet, ensuring controlled behavior of attack scenarios within the simulation environment. Additionally, we can deploy security devices such as IPS, IDS, and network probes on both virtual and physical switches to enable real-time monitoring of the network situation within the simulation platform. This setup provides a secure testing environment for subsequent crowdsourced testing activities.

4 Crowd-Sourced Testing of Electrical Substation Based on Network Emulation Platform

With the rise and development of the Crowd-soured testing, security testing has also entered this field. For small and medium-sized enterprises, using crowd-sourcing for security testing can save costs and improve efficiency compared to building a security team or purchasing security services on their own. At present, major security companies have developed their own crowd-sourced testing platforms, such as Uyun crowd-sourced testing, Baidu's crowd-sourced testing platform, Qi'anxin's Patian platform, Tencent Security Emergency Response Center (TSRC), and so on.

The main idea of the crowd-sourced testing method for substation emulation systems based on network emulation platforms is to install the crowd testing system to our network emulation platform. As an external portal of our network emulation platform, White hats connect to the crowd-sourced testing platform through the network to conduct penetration testing on the power company network in the network emulation platform.

In the previous section, we have deployed a virtualized power company network. We can publish this network on the crowd-sourced testing system and provide access to the power company network, such as the official website, or to the parts that we want to expose for white hats to penetrating. In order to attract more experienced white hats, we can invite students majoring in network security from well-known universities or network security penetration testing engineers from various security vendors to conduct comprehensive penetrating on this substation emulation system.

The crowd-sourcing testing process is mainly divided into three parts as shown in Fig. 2, including before crowd-sourced testing, during crowd-sourced testing, and after crowd-sourced testing.

1) **before crowd-sourced testing**

Step 1: Build a virtual network topology based on the real power company network in the network emulation platform to meet testing needs. The network topology includes the DMZ area, Substation secondary system and employee office area of the power company, the scale of the employee office area can be expanded using virtual nodes according to testing needs which is flexible and cost-effective.

Step 2: In order to emulate the PCs in the employee office network, DMZ and substation network, these desktop computers with specific software can be made into images and deployed to network emulation platforms, or corresponding software can be installed in virtual machines manually; In addition, the virtualization of physical objects such as switches, routers, and firewalls can be achieved in the network emulation platform resource database, the network emulation platform can conveniently and quickly build a topology structure of the real network environment by using the virtual switches, routers and firewalls. Besides, it is flexible to expand the scale by copying virtual machines or import topology template.

Step 3: Connect the physical equipments to virtual network by using Virtual Real Interconnection Method, configure the network according to the real network topology.

Step 4: Deploy network topology to make the virtual machines, routes and switches up and work. Make sure everything is connected in the network topology.

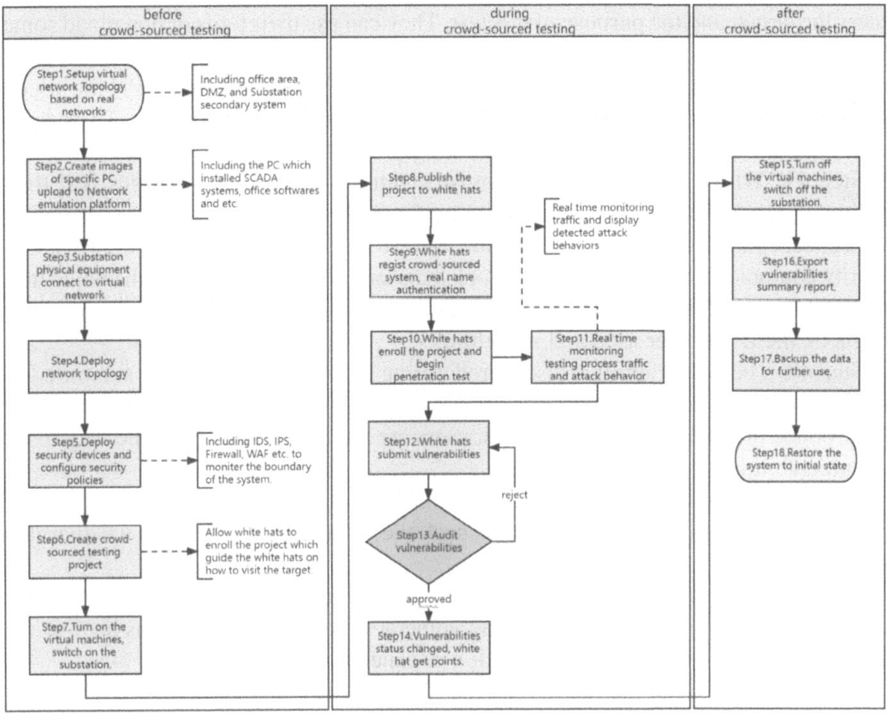

Fig. 2. The crowd-sourcing testing process.

Step 5: Deploy security devices on the network emulation platform, configure security policies, and monitor the traffic and logs of the network emulation platform. Such as firewall, WAF, IDS, IPS etc. They can also guarantee the penetrating testing is under control.

Step 6: After successful deployment, create a new project in the crowd-sourced testing system which contain the way to access the target, testing purpose, test duration, and how to log vulnerabilities. It allows the white hats to enroll the project, to get real name authentication, and enroll each project.

Step 7: Turn on the virtual machines, and switch on the substation physical equipment to make the substation emulation system ready to test. Now in the emulated power company, everything is up and working. You can adjust the corresponding control system software which is installed in virtual machine to connect the physical equipment, make sure the configuration is the same as the real one, and works well.

2) **during crowd-sourced testing**

Step 8: Publish crowd-sourced project on the internet, using rewards or bonus to attract more white hats, or invite the white hats from well-known security vendors to register in crowd-sourced testing system.

Step 9: White hats register in the crowd-sourced testing system, and after passing real name authentication, enroll the project. After that, white hats can found the way to access the target, and the purpose of this test. They can use their tools or download some tools to test.

Step 10: White hats begin their penetration test. They can download VPN file which allow them to remotely connect to the substation emulation network to test. There is no regional restrictions, and more flexible.

Step 11: While penetration test, Network emulation platform provides real time monitoring way to keep an eye on the attach traffic and behavior, to ensure the penetration test is under control. Once the white hats are trying to attack the emulation platform itself, security equipment, such as WAF will block the user's IP. Another case, if the white hats are trying to using VPN to attack some website on internet, we can detected the behavior and block the IP the user used and warn him.

Step 12: The vulnerabilities discovered during the testing process are submitted to the crowd-sourced testing system. User can also upload the write-up which explain how to reproduce the vulnerabilities. Besides, users are suggested to provide the way to fix the vulnerabilities.

Step 13: Vulnerability reviewers log in to the crowd-sourced testing system to verify and audit vulnerabilities. Reviewers can check the write-ups to determine whether it is a real vulnerability.

Step 14: Once the vulnerability reviewers needs more information or find the vulnerability is not a real one, they reject the vulnerabilities request, it returns back to the user who submit it to add more information or cancel.

Step 15: Once the vulnerability reviewers agree it is a real vulnerability.They will approved the request, users who submit it will get extra points or rewards. Usually, the vulnerabilities are separated into 3 degree, High, Medium, and low. Once high degree vulnerability is found, more points will be grant.

3) **after crowd-sourced testing**

Step 15: When the testing meets the end time, the project published in the crowd-sourced system is ended, the substation emulation system is shutdown by turn off the virtual machines and switch off the substation. White hats can no longer access the emulation power station and can't submit vulnerabilities.

Step 16: All approved vulnerability requests are gathered in Crowd-sourced testing system are ready to be exported. According to the vulnerabilities and the fix that white hats provide, we can make the power company network and system more secure. Besides, for the points the white hats get, we can grant rewards or dispatch bonus.

Step 17: Network emulation platform provides automatic data backup functionality. During the penetration test, white hats use all kinds of ways to penetrate the network and system which generated many kinds of testing data, such as traffic, configuration data, topology structure and etc. For those data, we can stored in database for further analysis. With the testing and traffic data, network emulation platform can also replay the test for further use.

Step 18: Network emulation platform provide automatic recover functionality. During the penetration test, the network and system itself might be destroyed. After testing, Network emulation platform need to restore the system and network to initial status.

5 Conclusion

In this paper, based on the environment and resources provided by the Pengcheng Cyber Range, we established a small-scale network emulation platform that aligns with the power system environment. We placed the entire power company's network on this emulation platform to monitor and safeguard the system boundaries of the entire power system. Then we proposed the crowd-sourced testing method for substation systems based on a network emulation platform. The experimental results indicated that our method is flexible, secure, and universal. In summary, the method we have proposed has the following advantages:

1) The substation system, including physical equipment and the network it is located, is emulated according to the real one to the virtual environment of the network emulation platform. Virtual environments have flexibility, scalability and low cost. It realized the security testing that couldn't performed in the product environment.
2) Integrating crowd-sourced testing systems into network emulation platforms, expands the scope of white hats and enables remote access testing.
3) The substation system is emulated in the Network emulation platform. The security equipment of the Network emulation platform was used to monitor and alarm the testing status, and the behavior of the attached behaviors was also effectively monitored.
4) Due to the security and effective isolation between the network emulation platform and the internet, it provides effective protection for the substation emulation system, ensuring the testing process is reliable and has no effect on the internet.

The network emulation platform enables low-cost, efficient, and high-precision emulation of substation systems, allowing us to emulate the power company's entire network where the substation system is located, effectively ensuring the safe operation of the substation network. By extension, we can connect devices from various industries to network emulation platforms through virtual and real interconnection techniques for network security crowd-sourced testing. With sufficient hardware resources, theoretically, all enterprise networks, even cities and country network, can be emulated in network emulation platforms to replicate real-world networks for security testing and drills. Its significance is even more profound.

References

1. Qi An Xin Technology Group Inc: Power system in Canada's second largest city under ransomme attack: forced to rebuild IT infrastructure. An Quan Nei Can, Aug (2023). https://www.secrss.com/articles/58380
2. 51CTO: Annual Report: 2022 Global Key Infrastructure Cybersecurity Events Overview, Dec (2022). https://www.51cto.com/article/742407.html
3. China Industry Research: Network security issues in the power industry are receiving attention, and frequent network attacks occur in the power industry, Mar (2022). https://www.chinairn.com/news/20220304/113317927.shtml
4. Sohu: The latest global cyber attack event highlights the energy industry Apr (2022). http://news.sohu.com/a/534539881_120205504

5. Salaxy security: Top Ten Cybersecurity Attacks in the Global Power Industry, July (2020). https://zhuanlan.zhihu.com/p/165567138
6. State Economic and Trade Commission of the People's Republic of China, Order No. 30 of the State Economic and Trade Commission of the People's Republic of China (2002)
7. State Electricity Regulatory Commission, Order No. 5 of the State Electricity Regulatory Commission, 2005
8. Ministry of Industry and Information Technology (2016)
9. State Administration for Market Regulation, National Standardization Administration (2018)
10. State Electricity Regulatory Commission (2006)
11. State Administration for Market Regulation, National Standardization Administration, Information Security Technology (2019)
12. State Administration for Market Regulation, National Standardization Administration (2019)
13. Cybersecurity and Infrastructure Security Agency (CISA) (2009)
14. National Institute of Standards and Technology (NIST)
15. Li, J., Zhang, X., Xie, X., Li, H.: J. Electr. Eng. **15**(4) (2020)
16. Shi, C., Tang, X.: Rural Electrification, no. 7 (2018)
17. You, X., Tang, Z., Hu, S., Lin, Q., Cao, X.: Electrical Technology, no. 12 (2018)
18. Chai, J.W., Liu, S.M.: IEEE PES Asia-Pacific Power and Energy Engineering Conference(APPEEC), pp. 1368–1373. IEEE, Xi'an, China (2016)
19. Noce, J., Lopes ,Y., Fernandes, N.C., et al.: IEEE 26th International Symposium on Industrial Electronics (ISIE), pp. 111–116. IEEE, Edinburgh, UK (2017)
20. Fovino, I.N., Masera, M., Guidi, L., Carpi, G.: IEEE, Poland (2010)
21. Elbez, G., Kellerh, B., Hagenmeyer, V.: IEEE International Conference on Communications, Control, and Computing Technologies for Smart Grids (SmartGridComm), pp. 1–6. IEEE, Aalborg, Denmark (2018)
22. Liu, S., Wang, S., Chai, J., et al.: J Chongqing Univ. **40**(7), 52–62 (2017)
23. Cao, Y., Yao, H., Huang, X., et al.: Electric Power Autom. Equip. **31**(6), 1–5 (2011)
24. Zhang, W., et al.: (2022)
25. Xiang ,W., Wang, H., Yang, S., Han, W., Qiu, J.: (2023)

Utilizing Digital Twin Simulation Technologies to Build ICS Cyber Ranges

Haifeng Zhou[1,2], Mohan Li[1,2], and Yanbin Sun[1,2](✉)

[1] Cyberspace Institute of Advanced Technology, GuangzhouUniversity, Guangzhou, China
sunyanbin@gzhu.edu.cn

[2] Guangdong Key Laboratory of Industrial Control System Security, Guangzhou, China

Abstract. With the integration of IT and OT, cyberattacks against ICS have been on the rise recently, and more and more advanced methods are resorted to by attackers to break down the physical system. To do further research in cybersecurity on ICS, building a cyber range is an economical, practical, and necessary approach considering the requirements of availability for the onsite production devices. Using all physical devices to build a cyber range is unrealistic and inflexible due to the high cost while pure software simulation lacks authenticity. Meanwhile, digital twin, a prevalent concept in recent years, shows its potential in ICS cyber ranges for its authenticity and flexibility due to the improvements in modeling methods and network communication. Therefore, the framework, implementation methods, and applications of using digital twin technology to build an ICS cyber range are discussed in this paper.

Keywords: digital twin · cyber range · ICS cyber security · machine learning

1 Introduction

Industry Control System *(ICS)* is an electronic control system and associated instrumentation used for industrial process control. ICSs can be implemented in the form of supervisory control and data acquisition (SCADA) or distributed control systems (DCS) with programmable logic controllers (PLC). It plays a so critical role in modern industries that tiny exceptions may lead to unbearable loss. With the integration of IT (Information Technology) and OT (Operation Technology), cyberattacks against ICS are more and more common and exposed, and more and more advanced methods are resorted to by attackers to break down the physical system. The well-known Stuxnet [1] destroyed Iranian physical nuclear facilities from workstations by changing the controlling parameters and misleading the operators to believe the onsite status was normal.

Ever since then, cyberattacks on ICS have been discovered in many types of industry systems, and a trend that the attacks were designed for a certain system and more and more sophisticated can be found. Cyberattacks on critical infrastructure can be catastrophic and fatal to a country, therefore, ICS security cannot be paid more attention for the sake of stability and safety. However, due to the requirements of production availability, which means the onsite devices should be online under most circumstances, analysis

Z. Gu et al. (Eds.): NSE 2023, CCIS 2063, pp. 283–294, 2024.
https://doi.org/10.1007/978-981-97-4519-7_20

of ICS cyberattacks and their corresponding policies cannot be performed directly on the real devices. Thus, building a testbed, or a cyber range, is a practical and necessary solution.

A **testbed** is built with physical devices to replicate the real devices and operation environment. To address the cybersecurity issues in energy delivery systems, the U.S. Department of Energy implemented a National SCADA Test Bed (NSTB) [2] based on all physical equipment. SWaT [3] is a physical testbed that simulates 6 sub-processes in the water treatment process based on physical tanks and pipes. Each sub-process has a physical PLC for process control. While these testbeds are of high fidelity, they are costly and inflexible for rapid deployment for educational use cases compared with cyber ranges. Therefore, building a physical testbed for operations in a cyber range is out of consideration.

A **cyber range** is a virtual environment that enables organizations to simulate cyber combat training, system/network development, testing, and benchmarking. And some attempts [4–6] have been made to build virtual ICS cyber ranges. MiniCPS [4] was developed purely based on simulations where the PLCs were emulated by scripts and physical layer interaction is based on file I/O. In [5], the cyber range is completed modularly by multiple simulation software. ICSrange [6] simulates a water process with emulated PLCs and real communication was established. Compared with traditional cyber ranges, ICS cyber ranges concentrate more on the OT domain instead of the IT domain where network technologies are relatively mature. Conventionally, the process in cyber range can be implemented by pure professional software simulation which holds certain thresholds for cyber security researchers who are not in this research field. In addition to this, although the purchase cost of related software is less than that of physical implementation, it is still not low, and the implementation of specific processes is often difficult to flexibly deploy according to the actual system.

These two approaches (physical and virtual) to build an ICS cyber range suffer from a tradeoff between authenticity and flexibility. Therefore, Building an ICS cyber range with both authenticity and flexibility is of research value. Meanwhile, a digital twin is a digital representation of the real corresponding object and serves as a data source ideally when the real object is unavailable or cannot be operated. Considering this feature, we believe that digital twin technologies can be utilized to build an ICS cyber range with both flexibility and authenticity.

The **digital twin** concept has been widely discussed in recent years. The twin concept of digital twin originates from NASA which built an identical aircraft on the ground in a certain environment to make decisions for the aircraft in outer space. And digital twin comes from Grieves in a PLM class in 2007 [7]. It was proposed to do car crash analyses in different virtual scenes in a virtual space.

Over the development of past years, digital twins mainly contributed to industries in predictive maintenance, virtual commissioning, and parallel scheduling, and in some research, it has been brought to the real factory to achieve production optimization. Based on historical data, AI and physical analysis methods can be used to build digital twin entities of field devices, and then constitute a virtual cyber range ideally. This can not only retain the flexibility and low cost of software simulation, but high-quality physical models and AI models can ensure overall authenticity to a certain extent. Using related technologies and concepts of digital twins is a method that is very consistent with the original intention of ICS cyber range, and this approach has been explored by other researchers [8, 9]. In [8], the benefits of digital twin for cybersecurity are investigated and concluded, a digital twin based-cyber range is used for Security Operations Centers (SOC) analysis. [9] uses Automation ML and expert knowledge to build a digital twin virtual environment and it can be used in different cybersecurity scenarios.

This paper proposes a framework for ICS cyber ranges based on digital twin simulation and models digital twin entities from different aspects according to a new connotation. The remainder of this article is organized as follows: Sect. 2 proposes the connotation of digital twins in the ICS cyber range; Sect. 3 explains the implementation methods of each sub-model of digital twins based on a scene; Potential applications of the ICS cyber range based on digital twin simulation is illustrated in Sect. 4.

Note: The digital twin concept, from the author's point of view, does not represent a technology, instead, it is a product of a combination of interdisciplinary technologies, like physics, graphics, AI, automation, mechanical engineering, and so on. In a virtual environment, there are a lot of digital twin entities (twin models) corresponding to the physical entities in the real world, and their modeling relies on these technologies.

2 The Connotation of Digital Twin in ICS Cyber Ranges

Based on the traditional IT cyber range, which is realized based on technologies like SDN and virtualization, the digital twin-based ICS cyber range focuses mainly on simulation in OT domain. There are two concerns to implement a digital twin entity in a cyber range, the first is how to build a digital twin entity itself and then is how to connect the digital twin entity with the real devices. Different connotations of digital twin have been proposed in different research and how to model a digital twin entity varies. In this paper, a digital twin entity can be modeled from five perspectives, i.e., geometry, behavior, logic, data, and service mode [10], as shown in Fig. 1. What should be mentioned is that while a digital entity can be modeled from these perspectives, all the perspectives are not necessary when it comes to implementation.

2.1 Models in One Digital Twin Entity

A geometry sub-model presents the object's appearance obviously, and the key point is to keep high fidelity to the physical part as operators observe them and give the first impression of the real devices. The geometry model provides a basis for other perspectives like collision detection in behavior models, and in a cyber range, it is a graphic instinct presentation which operators interact with. Feedback should be considered if specific data breaks through its threshold during modeling. To conclude, the more accurate this geometry model is, the more reliable the virtual data will be.

A behavior sub-model is the interaction between virtual models such as how and when a robot arm is supposed to rotate, grab, and move to a certain area. This sub-model represents the real-world logic, conforming with the physical laws, therefore, This can

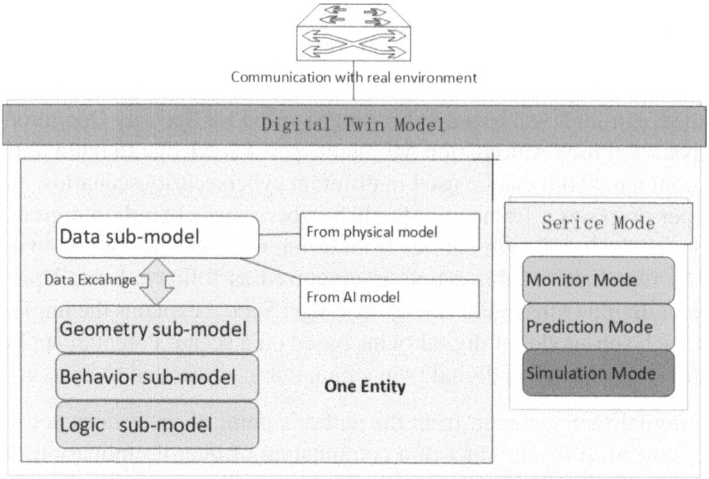

Fig. 1. The connotation of single digital twin entity

be implemented with code and script. In a virtual environment, a behavior model can change the surrounding parameters as it might interact with the environment.

A logic sub-model means that the digital twin should conform to the actual process logic and physical laws. For example, in an industrial process, the temperature rise must be due to the existence of a heat source. Therefore, when there is no heat source in the scene, the temperature change logic should be limited so it cannot rise. The implementation of the logic model depends on the natural laws involved in the overall system, which is often difficult, and the workload of implementing it from scratch is also very huge. Therefore, existing external tools can be used and based on the process-specific logic model can be realized and added.

A data sub-model refers to the data source of the digital twin. At the same time, the scientificity and rationality of the data need to be considered. The data source of the digital twin model is calculated from real-time data in the virtual environment and physical equations obtained after physical analysis. It can also be calculated based on virtual real-time data and AI models learned through artificial intelligence algorithms. The two correspond to the physical model and the AI model.

The **physical model** is a data source of the data model. Physical analysis of a system is required to get a corresponding sophisticated physical equation, which means expert knowledge in a certain field is inevitable. With an accurate physical model, the specific value of a system applied with external or internal effects can be calculated ideally at any time and under any situation, which also means it is almost impossible to establish such an equation for a large, complex system. As a replacement, simulating a physical process with approximation and calibration to get the most fitting equation with AI is also an alternative.

The **AI model** refers to models that are learned and trained through AI algorithms after collecting a large amount of historical data from the system through various sensors, analyzing the process, cleaning, selecting, and analyzing the data. AI algorithms can be machine learning algorithms or deep learning algorithms depending on the specific case and the complexity of the problem. The advantage of the AI model is that the generation of the model does not rely on the expertise of cyber security researchers in the physical field, and the model can be generated quickly. The disadvantage is that the accuracy of the model relies heavily on the diversity and accuracy of historical data. When historical data is missing in a certain situation, the reliability of the model will be greatly reduced compared to the physical model mentioned above.

Service Mode. The last one is from the service mode perspective(or service submodel [10]). Service mode is the usage purpose of the digital twin. Three service modes, which are monitor mode, prediction mode, and simulation mode, are proposed in this paper and which service mode to use depends on the operator's needs. **In monitor mode**, the digital twin is used as an HMI in an ICS, it collects data from the real world, monitors the whole process, and the entities' behavior changes according to the real-time data. Therefore, the behavior model is the most important sub-model in monitor mode, the behavior should be modeled to reflect the change of real-time data as a real device. **In prediction mode**, digital twins use collected history data in monitor mode and the digital model is used to predict future data with real-time communication with physical devices. **In simulation mode**, the reliability of the data model is fully evaluated. In simulation mode, there is no need for real-time data calibration as in prediction mode. That is to say, in simulation mode, digital twins can be in a cyber range to evaluate the harm of cyber attacks and the effectiveness of corresponding strategies. In prediction mode, models may show a better performance than the models in simulation mode as the input of real-time data.

2.2 Data Exchange

As for the data exchange inside the twin entity and between the entity and the real device, it needs to be realized according to the actual method of building the model. Different service modes and different implementation methods have different requirements for communication.

The purpose of external data communication is to obtain data and parameters corresponding to sensors, actuators, etc., so network communication with industrial field devices needs to be implemented in a virtual environment. This communication is based on industrial communication network protocols, such as Modbus, CNIP, OPC, and other common network protocols, as well as each vendor's internal proprietary protocols. Entities write the generated sensor data to the corresponding memory location inside the control device and accept the driver's device value from the control device to achieve data interaction.

Due to the property of the distribution of the data model, data communication is required within the twins. The data of the data model may come from models in different environments. Using a distributed approach can flexibly implement model iteration, thereby improving the reliability of the model. Of course, the data of the data model can also be synchronized in the same environment and reduce delays caused by communication, but this increases the coupling between models and the complexity of model replacement.

Fig. 2. The scenes of the simulating system, the metering station temperature control scene, and the corresponding real geometry model

3 Implementation

3.1 Implementation Methods

Take an oil metering station as an **example**, where temperature control logic is applied to keep the temperature of output oil within a certain range using filed devices in Fig. 2. To simulate a real system, a cooling fan running at a fixed rotation speed and two heat generators are used to control the temperature, and the physical entity is shown in Fig. 3.a. The objective of the digital twin in this scene is to predict the temperature under certain control parameters with realtime data, and based on this, if the digital model can run and generate data like the real one without exterior intervention or communication, then the digital twin can run in simulation mode and this scene is reproduced virtually and can be applied to cyber range consequently. Below is how to model such a digital twin entity from the perspectives mentioned above.

The Service Perspective. Service determines demand, so if the digital twin in the cyber range is only used in a scenario where real-time data is available and accuracy is not of the essence while the data trend is like in a honey pot, the digital model only needs to be generated in prediction degree. However, if it is in scenarios where real-time data is not available like emergency response validation, the digital model should include the simulation mode. In the metering station, to replace and reproduce the scene in a virtual environment, implementing simulation mode is necessary.

The Geometry Sub-model. The corresponding 3D models are used to build a geometry model as shown in the figure. The models can be built according to the real devices on CAD software in detail if the geometry property like inflation or deformation affects the process and cannot be ignored. In this scene, these 3D models just give an instinctive visual reflection on the process, the 3D model is downloaded from online sites to reduce the time consumption. The geometry model in this scene is shown in Fig. 3.b.

a. the temperature control scene b. the digital geometry model

Fig. 3. The temperature control scene and its geometry sub-model

The Behavior Sub-model. Based on reality, the fan should be spinning during cooling and stop when heating, while the opposite is the heat generators. Hence, the behavior model determines the reaction of the devices in response to the signal received from other models or the inner sub-models of the digital entity itself. This is usually implemented with the support of a virtual environment, Unity3D in this scene. In unity3D, the behavior model can be implemented in a state machine supported by component *animator* to simulate the changing rules of the model's behaviors, the corresponding actions can be implemented in geometric modeling.

The Logic Sub-model. The logic model represents the rules in the process, including operating rules and physical laws. In unity3D, physical laws can be implemented in conjunction with the physics engine and other components provided by Unity itself. For the operating rules in this scenario, after analysis, the operating logic mainly means that when the heater is started, the value of the temperature sensor will increase. After the radiator is turned on, the temperature rise will slow down; and when the temperature sensor is turned off, the temperature will drop rapidly. The state changes caused by the implementation of this running rule need to be controlled through scripts. The states in these rules are implemented in the behavioral model.

The Data Sub-model. All the data of the system is sensed and captured on Modbus/TCP once every second, and stored in a database. The history data needs to be screened to delete abnormal observations. The next time temperature minus the current one is the delta temperature observed and is our prediction target. The analysis of the correlation of the features should be performed as well to reduce the data dimensions.

The Physical Model. Limited by the author's professional knowledge, the derivation of physical equations, that is, the generation of physical models, may not be accurate, so this section only represents the process and method of building physical models. In this scene, a physical model is obtained based on heat transfer and thermodynamic analysis. In Eq. (1), Q_i is the heat received by the sensor, Q_o is the heat transferred out from the sensor, c is the Specific Heat Capacity (SHC), m is the mass of the sensed object, and Δ_t is the temperature changed during the interval, and this represents the heating process of the system.

$$Q_i - Q_o = cm\Delta T \tag{1}$$

In the stable condition of the heating process, $Q_i = Q_o$, $\Delta_t = 0$, the heat received equals the heat transferred out, according to heat transfer, the heat transferred out consists of heat conduction, heat convection, and heat radiation, as expressed in Eqs. (2), (3), (4) respectively:

$$\Phi_{cd} = \frac{\lambda(T_w - T_c)}{x} \tag{2}$$

$$\Phi_{cv} = h(T_w - T_c) \tag{3}$$

$$\Phi_r = \varepsilon A\sigma(T_w^4 - T_c^4) \tag{4}$$

$$Q_o = \Phi_{cd} + \Phi_{cv} + \Phi_r \tag{5}$$

where T_w, T_c are the stable temperature of the entity or the sensor and the surrounding temperature which is a consonant in this scene, λ, x, h, ε, A are thermal conductivity, the distance heat conducted in, convective heat transfer coefficient, emissivity, and radiant surface area, σ is the Stefan-Boltzmann constant with a constant value of 5.669×10^{-8} W/m^2 ·K4. Suppose that Q_i is a fixed value under a specific control parameter and equals to Q_o in stable conditions, and the thermophysical properties of the system are static, then the data change during the heating process under this control parameter can be obtained based on Eq. (1), (2), (3), (4), (5):

$$\Delta T = \frac{\frac{\lambda}{x}(T_w - T_w') + h(T_w - T_w') + \varepsilon A\sigma(T_w^4 - T_w'^4)}{cm} \tag{6}$$

where $T_w{}'$ denotes current temperature.

The AI Model. Compared with the physical model based on the physical equation, the performance of the AI model is related to the data's size and accuracy. Once the data in certain conditions is absent, the AI physical model may not be as accurate as the physical equation model would be.

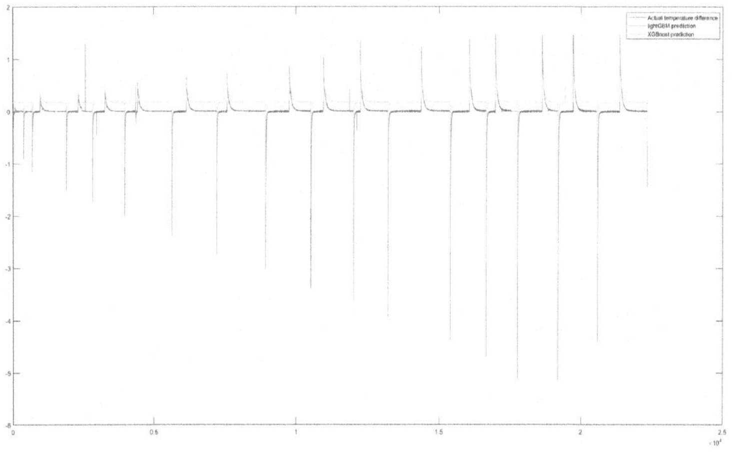

Fig. 4. The actual and predicted temperature difference

4 Result

Models Due to the limitation of expert knowledge as mentioned above and the lack of concrete physical property information of the system, this type of physical model does not meet the expectation, i.e. the equation does not predict the temperature accurately. As a result, the AI algorithms are resorted to implement the data model, i.e. an AI model. To simulate this process, XGBoost and lightGBM these machine learning algorithms, which are prevalent in industry data analysis, are used and the result is shown in Fig. 4. In Fig. 4, the abscissa represents the test cases, where the temperature control parameters of the test case gradually increase, and the ordinate represents the temperature difference. Blue is the actual difference, red is the difference predicted by lightgbm, and yellow is from XGBoost.

The result of Δ_t shows that lightGBM performs better prediction and fits with less error compared with XGBoost, and the fact is the RMSE of lightGBM(0.0120) is less than that of XGBoost(0.0397). It also can be seen that there are obvious errors in the temperature drop prediction values of the two algorithms when the temperature is high. However, when applied to the whole process, under the condition of periodical synchronization with physical equipment, both show good prediction trends and the curve of the actual temperature change under these controlling parameters as shown in Fig. 5, where the ordinate is the temperature and abscissa stay the same. However, the shortage is obvious. When the parameters change, the lack of corresponding data may lead to an unacceptable fitting error when accuracy matters in a scene. These two models can be integrated to comprise a whole digital twin model as a physical sub-model and therefore, a virtual data source in this scene but need an adequate amount of history data to train them.

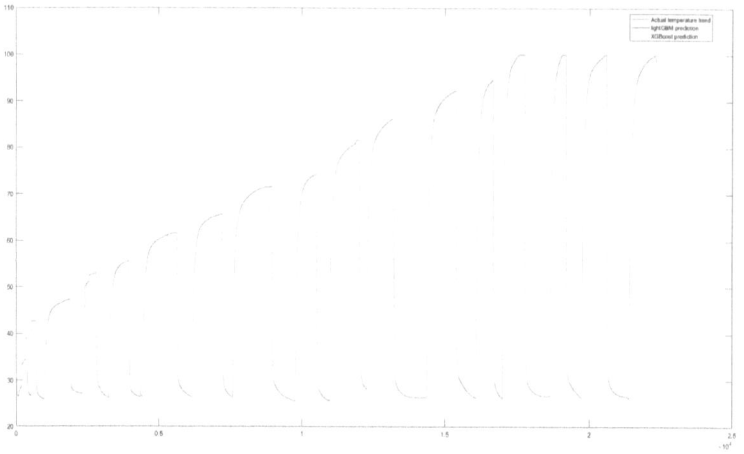

Fig. 5. The actual and predicted temperature changing curve

Communication. To connect the digital twin and physical twin devices, communication between the PLC and digital twin should be enabled. To read and write PLC, there are many industrial communication protocols to choose from. Modbus/TCP is a universal protocol that every PLC vendor supports and is still widely used in industries. However, since Modbus does not encrypt data nor either identify clients, there are a lot of cyberattacks targeting it. Therefore, it is a necessity to include Modbus as an alternative protocol in a cyber range. Meanwhile, OPC UA, a universal secure industry protocol, is also prevalent and gaining its place in the market, so OPC UA should be paid attention to as well. Besides, as OPC UA supports Subscribe-Publish mode, the parameter change in simulation mode can be published to the digital twin entity once subscribed, and this reduces the cost of communication due to fewer active data requests.

Apart from communications between virtual and real objects, the inner information and data exchange in a digital model is also considered. As the data model is an AI model, that is to say, the running environment of virtual entities is different from the one the predicted data generated, this data exchange is implemented using network protocols. To keep universality and compatibility, MQTT, a lightweight distributed and platform-independent IoT protocol, using the Pub/Sub messaging principle for machine-to-machine communication is realized. In our digital twin, it is used to achieve the model's inner information exchange.

5 Applications of Digital Twin-Based ICS Cyber Range

Cyber security use cases of digital twins rely on the mode of the digital twin. In different modes, the models that play a key role are also often different.

In the monitoring mode, all data originates from field devices, so the digital twin in this mode is mainly used for real-time observation of data. Based on this, rule-based anomaly detection can be implemented by defining a series of rules and combining them with the anomaly detection model. These rules can be derived based on historical data and process operation rules in the logical model.

In prediction mode, in addition to real-time field data, the virtual body itself also generates data, so in addition to rule-based anomaly detection [11] in monitoring mode, anomaly detection can also be performed by comparing actual data with predicted data. In addition, the prediction model can realize the prediction of data in advance and then can provide early warning based on the predicted value. In cyberattacks, it may be possible to avoid or reduce the losses and harm caused by the attack based on this early warning.

For digital twins **in simulation mode**, since the virtual environment does not obtain data from the real system, it cannot achieve anomaly detection of the real system. However, due to the training with a large amount of real historical data and verification under real-time connections, the virtual body can be used as a substitute for real devices in ICS cyber ranges, can be deployed as a honeypot for attack deception [12], and can be used as a substitute for real systems for attack testing[13, 14], evaluation and verification of the security strategies, etc. without affecting the real system.

6 Conclusion

In this paper, a framework of a cyber range utilizing digital twin technology is proposed. This cyber range is achieved with digital twin simulation technologies to keep both reproducibility and authenticity. Besides, the connotation of 5 general perspectives, geometry, behavior, logic, data, and service modes, to model a digital twin entity are also discussed. An example of a metering station is used to give the approach and concrete methods to model a digital twin entity. Finally, the application of digital twin-based ICS cyber range for cybersecurity is discussed. To build a comprehensive digital twin model fitting every condition in a scene, the pure AI model is not enough as there are always some conditions the history data did not explore. To solve this, expert knowledge and the physical entities' physical property information is inevitable and some research [15] on physical parameter calibration with deep reinforcement learning has been conducted, which is also our future research.

Acknowledgements. This work is funded by the National Key Research and Development Plan (Grant No. 2021YFB3101704), the National Natural Science Foundation of China (No. 62272119, 62072130, U20B2046), the Guangdong Basic and Applied Basic Research Foundation (No.2020A1515010450, 2021A1515012307), Guangdong Province Universities and Colleges Pearl River Scholar Funded Scheme (2019), and Guangdong Higher Education Innovation Group (No. 2020KCXTD007), Guangzhou Higher Education Innovation Group (No. 202032854), Consulting project of Chinese Academy of Engineering (2022-JB-04-05).

References

1. Langner, R.: Stuxnet: dissecting a cyberwarfare weapon. IEEE Secur. Priv. **9**(3), 49–51 (2011)
2. https://www.energy.gov/oe/national-scada-test-bed

3. Mathur, A.P., Tippenhauer, N.O.: SWaT: a water treatment testbed for research and training on ICS security. In: 2016 International Workshop on Cyber-Physical Systems for Smart Water Networks (CySWater), pp. 31–36. IEEE (2016)
4. Antonioli, D., Tippenhauer, N.O.: Minicps: a toolkit for security research on cps networks. In: Proceedings of the First ACM Workshop on Cyber-Physical Systems Security and/or Privacy, pp. 91–100 (2015)
5. Alves, T., Das, R., Werth, A., et al.: Virtualization of SCADA testbeds for cybersecurity research: a modular approach. Comput. Secur. **77**, 531–546 (2018)
6. Giuliano, V., Formicola, V.: ICSrange: a simulation-based cyber range platform for industrial control systems. arXiv preprint arXiv:1909.01910 (2019)
7. Grieves, M., Vickers, J.: Digital Twin: Mitigating Unpredictable, Undesirable Emergent Behavior in Complex Systems. Transdisciplinary Perspectives on Complex Systems, pp. 85–113. Springer, Cham (2017)
8. Vielberth, M., Glas, M., Dietz, M., et al.: A digital twin-based cyber range for SOC analysts. In: Data and Applications Security and Privacy XXXV: 35th Annual IFIP WG 11.3 Conference, pp. 293–311 (2021)
9. Eckhart, M., Ekelhart, A.: A specification-based state replication approach for digital twins. In: Proceedings of the 2018 Workshop on Cyber-Physical Systems Security and Privacy, pp. 36–47 (2018)
10. Dietz, M., Pernul, G.: Unleashing the digital twin's potential for ICS security. IEEE Secur. Priv. **18**(4), 20–27 (2020)
11. Eckhart, M., Ekelhart, A.: Towards security-aware virtual environments for digital twins. In: Proceedings of the 4th ACM Workshop on Cyber-Physical System Security, pp. 61–72 (2018)
12. Sun, Y., Tian, Z., Li, M., Su, S., Du, X., Guizani, M.: Honeypot identification in softwarized industrial cyber-physical systems. IEEE Trans. Industr. Inf. **17**(8), 5542–5551 (2021)
13. Luo, C., Tan, Z., Min, G., Gan, J., Shi, W., Tian, Z.: A novel web attack detection system for internet of things via ensemble classification. IEEE Trans. Industr. Inf. **17**(8), 5810–5818 (2021)
14. Tian, Z., Luo, C., Qiu, J., Du, X., Guizani, M.: A distributed deep learning system for web attack detection on edge devices. IEEE Trans. Industr. Inf. **16**(3), 1963–1971 (2020)
15. Tian, Y., Chao, M.A., Kulkarni, C., et al.: Real-time model calibration with deep reinforcement learning. Mech. Syst. Signal Process. **165**, 108284 (2022)

Cybersecurity Systems and Applications

Pairing-Free Certificateless Signature Scheme Based on SM2 Algorithm

Chuan He[1,2(✉)], Bo Zhang[1,2], Liang Zhang[3], Zesheng Xi[1,2], Yuan Fang[3], and Yunfan Wang[1,2]

[1] State Grid Smart Grid Research Institute Co., Ltd., Beijing, China
1255065000@qq.com, zhangbo@geiri.sgcc.com.cn
[2] State Grid Key Laboratory of Information and Network, Nanjing, China
[3] State Grid Anhui Electric Power Co., Ltd. Information and Telecommunication Branch, Hefei, China

Abstract. Certificateless signature scheme can be applied to ensure data integrity, where the public key infrastructure cannot be not easily deployed while the identity-based cryptography is not suitable either because of the key-escrow problem. In many applications, large amount of resource-limited equipment are installed and there is the demand of reducing the computational burden. On the other hand, the Chinese national cryptographic standards (SM algorithms) have been widely adopted in both domestic and abroad. The SM algorithms are preferred and even compulsory, in particular with the key infrastructure systems. For certificateless schemes, the national standard have not been selected and published. Meanwhile, traditional constructions extended from identity-based schemes share the same underlying mathematical structure. For example, the SM9 algorithm require bilinear pairing operations and a certificateless scheme extended from it also requires comparatively heavy computations. In this paper, we study how to achieve pairing-free certificateless signature scheme and propose a concrete construction based on SM2 algorithm. The core idea of the construction is similar to the general transformation from traditional signature to identity-based signature by Bellare et al. and it does not resort to bilinear pairing.

Keywords: Certificateless signature · National cryptographic algorithm · Pairing-free · SM2 algorithm

1 Introduction

1.1 Certificate-Based Signature and Certificateless Signature

Digital signature is one of the most fundamental cryptographic primitives to protect security in information systems. It can be applied to guarantee the data integrity and non-repudiation. In many other scenarios, such as the identity authentication, signature technique is also a candidate. More applications of digital signature can be found in [1]. As a traditional public-key cryptographic component, the user has a pair of keys,

Z. Gu et al. (Eds.): NSE 2023, CCIS 2063, pp. 297–308, 2024.
https://doi.org/10.1007/978-981-97-4519-7_21

namely public key and secret key. To ensure the authenticity of a user's identity, certificate is issued to the user to prove the validity of its public key. However, this requires the implementation of public-key infrastructure (PKI). In such a system, the certificate management is complex and there needs a trusted third party named certificate authority (CA) to maintain the daily operations. The costs of a CA and all the related issues, such as certificate distribution, revoked certificate list, certificate application and update, etc. are huge. Therefore, to deploy a PKI system is only feasible for large amount of users and big industry systems. For other smaller environments, implementing PKI is not suitable.

To simplify the certificate management complexity in traditional public-key systems, identity-based cryptography [2] was introduced. In such a system, the motivation is directly using the user's identity as its public key. As a result, all the complex issues with respect to certificate management no longer exist. To achieve this goal, there needs an entity, which generates and distributes user's private keys. The private key can only be generated by this party and cannot be computed by the user. This solves the certificate complexity problem. However, in identity-based systems, there needs a fully trusted third party called private key generator (PKG). The PKG is responsible for generating and distributing all the users' private keys and hence the key-escrow problem is inevitable. There are two main approaches that solves the key-escrow problem while still maintain the inherent authenticity of user's identity, namely the certificateless cryptography [3] and the certificate-based cryptography [4]. The core idea of these two primitives is similar but the detailed procedures are different.

In certificateless cryptography (CL-PKC), the private key of a user consists of two parts, namely partial private key and secret value. The partial private key is generated by a third-party called key generation center (KGC). This phase is similar to the identity-based system. The secret value is selected by the user itself and then used to compute the corresponding public key. This phase is similar to the traditional public-key system. Therefore, it enjoys both the advantages of simplified certificate management and no key-escrow problem. In certificate based cryptography (CB-PKC), the private key is also composed of two parts. The main difference is the sequence of the two parts generation. In detail, the user first selects the secret key and computes the corresponding public key. This phase is similar to the PKI setting. Then, the user uploads the public key to the third party KGC and verifies the identity with the public key. Next, the KGC generates a certificate for the public key and transmits it to the user. Please note that this certificate does not need to by verified or authenticated to others later and hence there is no need to maintain the complex certificate management systems like the PKI setting. The certificate is only used during the decryption or signing phases. Consequently, the certificate management is simple and there does not exist key-escrow problem either in a CB-PKC system.

As for signature schemes, the main difference between certificateless signature (CLS) and certificate-based signature (CBS) is the sequence of secret value generation algorithm and partial private key generation algorithm. Both of these two primitives have been extensively studied recently and many constructions are available in the literature [5–7]. However, many of the aforementioned schemes are proposed by individual researchers, foreign organizations, or published as international cryptographic standards. To promote the development of domestic cryptography research and achieve the independent intellectual property rights, the Chinese national cryptographic standard has been proposed and published. The SM (abbreviation of "Shang Mi", commercial cryptography in Chinese) series algorithms are currently the standard. Among these algorithms, SM2 is the traditional public-key system and SM9 is the identity-based system.

1.2 General Constructions from Traditional Public-Key Schemes

There are many concrete constructions of various cryptographic systems. Furthermore, there are also some general constructions for a particular type system from traditional public-key systems. This provides us more possibilities for achieving more constructions. For example, Bellare et al. [8] proposed a general construction (a.k.a. BNN transformation) of identity-based signature (IBS) from a traditional signature scheme. Subsequently, Galindo et al. extended the general construction for an identity-based signature with additional properties, such as proxy signature, threshold signature, blind signature etc.

With respect to certificateless signature and certificate-based signature, how to achieve general constructions lacks systematic researches. Since the inherent structure of certificate-based signature is more compatible with the BNN transformation, it is easier to achieve certificate-based signature from a traditional signature scheme in a more straightforward way. For certificateless signature, it requires more complex interactions between the key generation center (KGC) and the user. The detailed constructions are presented in Sect. 3 and Sect. 4, respectively.

In addition to general constructions, another alternative approach is to extend the identity-based scheme. As for the Chinese national cryptographic standard, the only candidate of identity-based scheme is SM9. The construction is built upon the pairing-friendly group, which involves the complex computation called bilinear pairing. As a result, the CLS or CBS schemes based on the SM9 algorithm also require the bilinear pairing operations. Whether we can achieve efficient constructions based on pairing-free national cryptographic standards, such as SM2 algorithm, remains unknown.

1.3 Motivation and Contribution

From the above discussions, we can see that the design and analysis of CLS and CBS schemes have been extensively studied. However, there still have some research gaps. For example, how to achieve efficiently general constructions from traditional public-key schemes. In addition, how to achieve pairing-free constructions based on SM series standards. In many particular systems, there exists the requirement of applying national cryptographic standards and lightweight schemes are preferred, especially if the system has large amount of resource limited equipment.

Based on the above observations, we study the constructions of CLS and CBS schemes based on Chinese national cryptographic standards, in particular with the SM2 algorithm. The main contributions of this paper are three-fold and presented as follows.

1. We propose the concrete constructions of a certificate-based signature scheme and a certificateless signature scheme based on SM2 algorithm, respectively.

2. The constructions can be seen as an extension of the Bellare et al.'s general construction (BNN transformation) of IBS schemes. This also provides an answer to the question of how to achieve CLS and CBS schemes from traditional public-key signatures.
3. The schemes proposed in this paper does not resort to the bilinear pairing operations, which is more cost-efficient in practical applications.

1.4 Roadmap

The rest of this paper is organized as follows. In Sect. 2, we review the formal definitions of related concepts and the building block of this paper. Then, we present in Sect. 3 the construction of CBS scheme based on the SM2 algorithm. Next, we show how to extend the construction to achieve CLS scheme in Sect. 4. Finally, we conclude this paper in Sect. 5.

2 Preliminaries

2.1 Related Concepts

In this section, we review the related concepts including signature, certificate based signature, and certificateless signature schemes.

Definition 1 (Digital Signature). *A signature scheme is composed of four algorithms, namely the* setup *algorithm* Setup, *key generation algorithm* Key Gen, *signing algorithm* Sign, *and verification algorithm* Verify. *Each algorithm's description of input and output is depicted as follows.*

- Setup(1^λ): *The setup algorithm inputs the security parameter* 1^λ *and outputs the system parameter* SP.
- KeyGen(SP): *The key generation algorithm inputs the system parameter* SP *and outputs a pair of public key and secret key* (pk, sk).
- Sign(SP, sk, m): *The signing algorithm inputs the system parameter* SP, *the signer's secret key sk, and a message m and outputs a signature* σ_m *on the message m.*
- Verify(SP, pk, m, σ_m): *The verification algorithm inputs the system parameter* SP, *public key pk, message m, and a signature* σ_m *and outputs the verification result. If it returns "1", it means that the signature is valid; and it means the signature is invalid if it returns "0", otherwise.*

Definition 2 (Certificate-Based Signature). *A certificate-based signature (CBS) scheme is composed of six algorithms, namely the system setup algorithm Setup, secret value generation algorithm* ScretValue, *public key generation algorithm* PublicKey, *certificate generation algorithm* Cert, *signing algorithm* Sign, *and verification algorithm* Verify. *Each algorithm's description of input and output is depicted as follows.*

- Setup(1^λ): *The system setup algorithm inputs the security parameter* 1^λ *and outputs the master public and secret key pair (mpk, msk).*

- ScretValue(*mpk*, ID): *The secret value generation algorithm inputs the master public key mpk and a user's identity* ID *and outputs the secret value* sv_{ID} *for the user.*
- PublicKey(*mpk*, ID, sv_{ID}): *The public key generation algorithm inputs the master public key mpk, a user's identity* ID, *and the secret value* sv_{ID} *of the user and outputs the public key* pk_{ID} *for the user.*
- Cert(*mpk*, *msk*, ID, pk_{ID}): *The certificate generation algorithm inputs the master public and secret key pair (mpk, msk), a user's identity* ID, *and the user's public key* pk_{ID} *and outputs the certificate* c_{ID} *for the user.*
- Sign(*mpk*, sv_{ID}, c_{ID}, *m*): *The signing algorithm inputs the master public key mpk, a user's secret value* sv_{ID}, *certificate* c_{ID}, *and a message m and outputs a signature* σ_m *on message m.*
- Verify(*mpk*, ID, pk_{ID}, m, σ_{m}): *The verification algorithm inputs the master public key mpk, a user's identity* ID, *public key* pk_{ID}, *message m, and a signature* σ_m *and outputs the verification result. If it returns "1", it means that the signature is valid; and it means the signature is invalid if it returns "0", otherwise.*

Definition 3 (Certificateless Signature). *A certificateless signature (CLS) scheme is composed of six algorithms, namely the system setup algorithm* Setup, *partial private key generation algorithm* KeyExt, *secret value generation algorithm* ScretValue, *public key generation algorithm* PublicKey, *signing algorithm Sign, and verification algorithm* Verify. *Each algorithm's description of input and output is depicted as follows.*

- Setup(1^λ): *The system setup algorithm inputs the security parameter* 1^λ *and outputs the master public and secret key pair (mpk, msk).*
- KeyExt(*mpk*, *msk*, ID): The *partial private key generation algorithm inputs the master public and secret key pair (mpk, msk) and a user's identity ID and outputs the partial private key* d_{ID} *for the user.*
- ScretValue(*mpk*, ID): *The secret value generation algorithm inputs the master public key mpk and a user's identity* ID *and outputs the secret value* sv_{ID} *for the user.*
- PublicKey(*mpk*, ID, sv_{ID}): *The public key generation algorithm inputs the master public key mpk, a user's identity* ID, *and the secret value* sv_{ID} *of the user and outputs the public key* pk_{ID} *for the user.*
- Sign(*mpk*, d_{ID}, sv_{ID}, *m*): *The signing algorithm inputs the master public key mpk, a user's partial private key* d_{ID}, *secret value* sv_{ID}, *and a message m and outputs a signature* σ_m *on message m.*
- Verify(*mpk*, ID, *m*, σ_m): *The verification algorithm inputs the master public key mpk, a user's identity* ID, *public key* pk_{ID}, *message m, and a signature* σ_m *and outputs the verification result. If it returns "1", it means that the signature is valid; and it means the signature is invalid if it returns "0", otherwise.*

From the above definitions, we can see that the main difference between CBS and CLS schemes is the sequence of secret value generation and the KGC's computation. In a CBS scheme, the user generates its public and secret key pair first and then transmits the key pair to the KGC. The KGC computes the corresponding certificate for the user subsequently. In a CLS scheme, the KGC first computes the partial private key for the user before the user generates its public and secret key pair.

2.2 General Construction of Identity-Based Signature (BNN Transformation)

In [8], Bellare et al. proposed a general construction (BNN Transformation) of IBS signature from a traditional signature in public-key system. The core idea of BNN transformation is to let the PKG run the signing algorithm to generate a signature on a user's identity concatenated with another public key selected by the PKG. The corresponding secret key is assigned to the user as its private key that will be used to sign messages later. The signature consists of two parts, namely a signature signed by the user using its private key and the signature on its identity concatenated with a public key generated by the PKG. The verification of the user's signature also has two phases, namely verifying the above two signatures, respectively. The high-level description of the transformation is further summarized in [9] as follows.

2.3 SM2 Cryptographic Standard

The SM2 algorithm is the published national standard of public-key cryptography. It consists of encryption, signature, and key exchange schemes. The building block of this paper is the SM2 signature scheme and the details of the scheme is depicted as follows.

- Setup(1^λ): The setup algorithm inputs the security parameter 1^λ and chooses an elliptic curve group ($\mathbb{G}$, p, P) of prime order p, where P is a generator of the group. In addition, the description of group $\mathbb{G}$ also includes the parameters a, b in defining the curve and x_P, y_P, which are the coordinates of the base point P. It then selects a hash function H, such as the SM3 algorithm. The system parameter SP = ($\mathbb{G}$, p, P, H) (Fig. 1)

- KeyGen(SP): The key generation algorithm inputs the system parameter SP and selects a random $d \in \mathbb{Z}_p$. Then, it computes $[d]P$ and the point $[d]P$ also includes the coordinates x_{pk}, y_{pk}. Next, it outputs the public and secret key pair $(pk, sk) = ([d]P, d)$.
- Sign(SP, sk, m): The signing algorithm inputs the system parameter SP, the signer's secret key sk, and a message m. It first computes the hash value $Z = H(ENTL \parallel ID \parallel a \parallel b \parallel x_P \parallel y_P \parallel x_{pk} \parallel y_{pk})$. Then, it computes the hash value $e = H(Z \parallel m)$ and selects a random $k \in \mathbb{Z}_p$. . Next, it computes $[k]P = (x_1, y_1)$ and $r = e + x_1 \bmod p$. At last, it computes and outputs $\sigma_m = (r, s)$ as the signature on the message m.

$$s = \left((1+d)^{-1} \cdot (k - r \cdot d)\right) \bmod p$$

- Verify(SP, pk, m, σ_m): The verification algorithm inputs the system parameter SP, public key pk, message m, and a signature $\sigma_m = (r, s)$. It first computes the hash value Z from the system parameter and the signer's public key. Then, it computes the hash value $e' = H(Z \parallel m)$ and $t = r + s \bmod p$.

Next it computes $[s]P + [t]pk = (x_1', y')_1$ and $R = e' + x_1' \bmod p$. At last, it outputs "1", if r = R; otherwise, it outputs"0".

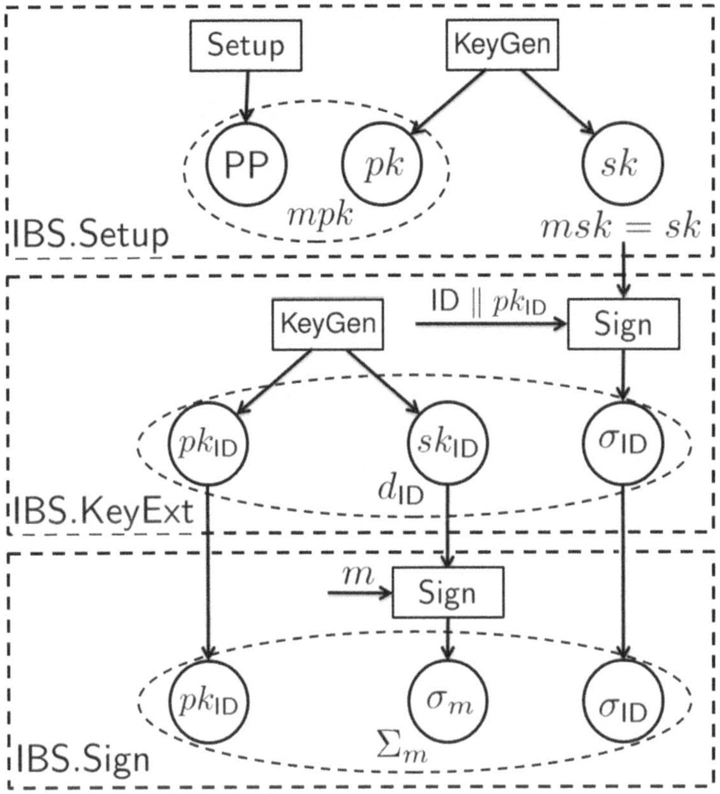

Fig. 1. The BNN Transformation.

3 Certificate-Based Signature Based on SM2

In this section, we extend the BNN transformation to achieve the certificate based signature from a public-key scheme. The core idea is as follows.

From Fig. 2, we can see that the setup phase is similar to the BNN transformation. For CBS scheme, the secret value is selected by the user itself and the detailed process is the same as the key generation algorithm of the traditional signature scheme. After the secret value and public key are generated, the KGC runs the certificate generation algorithm with its master secret key and generates the certificate for the user. The signing algorithm is also similar to the BNN transformation, where the signature consists of a part signed by the secret value and the certificate.

If we use the SM2 algorithm as the concrete instance in the general construction, it immediately yields a CBS signature scheme based on national cryptographic standard without bilinear pairing. The details come as follows.

- Setup(1^λ): The system setup algorithm inputs the security parameter 1^λ and runs the SM2 setup algorithm. It chooses an elliptic curve group $(\mathbb{G}, p, P)$ with parameters a, b and coordinates x_P, y_P. Then, it selects a random $\alpha \in \mathbb{Z}_p$ and computes $P_{pub} =$

$[\alpha]P$. Next it selects a hash function H, such as the SM3 algorithm and outputs the public and secret key pair as

$$mpk = \left(\mathrm{G}, p, P, P_{pub}, H\right), msk = \alpha.$$

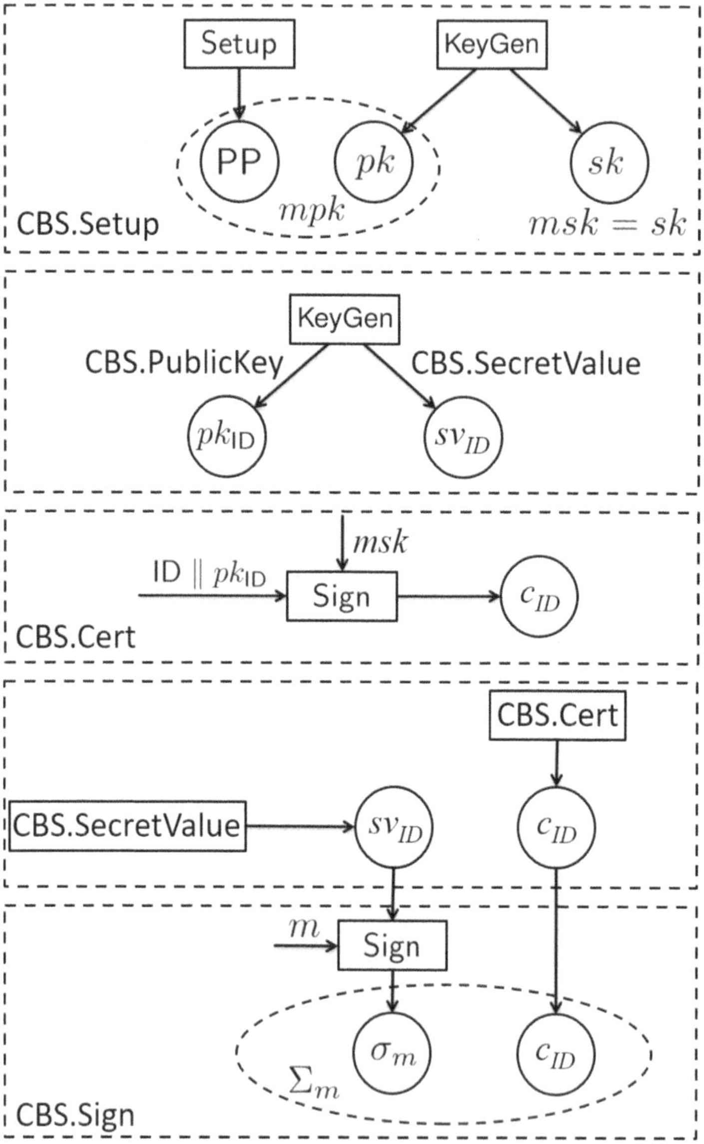

Fig. 2. The CBS Construction.

- ScretValue(*mpk*, ID): The secret value generation algorithm inputs the master public key *mpk* and a user's identity ID. Then, it runs the SM2 key generation algorithm. It selects a random $sv_{ID} \in \mathbb{Z}_p$ and outputs the secret value sv_{ID}.
- PublicKey(*mpk*, ID, sv_{ID}): The public key generation algorithm inputs the master public key *mpk*, a user's identity ID, and the secret value sv_{ID} of the user. Then, it computes $pk_{\mathrm{ID}} = [sv_{ID}]P$ and outputs the public key pk_{ID}.
- Cert(*mpk*, *msk*, ID, pk_{ID}): The certificate generation algorithm inputs the master public and secret key pair (*mpk*, *msk*), a user's identity ID, and the user's public key pk_{ID}. It first concatenates user's identity ID and its public key pk_{ID}. Then, it runs the SM2 signing algorithm with the master secret key to generate the certificate. In detail,

  1. Compute $e = H(\mathrm{ID} \parallel pk_{\mathrm{ID}})$;
  2. Pick a random $k \in \mathbb{Z}_p$ and compute $[k]P = (x_1, y_1)$, $r = (e + x_1) \bmod p$;
  3. Compute $s = ((1 + \alpha)^{-1} \cdot (k - r \cdot \alpha)) \bmod p$. Next, it outputs the certificate $c_{\mathrm{ID}} = (r, s)$.

- Sign(*mpk*, sv_{ID}, c_{ID}, *m*): The signing algorithm inputs the master public key *mpk*, a user's secret value sv_{ID}, certificate c_{ID}, and a message *m*. It first concatenates user's identity ID and the message *m*. Then, it runs the SM2 signing algorithm with the secret value to generate the part signature. In detail,

  1. Compute $e = H(ID\|m)$;
  2. Pick a random $k \in \mathbb{Z}_p$ and compute $[k]P = (x_1, y_1)$, $r^{'} = (e + x_1) \bmod p$;
  3. Compute $s^{'} = ((1 + sv_{\mathrm{ID}})^{-1} \cdot (k - r^{'} \cdot sv_{\mathrm{ID}})) \bmod p$.

  Next, it outputs the signature $\sigma_m = (c_{\mathrm{ID}}, r^{'}, s^{'})$.

- Verify(*mpk*, ID, pk_{ID}, *m*, σ_m): The verification algorithm inputs the master public key *mpk*, a user's identity ID, public key pk_{ID}, message *m*, and a signature $\sigma_m = (c_{\mathrm{ID}}, r^{'}, s^{'})$. It first parses $c_{\mathrm{ID}} = (r, s)$ and runs the SM2 verification algorithm to check the validity of (r, s) and $(r^{'}, s^{'})$, respectively. In detail,

  1. Compute $e_1' = H(ID\|pK_{ID})$, $e_2' = H(ID\|m)$;
  2. Compute $t_1 = (r + s) \bmod p$, $t_2 = (r' + s') \bmod p$;
  3. Compute $[s]P + [t_1]P_{pub} = (x_1', y_1')$, $[s']P + [t_2]pk_{ID} = (x_2', y_2')$;
  4. Compute $R = (e_1' + x_1')$, $R' = (e_2' + x_2')$.

  Next, it outputs "1", if $r = R$ and $r^{'} = R^{'}$; otherwise, it outputs "0".

4 Certificateless Signature Based on SM2

In CBS schemes, the user first generates the secret value and public key, then the KGC generates the certificate for the user. Unlike CBS signature, CLS signature generates partial private key and selects secret value independently. Therefore, the BNN transformation cannot be directly extended in a straightforward way. In practice, there is the possibility that the KGC generates partial private key for the user first, and then the user generates its secret value and public key. In this case, how to bond the user's identity and its public key is the core of the CLS signature construction from a traditional public-key scheme.

In this paper, we propose the following technique to overcome the aforementioned problem. The KGC first uses the same procedure as BNN transformation to generate the partial private key with a selected (partial) secret value and public key. Then, the user generates its secret value and public key from the received (partial) secret value and public key. This phase requires the user to combine the partial result from the KGC and the value generated by itself together. The core idea is summarized as Fig. 3.

The detailed construction is depicted as follows.

- Setup(1^λ): The system setup algorithm inputs the security parameter 1^λ and runs the SM2 setup algorithm. It chooses an elliptic curve group ($\mathbb{G}$, p, P) with parameters a, b and coordinates x_P, y_P. Then, it selects a random $\alpha \in \mathbb{Z}_p$ and computes $P_{pub} = [\alpha]P$. Next it selects a hash function H, such as the SM3 algorithm and outputs the public and secret key pair as

$$mpk = \left(\mathbb{G}, p, P, P_{pub}, H\right), msk = \alpha$$

- KeyExt(mpk, msk, ID): The partial private key generation algorithm inputs the master public and secret key pair (mpk, msk) and a user's identity ID. It first picks a random $x \in Z_p$ and computes $ppk_{\text{ID}} = [x]P$. Then, it concatenates the user's identity ID and the partial public key ppk_{ID}. Next, it runs the SM2 signing algorithm with the master secret key to generate the partial private key.

  1. Compute $e = H(ID\|ppk_{ID})$;
  2. Pick a random $k \in Z_p$ and compute $[k]P = (x_1, y_1), r = (e + x_1)\text{mod}p$;
  3. Compute $s = \left((1+\alpha)^{-1} \cdot (k - r \cdot \alpha)\right) \bmod p$

  It outputs the partial private key $d_{\text{ID}} = (r, s, x, ppk_{\text{ID}})$.

- ScretValue(mpk, ID): The secret value generation algorithm inputs the master public key mpk and a user's identity ID. Then, it runs the SM2 key generation algorithm. It selects a random $y \in \mathbb{Z}_p$ and sets $sv_{\text{ID}} = (x, y)$ with the random value x received from KGC. Next, it outputs the secret value sv_{ID}.
- PublicKey(mpk, ID, sv_{ID}): The public key generation algorithm inputs the master public key mpk, a user's identity ID, and the secret value $sv_{\text{ID}} = (x, y)$ of the user. Then, it computes $[y]P$ and generates a non-interactive zero knowledge proof π of holding the corresponding y with respect to $[y]P$. Next, it sets $pk_{\text{ID}} = (ppk_{\text{ID}}, [y]P, \pi)$ and outputs the public key pk_{ID}.
- Sign(mpk, d_{ID}, sv_{ID}, m): The signing algorithm inputs the master public key mpk, a user's partial private key $d_{\text{ID}} = (r, s, x, ppk_{\text{ID}})$, secret value $sv_{\text{ID}} = (x, y)$, and a message m. It first concatenates user's identity ID and the message m. Then, it computes $(x + y) \bmod p$ and runs the SM2 signing algorithm with $(x + y)$ to generate the part signature. In detail,

  1. Compute $e = H(ID\|m)$;
  2. Pick a random $k \in \mathbb{Z}_p$ and compute $[k]P = (x_1, y_1)$, $r' = (e + x_1) \bmod p$;
  3. Compute $s' = (1 + (x + y))^{-1} \cdot \left(k - r' \cdot (x + y)\right) \bmod p$. Next, it outputs the signature $\sigma_m = (r, s, r', s')$.

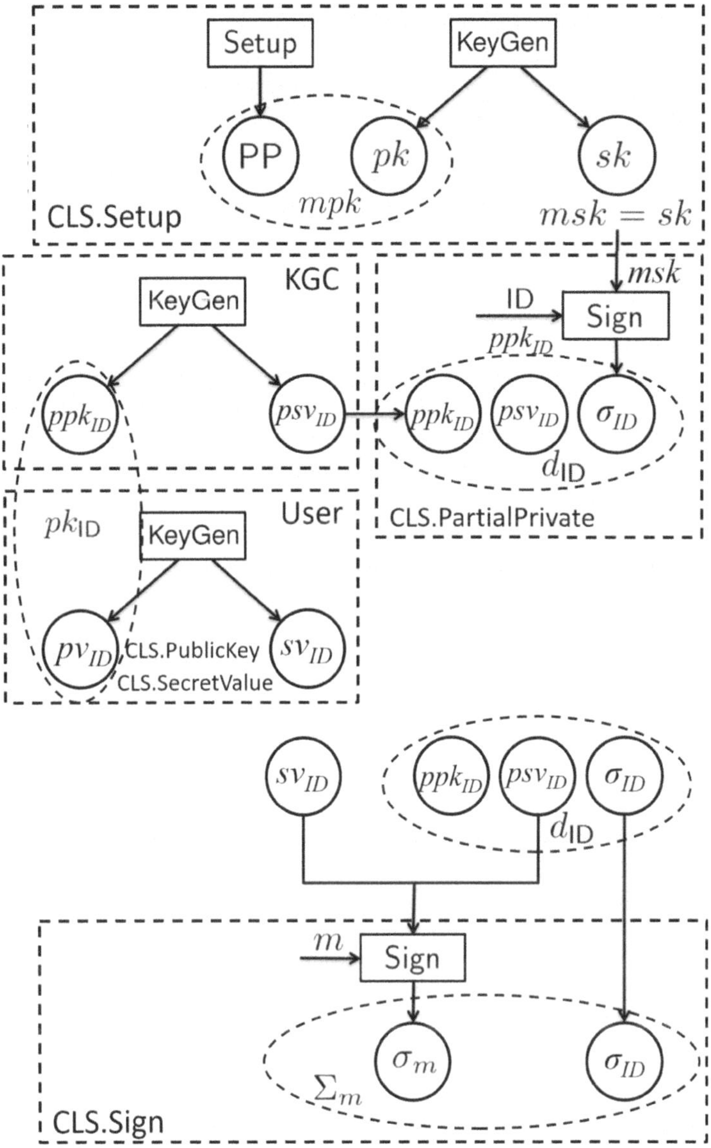

Fig. 3. The CLS Construction.

– Verify(mpk, ID, pk_{ID}, m, σ_m): The verification algorithm inputs the master public key mpk, a user's identity ID, public key $pk_{\mathrm{ID}} = (ppk_{\mathrm{ID}}, [y]P, \pi)$, message m, and a signature $\sigma_m = (r, s, r', s')$. It first checks the validity of the non-interactive zero-knowledge proof π with respect to $[y]P$. If the proof is valid, then computes $ppk_{\mathrm{ID}} + [y]P$ and runs the SM2 verification algorithm to check the validity of (r, s) and (r', s'), respectively. In detail,

1. Compute $e_1' = H(\mathrm{ID}\|pk_{ID})$, $e_2' = H(\mathrm{ID}\|m)$;
2. Compute $t_1 = (r + s) \bmod p$, $t_2 = (r' + s') \bmod p$;
3. Compute $[s]P + [t_1]P_{pub} = (x_1', y_1')$, $[s']P + [t_2]pk_{ID} = (x_2', y_2')$;
4. Compute $R = (e_1' + x_1')$, $R' = (e_2' + x_2')$.

Next, it outputs "1", if the proof π is valid, $r = R$, and $r^{'} = R^{'}$; otherwise, it outputs "0".

5 Conclusion

In this paper, we focused on the construction of certificate-based signature (CBS) and certificateless signature (CLS). For the constructions based on Chinese cryptographic standard, currently there lacks the systematic research. The identity based scheme SM9 algorithm relies on the bilinear pairing. To fill this gap, we extend the BNN transformation to achieve CBS scheme based on SM2 algorithm. As for CLS scheme, the BNN transformation cannot be directly applied. We proposed a method to achieve CLS scheme which is also based on the SM2 algorithm. The most distinctive feature of our schemes is that they do not involve the bilinear pairing operations.

Acknowledgements. This work is supported by the science and technology project of State Grid Corporation of China "Research on lightweight cryptographic technology for power IOT terminal"(Grand No. 5700-202255186A-1-1ZN).

References

1. Katz, J., Lindell, Y.: Digital Signature Schemes, 1st edn. Springer, Berlin (2010)
2. Shamir, A.: Identity-based cryptosystems and signature schemes. In: Blakley, G.R., Chaum, D. (eds.) CRYPTO 1984, LNCS, vol. 196, pp. 47–53. Springer, Heidelberg (1984). https://doi.org/10.1007/3-540-39568-7_5
3. Al-Riyami, S., Paterson K.: Certificateless public key cryptography. In: Laih, C. (eds.) ASIACRYPT 2003, LNCS, vol. 2894, pp. 452–473. Springer, Heidelberg (2003). https://doi.org/10.1007/978-3-540-40061-5_29
4. Gentry, C.: Certificate-based encryption and the certificate revocation problem. In: Biham, E. (eds.) EUROCRYPT 2003, LNCS, vol. 2656, pp. 272–293. Springer, Heidelberg (2003). https://doi.org/10.1007/3-540-39200-9_17
5. Wu, G., Zhang, F., Shen, L., Guo, F., Susilo, W.: Certificateless aggregate signature scheme secure against fully chosen-key attacks. Inf. Sci. **514**, 288–301 (2020)
6. Hussain, S., Ullah, S., Ali, I., Xie, J., Inukollu, V.: Certificateless signature schemes in Industrial Internet of Things: a comparative survey. Comput. Commun. **181**, 116–131 (2022)
7. Qiao, Z., Yang, Q., Zhou, Y., Yang, B., Xia, Z., Zhang, M., Wang, T.: An efficient certificate-based aggregate signature scheme with provable security for industrial internet of things. IEEE Syst. J. **17**(1), 72–82 (2023)
8. Bellare, M., Namprempre, C., Neven, G.: Security proofs for identity-based identification and signature schemes. In: Cachin, C., Camenisch, J. (eds.) EUROCRYPT 2004, LNCS, vol. 3027, pp. 268–286. Springer, Heidelberg (2004). https://doi.org/10.1007/978-3-540-24676-3_17
9. Wu, G., Zhao, Z., Guo, F., Susilo, W., Zhang, F.: On the general construction of tightly secure identity-based signature schemes. Comput. J. **63**(12), 1835–1848 (2020)

A Quantitative Assessment Method of System Defense Capability Based on Attack Surface

Ziqi Yin[1], Shudong Li[1,2](✉), Weihong Han[2], Xiaobo Wu[3], and Shumei Li[4]

[1] Cyberspace Institute of Advanced Technology, Guangzhou University, Guangzhou 510006, China
lishudong@gzhu.edu.cn

[2] Department of New Networks, Peng Cheng Laboratory, Shenzhen 518000, China
hanwh@pcl.ac.cn

[3] School of Computer Science and Cyber Engineering, Guangzhou University, Guangzhou 510006, China
happywxb@gzhu.edu.cn

[4] Computer College, Jilin Normal University, Siping 136000, China
lishumeiwei@sina.com

Abstract. A multitude of incidents, such as data leakage, cyber fraud, phishing, Trojan viruses, and DDoS attacks, present grave threats to the privacy, property, and security of individuals and organizations. To address these challenges effectively, it is imperative to employ scientific and efficient techniques for network security assessment, which play a vital role in augmenting the security performance of network systems and bolstering their resilience against attacks. In this paper, we propose a quantitative assessment methodology for evaluating the defense capability of network security, based on the concept of attack surfaces. Through the construction of a hierarchical model to represent attacks, we successfully mitigate the intricacies and scale-related issues associated with attack graphs. Additionally, our approach leverages the wealth of system information available to the assessor by constructing attack surfaces, enabling the simulation of attacks and the analysis of the defense strategy from an adversarial perspective. This methodology facilitates a more objective and comprehensive assessment of the security status of a system, ensuring a proactive and robust defense against potential threats.

Keywords: quantitative assessment · attack surface · hierarchical attack representation model

1 Introduction

Network security assessment technology refers to the technology that detects, analyzes and evaluates the security of the network system. This technology aims to identify the hidden vulnerabilities existing in the network system and improve the security performance of the network system and its ability to resist attacks. Scientific and objective network security assessment technology is a necessary condition to provide protection for the system. Previously, the traditional network security assessment methods were

Z. Gu et al. (Eds.): NSE 2023, CCIS 2063, pp. 309–321, 2024.
https://doi.org/10.1007/978-981-97-4519-7_22

based on mathematical logic and established a comprehensive and cumbersome index system, which can integrate the overall situation of the network to a certain extent and provide decision making and suggestions for network managers, but the whole process mainly relies on expert assessment, which makes the assessment inefficient and the improvement cannot be carried out in a timely manner. With the advent of the big data era, some traditional methods can no longer meet the requirements of dealing with multiple network attacks, and it is difficult to accurately assess the security status of the network system. In terms of the assessment perspective, the existing assessment methods are more targeted at the attacker's ability, and fewer of them assess the defender's own ability to resist.

Throughout the development of the filed of network security assessment, the mainstream research on related technologies is roughly divided into two categories. One is the assessment method based on mathematical models, mainly hierarchical analysis, integrated analysis, etc. This type of method is able to comprehensively consider the factors of network security situational awareness, but the evaluation model constructed by this method and the definition of its variables involves more subjective factors, and lacks objective uniformity. In recent years, machine learning has emerged at home and abroad, and some researchers have introduced the knowledge of artificial intelligence into the filed of security, bur from the point of view of practical application, this type of method to obtain knowledge is still relatively single, mainly relying on machine learning and expert knowledge base, relying on the accumulation of experience, and at the same time, there are operational difficulties. Another category is knowledge graph technology, which as an important branch of machine learning is also used in assessment techniques, but the construction, maintenance and reasoning ability of knowledge graph is limited by the complexity and expression ability of the ontology model, so the simple attack graph and attack tree methods have the problems of poor scalability and low assessment efficiency. Finally current research on network assessment methods as a defender mostly focuses on system vulnerabilities, and we believe that the defender should take advantage of its comprehensive knowledge of the network system to build attack surface management, join the perspective of attackers, and rehearse attack scenarios by integrating attack information, so as to use this information to support the automatic assessment of network security. In this way, the defender can have a comprehensive understating of the network weaknesses and assess the defense available.

In response to the above problems, the present invention proposes a quantitative assessment method of network security defense capabilities based on attack surface. This method considers all possible attack vectors that system administrators need to detect while evaluating the security performance of the system, including vulnerabilities, entry points, and paths that attackers can exploit. We also introduce the method of attack path enumeration, leveraging the defender's advantage in system information. From an adversarial perspective, we simulate attacks on the network system and analyze defense strategies. This approach enables us to obtain all possible attack paths in the network system, along with their probabilities, costs, benefits, and other parameters. As a result, we can quantitatively analyze the overall and local security risks of the network system, providing a basis for the formulation of defense strategies and a more objective and comprehensive assessment of the system's security status.

This paper is structured as follows: in Sect. 2, we analyze the current mainstream methods for network security assessment and analyze their advantages and shortcomings. Section 3 briefly introduces the main modules of the model, including the security graphical model construction module, the attack path planning module, and the assessment module. Next, in Sect. 4, we introduce the main process of defense capability assessment using the security model. Finally, in Sect. 5, we summarize the paper and present our outlook for future work.

2 Related Work

2.1 Attack Map Evaluation

The graph-based approach is one of the most commonly used security modeling techniques, which can graphically represent vulnerabilities, hosts, connections, and other information in a network system in order to analyze and assess the security risk of the system. K. Ingols et al [1] proposed a scalable AG construct where they used Multiple Precondition (MP) graphs to construct attack models of polynomial complexity. X. Ou et al [2] proposed a logical AG with polynomial constructive complexity. K. Ingols et al [3] used a Network Security Planning Architecture (NetSPA), a Network Security Analysis Tool to analyze network security. To address the problem that existing attack graph generation and analysis methods cannot accurately reflect the real risk of nodes and attack paths, Ye et al [4] designed a knowledge graph based on atomic attack ontology and proposed an extended knowledge graph-based attack graph generation framework. Based on this, an attack graph generation algorithm and a calculation method for attack success rate and attack profit are proposed. However, the calculation method of attack success rate and attack profit of this method is relatively simple, and only verifies the feasibility of this method, and does not give the effect of the application of this method in large-scale networks.

The following drawbacks exist in the attack models of modeling techniques based on the graph approach: 1) modeling based on the attack graph approach tends to incorporate too much information in a single attack graph, resulting in an attack graph that is too complex and large to deal with large-scale and highly dynamic networked systems; 2) the complexity of calculating all the possible attack paths in a single attack graph is exponential, and the solution only considers a subset of the complete attack paths, which poses a security modeling a huge challenge. In contrast, the layered security graph model proposed in this approach separates the attack graph functionality into two layers: the upper layer is the vulnerability dependency graph, which focuses on describing the vulnerabilities and their inter-dependencies that exist in the network system; and the lower layer is the host information tree, which focuses on describing hosts in the network system and their configurations, services, connections, and other information. This layered design effectively reduces the complexity and scale of the attack graph, improves the scalability and adaptability of security modeling, effectively solves the problems of excessive complexity and resource consumption faced by the traditional attack graph model in practical applications, and is more suitable for application in large and complex network scenarios, and avoids duplication by grouping the interactions between independent components in different layers in the construction phase.

2.2 Machine Learning

Aiming at the problems of poor performance and flexibility of traditional cybersecurity assessment methods when dealing with a large amount of cyberattack data, Yang et al [5] proposed a cybersecurity posture assessment method based on adversarial deep learning, where feature learning is performed on DAE networks and DNN networks are used as cyberattack classifiers. Wang et al [6] developed a more flexible alternative FAIR-BN in order to address the limitations of the fair model in order to implement the FAIR model using Bayesian networks (BNs). Containing the features of existing FAIR models while 1) allowing a wider set of distributions to represent and process input variables; and 2) supporting deeper modeling of cyber-attack-defense processes as well as decision-making and evaluation. Wang Jinsong et al [7] analyzed the cyberspace defense capability and its influencing factors, constructed the defense capability assessment index system, used the AHP method to determine the weights of the indexes, and assessed the capability indexes using the successful evaluation data obtained by applying the fuzzy comprehensive evaluation method as an assessment model of Hopfield neural network. However, this type of methodology suffers from the problem of sample bias in data collection and lacks innovation in assessment.

2.3 Security Metrics

Metrics can support objective, repeatable, and quantifiable measurements and ratings, or allow for the calculation of overall system properties. W Moret et al [8] developed a methodology for obtaining metrics with aggregated values from results reported by multiple vulnerability scanners. Arabsorkhi and Ghaffari [9] discuss the literature on metrics that can be used in cybersecurity assessments. In addition, they provide useful metric choices (mainly quantitative) and metric categorization. Due to the development of industrial automation and the application of information technology and intelligent systems in different areas of human life, more and more solutions combine information technology with control elements of physical processes, these systems are called cyber-physical systems, Zegzhda et al [10] describe cybersecurity metrics for cyber-physical systems. The indicator-only approach has some drawbacks, such as the selection of indicators and the determination of weights may be subjective and biased, the collection and matching of indicators may face incomplete or inaccurate data, and the assessment of indicators may ignore the dynamics and complexity of the network system.

3 Model

This paper introduces a quantitative assessment method for network security defense capability based on the attack surface. It constructs a hierarchical attack representation model using open-source scanning tools and system information. The model consists of an upper layer vulnerability dependency graph and a lower layer vulnerability-host information tree, allowing the identification of attack paths or surfaces. The method combines CVSS scoring to obtain attack cost values for the paths, calculates an attacker's score after exploiting different paths, and supports automatic network security assessment.

Security graphical model construction module. This module utilizes publicly disclosed vulnerabilities and network information known to the administrator to create a hierarchical security graphical model with an upper vulnerability dependency graph and a lower vulnerability host information tree. This module accepts information including the host corresponding to the vulnerability CVSS number, vulnerability reachability information, host information, host user privilege information, services, host operating system type, service ports opened by the host, configuration information of the network system, topology information of the network system, and vulnerability data, etc., and builds a hierarchical security graphical model based on the metric data.

Attack path planning module. This module utilizes the vulnerability dependency graph constructed in the hierarchical attack representation model to derive all the existent attack paths by graph traversal method and find the shortest paths by depth-first search algorithm.

Evaluation module. Combines the AC value (Attack Complexity Indicator) of CVSS benchmark metrics and the number of hosts on the path to derive the attack path's cost value. Scores the threat value of the attack path based on asset importance, attack type, and other vectors in the CVSS benchmark metrics. Finally, synthesizes the cost value and threat value to obtain composite scores for the attack paths, allowing the grading of defense capability based on the scores. If undiscovered paths are detected, the defense capability is assigned the lowest level.

4 Procedure

The main steps of the methodology include data collection preprocessing, generation and construction of a hierarchical attack representation model, and evaluation based on each metric of the tool path. The main flow is shown in Fig. 1.

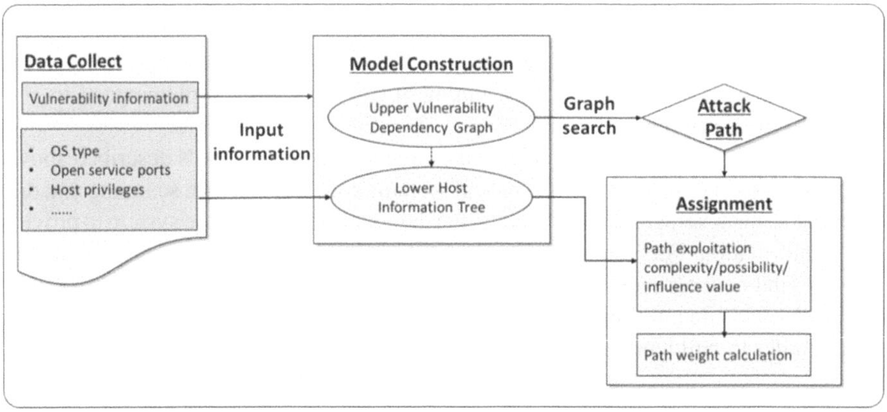

Fig. 1. Main steps in conducting evaluating attack capability based on attack representation model.

4.1 Data Collection and Processing

This step uses the following open source tools: Nmap, OpenVas, and Nessus. Nmap is a common network scanning and enumeration tool that helps detect hosts and their state of being, services, ports, and operating systems in a target network. OpenVas is an open vulnerability assessment system that captures information about security weaknesses and vulnerabilities in systems and software, the severity and impact of the vulnerabilities, and the vulnerability's security score. Nessus is a vulnerability scanning tool that informs what known or unknown security vulnerabilities and misconfigurations are present in a host, their severity and solutions, in addition to first-hand knowledge of the host's compliance status, baseline status and asset information.

Configure the host IP in Nmap, specify the TCP full-connection scanning method to detect which hosts are online, and use the "-0" parameter to get the operating system version, which can be integrated with the operating system information provided by the administrator and compared. You also need to use Nmap's host discovery option to set the parameter "-sn" to Ping scan mode for host discovery, and then use the parameter "-PR" to set the parameter to ARP Ping scan mode for discovering active hosts in the local network, thus generating a network topology. Use OpenVas to set up the scanning host information and scanning target, fill in the name and target IP, configure the corresponding scanning policy, because the defense side of the test, so you can turn off the firewall and other protective equipment, set the tool to full scan. The results of the scan are generated with a variety of information, including the vulnerability criteria score, the percentage of the vulnerability, the description of the vulnerability, the vulnerability number and link. In our method we emphasize on the need for the description, number and link of the vulnerability, the expert classifies the vulnerability into one of four categories based on the relevant description of the vulnerability, namely, theft of user information, destruction of data integrity, destruction of data validity, and destruction of system availability, and thus assigns different scores to the vulnerability. According to the vulnerability of the CVSS number to obtain the vulnerability of the AC and AV value to facilitate the subsequent calculation. The function of Nessus and OpenVas similar to the configuration of the host information can be obtained after the vulnerability of the information, we also extracted the description of the vulnerability and vulnerability number of the information as a comparison of the results of the scanning of OpenVas and complementary, to give the experts more reference material.

In addition to the information collected using the open source tools described above, the framework is not limited to information systems collected through scanning and discovery tools, but leverages the knowledge that defenders have about the system to provide known information. For example, a network administrator may have a network map of devices found on their network, including their configuration information, which can be combined with vulnerability information from scanning tools to link vulnerabilities to hosts in order to build an upper-level vulnerability dependency graph The information collected above is filtered and finally the following metrics are extracted: vulnerability information of the host, operating system type, open service ports, and host privileges.

The information collected above is filtered and the following metrics are finally extracted: host corresponding to the vulnerability CVSS number, vulnerability reachability information, host information, host user privilege information, services, host operating system type, service ports opened by the host, configuration information of the network system, topology information of the network system and vulnerability data.

4.2 Generation and Construction of Model

Vulnerability Dependency Graph Construction. In order to analyze the dependencies between vulnerabilities, we can use the vulnerability dependency graph, a security model that defines all the vulnerabilities and their interactions that exist in devices in the network based on the topology information and vulnerability data of the system. Vulnerability dependency graph, also known as exploit dependency graph, is a kind of attack graph constructed based on the topology information and vulnerability data of the system. The upper vulnerability dependency graph is denoted as $VDG = (V, E)$, where V is the set of points representing all vulnerabilities existing in the devices in the network, and E is the set of directed edges representing the dependencies between vulnerabilities. If vulnerability Y is a successor node of vulnerability X, then it means that in order to exploit vulnerability Y , vulnerability X must be exploited first.

For example, suppose there is a web application with two vulnerabilities: vulnerability X and vulnerability Y. Vulnerability X could be a code execution vulnerability caused by insufficient input validation, while vulnerability Y could be an administrator privilege gained by executing a malicious operation through vulnerability X. In this case, an attacker would need to first use vulnerability X to execute malicious code, and then use vulnerability Y to elevate privileges to take full control of the application. The vulnerability dependency graph in this model focuses only on the vulnerability factors in the network environment and does not involve a lot of state information and host information contained in other types of attack graphs (e.g., state attack graphs and attribute attack graphs), and is therefore relatively simple to analyze. When new nodes appear in the graph, the complexity of dynamic changes is also low, which is conducive to improving the efficiency of analysis.

Vulnerability dependency graphs can be generated using open-source tools for attack graph generation such as MulVal. Input vulnerability information and system configuration information as well as topology, etc., and write them as Datalog rules.

Exploit Path Generation. This method generates attack paths based on the upper vulnerability dependency graph, and employs graph algorithms to measure and analyze the nodes, edges, and paths in the attack graph. Using some concepts and algorithms in graph theory, such as shortest path, average path length, and node importance, the advantages and disadvantages of different attack schemes can be evaluated. Since the vulnerability dependency graph is a directed graph, a vulnerability node with no entry indicates that it can be exploited directly, while a vulnerability node with an entry indicates that its predecessor node needs to be exploited before it can be used, a depth-first search algorithm can be used directly to find out all the paths that can reach the target hosts, and the order of nodes on each path is the order in which vulnerabilities are exploited in the attack.

Host Information Tree Construction. However, the vulnerability dependency graph does not fully reflect the complexity of the actual network environment because it ignores state information of the hosts, such as whether the hosts are alive or not, which ports are open, and what kind of importance they have. If we incorporate both this state information and vulnerability information into the attack graph, the size and complexity of the attack graph will increase dramatically, leading to scalability problems. In order to solve this problem, this method adopts the technique of layered attack model construction, which divides the attack graph into two layers: the upper layer is the vulnerability dependency graph, which represents the dependency relationship between vulnerabilities and is used to analyze the vulnerabilities and risks in the network; the lower layer is the vulnerability-associated host information graph, which represents the multiple sub-targets corresponding to each vulnerability node, including the host information, the importance level of the network device (human-defined), the type of vulnerability (Stealing user information category, Destroying data integrity category, Destroying data validity category, Destroying system availability category) and so on. By doing so, we can reduce the size and complexity of the attack graph while preserving the core information of the attack graph, and improve the scalability of using the security model, especially for large network systems. Where the left sub-tree keeps the CVSS score information and the right sub-tree keeps the corresponding asset information. The attack tree node information is shown in Table 1. And the sample of the graph is shown in Fig. 2.

4.3 Evaluation

Path utilization complexity calculation. First we have obtained the CVE number corresponding to each vulnerability node. Based on the CVE number we can query the CVSS value of the vulnerability, where the AC (Access Complexity) value in the benchmark metrics reflects the difficulty of the conditions required to exploit the vulnerability. In order to calculate the attack cost of this path, we use the AC value as one of the reference values. There are two values for the AC value: Low(L) low complexity, which indicates that the attacker can attack at will and there is no penalty mechanism, and High(H) high complexity, which indicates that the attack can not be accomplished at will, and the attacker needs to invest a lot of preparations on the vulnerable components before the attack succeeds. We also count the number of hosts involved in the path V. In actual attacks, rational attackers tend to choose a set of machines that are closer to the target, less costly, and less risky to compromise, so a higher number of hosts implies a higher attack cost. Taking the above factors into consideration, we define the attack cost formula of the path using formula (1).

$$\cos t = \frac{\sum_{i=1}^{V} ac_i}{V} \tag{1}$$

Calculation of path utilization possibilities. This step uses the AV vector and the number of hosts V from the CVSS benchmark metrics, the AV value is the value of the Attack (Logical/Physical) Distance metric, the farther the distance (the more potential attackers) the greater the value, reacting to the likelihood that the vulnerability will be exploited, and the possible values include Network (N), Neighborhood (A), Local (L), and Physical (P). For example, assume that the AV value in the CVSS benchmark metrics

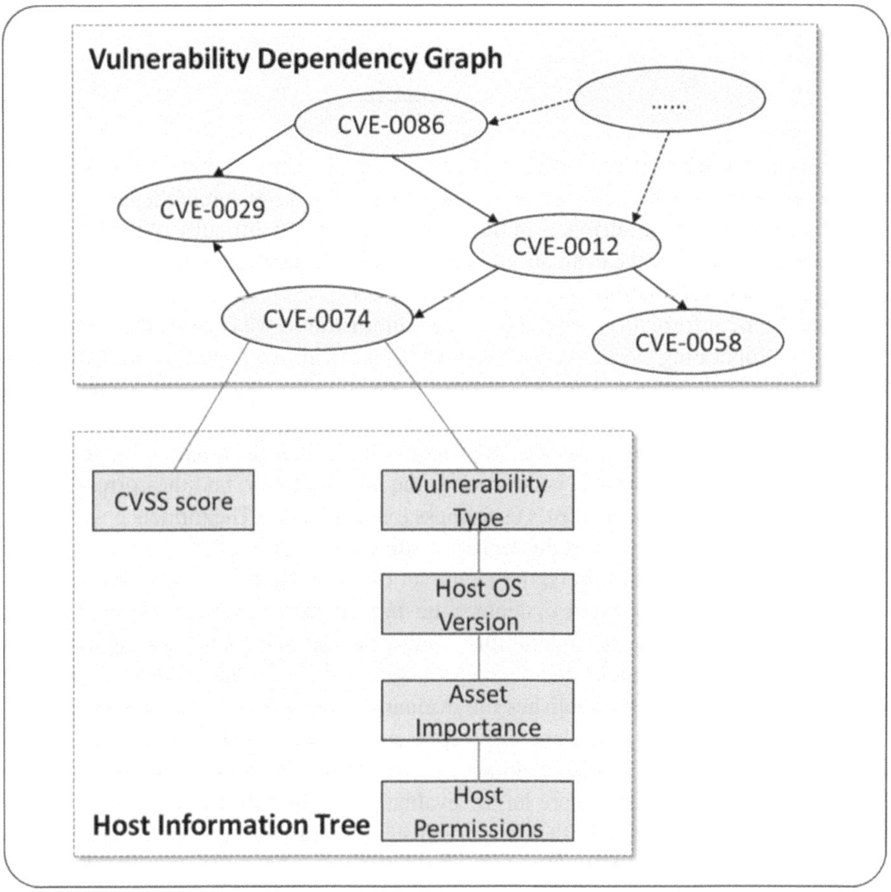

Fig. 2. A sample hierarchical security graph.

Table 1. Attack tree nodes information.

Element Name	Description
CVSS correlation scores	AC/AV
Vulnerability Type	Steal information/Corrupt I/Corrupt V/Corrupt A
Host operating system version	Windows/Linux
Asset importance	A/B/C/D
Host user rights	Administrator rights/general rights

for a particular vulnerability is N (network). This means that the vulnerability can be attacked over a network and the attacker does not need to have direct access to the target host to exploit the vulnerability. With the same idea of complexity, the path likelihood *pro* is calculated according to formula (2).

$$pro = \frac{\sum_{i=1}^{V} av_i}{V} \tag{2}$$

Calculation of path impact values. The above method this method intuitively and simply considers the complexity of the attack path use, but is not comprehensive enough, for example, ignoring the attributes of the nodes, the cost and difficulty of the attack and other factors. Therefore, the evaluation process also considers the lower level values for path impact value calculation.

Based on the information provided by the administrator, we classify the vulnerability type, host operating system type, host operating system version, host user privilege and host device importance. The vulnerability type contains four aspects, namely, theft level, modification level, failure level, and damage level; the host type, e.g., Windows or Linux operating system, the newer the version, the higher the quantitative score; the corresponding asset importance is defined by the administrator; and the corresponding host user privilege reflects the host's usefulness to the attacker. The influence value contains more factors and has a certain degree of subjectivity, so the fuzzy comprehensive evaluation method is used. Firstly, the factor set of comprehensive evaluation is established $U = (u_1, u_2, u_3, u_4)$, where u_i denotes the *ith* factor affecting the evaluation object, where u_1 denotes the type of vulnerability, and u_2 denotes the corresponding operating system version, and u_3 denotes the degree of asset importance, and u_4 denotes the user privilege. The second step establishes the evaluation set for comprehensive evaluation $V = (v_1, v_2, \ldots, v_n)$, where the element v_j represents the *jth* evaluation result, which can be expressed by different grades, comments or numbers according to the needs of the actual situation. For example, here let the evaluation set of vulnerability types be $V = (v_1, v_2, v_3, v_4)$, which represents the category of stealing user information/damaging data integrity/damaging data validity/damaging system availability, respectively. Thirdly, if the affiliation degree of the *ith* element in the factor set U to the first element in the evaluation set V is r_{i1}, then the result of the single factor evaluation of the *ith* element is expressed in fuzzy sets as: $R_i = (r_{i1}, r_{i2}, \ldots, r_{in})$, with m single-factor evaluation sets $R_1, R_2, \ldots, R_m$ as the rows to form a matrix $R_{(m*n)}$, is called the fuzzy comprehensive evaluation matrix. In the fourth step of the evaluation process, the importance of each factor is different, and for this reason, each factor is given u_i a weight a_i, the fuzzy set of the set of weights for each factor, denoted by $A = (a_1, a_2, \ldots, a_m)$. After determining the single factor evaluation matrix R and the factor weight vector A, the fuzzy change from fuzzy vector A on U to fuzzy vector B on V is calculated as in equation (3).

$$B = A_{1*m} * R_{m*n} = (b_1, b_2, \ldots, b_n) \tag{3}$$

After the comprehensive evaluation model is determined, the system score is determined as equation (4).

$$F = B_{1*n} * S^T \tag{4}$$

where F is the path influence value score and S is the rating score of the corresponding factor in V.

Path weight calculation. Combining using complexity, likelihood and impact value, we can derive the path composite value, i.e., weight, by multiplying and normalizing the above three values. The integrated value of the *ith* path sum_i is calculated in formula (5).

$$sum_i = cost_i * pro_i * F_i \tag{5}$$

There are n paths in total, and the weight of the *ith* path is calculated according to formula (6).

$$weight_i = \frac{sum_i}{\sum_n^1 sum_i} \tag{6}$$

Path weights are used in the evaluation as follows: the defender pre-calculates all potential attack paths and their corresponding weights, i.e., the complete definition of the attack surface is obtained. When the attacker invades, the attack paths used by the attacker are organized, and the path weights are used to assign different scores to the attack effects. The score corresponding to each attack path utilized by the attacker is the path weight, using multiple paths i.e., the weights can be added together. Thus, the defense capability score *Def* is calculated according to formula (7).

$$Def = 1 - \sum_{i=1}^{m} weight_i \tag{7}$$

$\sum_{i=1}^{m} weight_i$ is the sum of the attacker's usage path weights. For the defense capability score we make the following Table 2:

Table 2. Mapping of Defense Rating to Composite Score.

Defense Rating	Aggregate Score	Description
D	0.0–0.29	Weak defense, needs to be rectified
C	0.3–0.49	Average defense, needs to be enhanced
B	0.5.0–0.79	Strong defense, needs to be reinforced
A	0.8–1	Stronger defense, keep level

5 Conclusions

Aiming at the lack of existing security assessment methods in the attack surface construction, as well as the problems of network security model complexity is too high and not flexible enough, a quantitative assessment method of network security defense capability based on attack surface is proposed from the defense perspective. The main contributions of this paper are as follows: 1) make full use of the existing open source tools and the knowledge of the defender for the system to collect specific indicators, the operability is strong, and the comparison of the results of multiple tools also ensures that the results of the scientific nature; 2) put forward a hierarchical security representation model, the upper layer of which is the vulnerability dependency graph, the lower layer of which is the associated host information tree, which effectively reduces the scale of the complexity of the attack graph, and connects vulnerabilities with assets, making the results more realistic for the defender. The results are more realistic for the defender; 3) The defender can get the system attack surface according to the attack path through the vulnerability dependency graph, and get the system defense capability through the proportion of the attack surface defeat.

Future research will consider the following three areas. One is to combine the existing assessment methodology with the MITRE ATT&CK attack technique and tactics system, the framework has become the common language of the industry, how to give appropriate weight to the techniques and tactics and make dynamic adjustments is to make the assessment methodology more universal need to think about the problem. Second, the state has introduced a large number of cybersecurity-related standards, and we may add the compliance requirements to the assessment indicators in the future. Finally, the indicators for constructing the hierarchical attack representation model need to be further polished, and more open source tools should be integrated into the security graphical model building module.

Acknowledgements. This research was funded by NSFC (No. 62072131, 62372127, 61972106), the Major Key Project of PCL (Grant No. PCL2022A03-3), Key R&D Program of Guangdong Province (No. 2019B010136003), Guangdong Basic and Applied Basic Research Foundation (No. 2022A1515011401), DongGuan Innovative Research Team Program (No. 2018607201008).

References

1. Ingols, K., Chu, M., Lippmann, R., et al.: Modeling modern network attacks and countermeasures using attack graphs. In: 2009 Annual Computer Security Applications Conference, pp. 117–126. IEEE (2009).
2. Ou, X, Boyer, W.F., McQueen, M.A.: A scalable approach to attack graph generation. In: Proceedings of the 13th ACM Conference on Computer and Communications Security, pp. 336–345 (2006)
3. Lippmann, R., Ingols, K., Scott, C., et al.: Validating and restoring defense in depth using attack graphs. In: MILCOM 2006–2006 IEEE Military Communications Conference, pp. 1–10. IEEE (2006)
4. Ye, Z., Guo, Y., Li, T., et al.: An extended attack graph generation method based on knowledge graph. Comput. Sci. **46**, 165–173 (2019)
5. Yang, H., Zeng, R., Xu, G., et al.: A network security situation assessment method based on adversarial deep learning. Appl. Soft Comput. **102**(8), 107096 (2021)
6. Wang, J., Neil, M., Fenton, N.: A Bayesian Network Approach for Cybersecurity Risk Assessment Implementing and Extending the FAI
7. Wang, J., Li, Z., Xu, Y.: Evaluation of cyberspace defense capabilities based on Hopfield network. Control. Eng. **24**(2), 263–270 (2017)
8. Moret, W.: Vulnerability Assessment Methodologies: A Review of the Literature, p. 360. FHI, Washington, DC (2014)

9. Arabsorkhi, A., Ghaffari, F.: Security metrics: principles and security assessment methods. In: 2018 9th International Symposium on Telecommunications (IST), pp. 305–310. IEEE (2018)
10. Zegzhda, D.P., Poltavtseva, M.A., Lavrova, D.S.: Systematization and security assessment of cyber-physical systems. Autom. Control. Comput. Sci. **51**, 835–843 (2017)

A Link Simulation Technique for LEO Satellite Constellations in the Electromagnetic Environment

Yi Zhang[1], Xiaofeng Wang[1,2](✉), Taiqian Shen[1], and Guoxiu Zhang[1]

[1] School of Artificial Intelligence and Computer Science, Jiangnan University, Wuxi, China
wangxf@jiangnan.edu.cn
[2] Peng Cheng Laboratory, Shenzhen, China

Abstract. Link simulation can provide strong support for the verification and evaluation of new techniques for low earth orbit (LEO) satellite constellations. Aiming at the feature that communication links of LEO satellite constellations are susceptible to electromagnetic environment interference (EEI), we propose a link simulation architecture named LEOLS-Stack. To improve usability, scalability and fidelity, the proposed architecture integrates three key techniques: (1) models construction of natural electromagnetic interference, interstellar electromagnetic interference and artificial electromagnetic interference, which can quickly construct scenarios containing different electromagnetic environments; (2) automatic extraction of interference links, which can automatically divide interference links and calculate link performance data; (3) link simulation combining performance data with interference signals, which can identify the characteristics of interference signals and simulate interference links. Experiments show that constellations in different electromagnetic environments can be quickly built by the interference construction and simulation process. At the same time, the fidelity of link simulation under electromagnetic interference is improved.

Keywords: LEO satellite constellations · electromagnetic environment · network simulation · link simulation

1 Introduction

Low earth orbit (LEO) satellite constellations generally consist of hundreds to tens of thousands of satellites. Thus, they naturally have the characteristics of heterogeneous node types and dynamic changes in network topology [15]. The spatial geometry of LEO satellite constellations changes rapidly, offering the ability to enhance positioning time and range, provide global coverage and improve satellite remote sensing. Due to the short transmission distance, the quality and accuracy of the broadcast signals can be increased. As an effective complement to current global navigation satellite systems, LEO satellite constellations can play an important role in realizing navigation enhancement [5].

Supported by the Major Key Project of PCL (Grant No. PCL2022A03) and the National Natural Science Foundation of China (Grant Nos. 62172191 and 61972182).

Z. Gu et al. (Eds.): NSE 2023, CCIS 2063, pp. 322–335, 2024.
https://doi.org/10.1007/978-981-97-4519-7_23

With the development of satellite manufacturing technology, rocket launching technology, integrated circuit technology and communication technology, the costs for the manufacturing, launching and maintenance of LEO satellites have decreased. However, compared with the construction cost of base stations in ground networks, the construction cost of LEO satellite constellations is still too high. Therefore, to reduce additional construction costs incurred due to technical and system risks [10], there is a need to design an experimental network simulation platform that can meet the requirements of rolling construction and have high fidelity for LEO satellite constellations.

Traditional simulation tools based on discrete event simulation technology, such as commercial software OPNET [17], QualNet/Exata [1], Systems Tool Kit(STK) [16] and open source software NS2/NS3 [14], OMNeT++ [11], Mininet [9]. Although the subsequent development of related modules can meet the needs of link simulation, digital simulation cannot carry real business systems and traffic, so it is not the best choice to build a link simulation platform. However, virtualization technology [12] and cloud computing technology [2] provide new ideas for link simulation scheme design by logically integrating and managing resources, so as to have better simulation capabilities. Link simulation schemes based on virtualization and cloud computing are shown in reference [13]. However, electromagnetic waves are transmitted without relying on physical equipment during the process of satellite communication. Consequently, the links are extremely susceptible to both intentional and unintentional interference. Regardless of the type of electromagnetic interference, it will affect the communication quality of the link, which is particularly important to study link simulation in the electromagnetic environment.

Therefore, we propose a link simulation platform named LEOLS-Stack (Link Simulation Stack for LEO satellite constellations) based on virtualization and cloud computing technology. LEOLS-Stack can improve the fidelity of link simulation under electromagnetic interference. First, we propose the models construction technique of nature electromagnetic interference, interstellar electromagnetic interference and artificial electromagnetic interference, which can quickly construct scenarios containing different electromagnetic environments. Then, we propose an automatic extraction algorithm of interference links, which can automatically divide interference links and calculate link performance data. Finally, a link simulation technique combining performance data with interference signals is presented, which can identify the characteristics of interference signals and simulate interference links.

Based on the above considerations, the main contributions of this paper can be summarized as follows:

(1) A link simulation platform LEOLS-Stack based on virtualization and cloud computing technology is proposed.
(2) By virtue of the models construction technique of natural electromagnetic interference, interstellar electromagnetic interference and artificial electromagnetic interference, automatic extraction algorithm of interference links, and link simulation technique combining performance data with interference signals, LEOLS-Stack can construct scenarios containing different electromagnetic environments and simulate links.

(3) LEOLS-Stack improves the usability, scalability and fidelity of the link simulation platform.

2 Related Work

The link simulation solution based on virtualization [12] and cloud computing [2] has become a trend currently, which can carry real business systems and traffic and become the main solution in the simulation platform for LEO satellite constellations.

For link simulation, the celestial body integration network proposed in [18] used software-defined networking (SDN) technology. A flow table isolation strategy and a flow table forwarding strategy were designed to deliver flow tables so as to realize time-varying topology simulation. However, a dynamic time-varying topology leads to insufficient real-time dynamics and insufficient accuracy. Wang et al. [7, 8, 13] used virtualization, SDN, traffic control (TC), Docker and KVM to simulate network nodes. Their method improves the simulation performance of a platform but cannot simulate electromagnetic environments in the scenario of constellations.

The above work has certain advantages in link simulation but it does not consider electromagnetic interference. Dong et al. [4] investigated the electro-magnetic interference mechanism between the automatic identification system (AIS) and the satellite sensor, which can realize the analysis and suppression of artificial electromagnetic interference but only artificial electromagnetic interference can be simulated. Cao et al. [3] carried out the application of space physical collision theory in electromagnetic interference of satellite equipment through numerical simulation and analysis based on the theoretical model of physical collisions of space objects, but these authors were unable to build a model of artificial electromagnetic interference. Kim et al. [6] used the simplified Appleton-Hartree equation to construct the ionospheric environment and calculated the atmospheric attenuation of a satellite dish from an LEO satellite to a ground station under different conditions; however, these authors did not achieve satisfactory results in constructing models of interstellar electromagnetic interference or artificial electromagnetic interference.

In summary, through various techniques, the fidelity and accuracy of simulations in this field have been improved to a certain extent. However, the electro-magnetic interference models and the existing methods for the correct extraction and simulation of interference links still suffer from deficiencies of varying degrees, which limit the ease of use, scalability and fidelity of simulation platforms.

3 Architecture of LEOLS-Stack

Figure 1 shows the overall architecture of LEOLS-Stack, which is based on the OpenStack cloud platform. SAGINHTE-Stack mainly includes a physical layer, a device layer and a service layer, which can be used to construct scenarios containing different electromagnetic environments and simulate links.

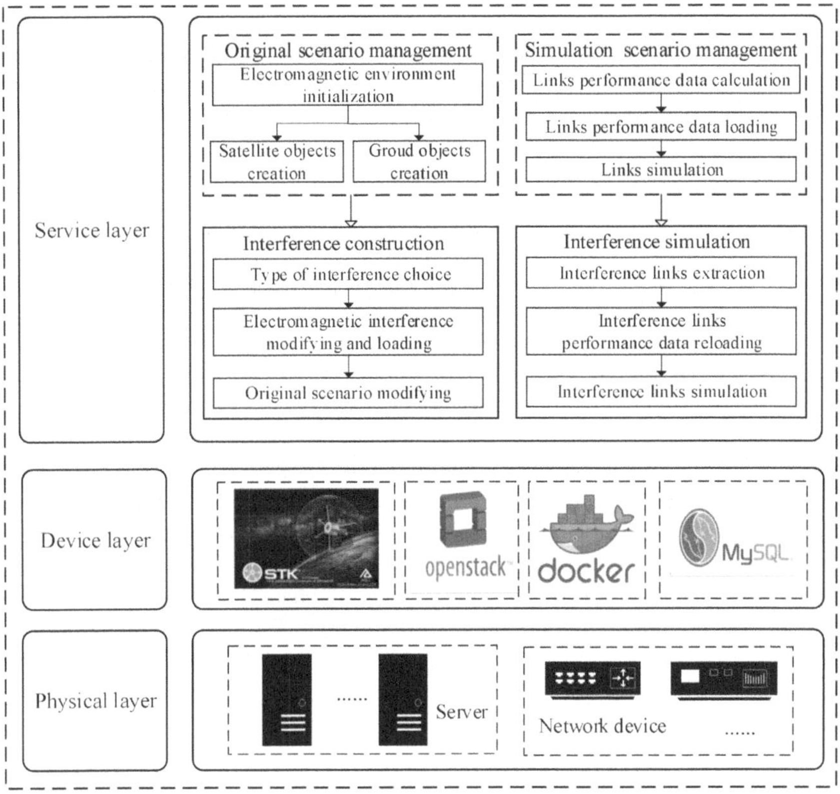

Fig. 1. Architecture of LEOLS-Stack.

3.1 Physical Layer

The role of the physical layer is to provide physical devices for the entire system, including physical resources such as servers and network devices, and to manage and maintain these physical resources. At the same time, the physical layer can scale the system by increasing or decreasing the physical resources.

3.2 Device Layer

The role of the device layer is to build the software devices required by the entire system, including STK, OpenStack, Docker and Mysql, and provide these software devices to the service layer.

OpenStack is the foundation for simulation environment deployment, providing virtualization and cloud computing technology for simulation. STK is a management tool that provides link performance data accuration for link simulation in scenarios. As a containerization technology, Docker has the advantages of faster startup and less resources than traditional virtual machines. Mysql manages all kinds of data in an orderly manner and provides unified interfaces. Detailed application of the device layer is shown in reference [13].

3.3 Service Layer

The role of the service layer is to further encapsulate the functionality and interfaces provided by the physical and device layer. The function of the model of original scenario management is to build the scenario of the satellite and ground object, and the interference construction module is mainly responsible for constructing electromagnetic interference after the scenario is built. The function of the model of simulation scenario management is to simulate all links in real time, and the interference simulation module is mainly responsible for extracting and simulating interference links.

Interference types of natural electromagnetic interference, interstellar electromagnetic interference and artificial electromagnetic interference are different, so it is necessary to select models of three different electromagnetic interference to construct different scenarios in the interference construction module, of which three different electromagnetic interference models are specifically shown in Sect. 4.1. After we realize the interference links extraction in the interference simulation module, which is used to finally complete the link simulation in various electromagnetic environments. Interference links analysis and simulation are exhaustively shown in Sect. 4.2.

4 Implementation

This section describes the core methods and shows how the architecture constructs electromagnetic interference models and completes the correct extraction and simulation of interference links in various electromagnetic environments.

4.1 Models Construction of Electromagnetic Interference

Natural Electromagnetic Interference. The natural electromagnetic environment affects communication links through attenuation of different types and intensities. Three types of attenuation are supported by STK: a rain attenuation, a cloud and fog attenuation and an atmosphere attenuation. The construction of these models is relatively complicated, and STK does not directly provide a way to customize their construction but instead uses a combination of attenuation models and other parameters provided by the International Telecommunication Union Radiocommunication Sector (ITU-R) to construct the natural electromagnetic environment. The model is summarized in Table 1.

Table 1. Model of natural electromagnetic interference.

Type	Model	Parameter
Rain	Crane 1985 ITU-R P618-12 CCIR 1983 Crane 1982 ITU-R P618-10 ITU-R P618-5 TU-R P618-8 ITU-R P618-9	surface temperature
cloud and fog	ITU-R P840-6 ITU-R P840-3	cloud ceiling cloud layer thickness cloud temperature cloud liquid water density
atmosphere	ITU-R P676-9 Simple Sactom TIREM 3.31 ITU-R P676-3 ITU-R P676-5	water vapor surface temperature

Three attenuation types are documented in the table, of which there are eight models of rain attenuation, two models of cloud and fog attenuation, and five models of atmosphere attenuation. The various parameters under the model are recorded in "Parameter".

Interstellar Electromagnetic Interference. When STK is directly used to construct the links between LEO satellites and ground stations, STK will analyze and calculate the communication links from two different satellites to the same ground station dynamically, neglecting any possible interference between the two links. However, the "CommSystem" object of STK can be used to obtain link performance data between a transmitter and a receiver when a interferer interferes. Therefore, it is necessary to study the "CommSystem" object of STK to construct a model of interstellar electromagnetic interference.

The main function of the "CommSystem" object is to analyze the communication performance and anti-interference performance of links. The "CommSystem" object can be associated with various "Constellation" objects as its subobjects, which consists of transceivers from different satellites and these subobjects can be set as transmitters, receivers and interferers. Therefore, the model of interstellar electromagnetic interference can be designed hierarchically, as shown in Table 2.

Where "CommSystem" is a list of all interstellar electromagnetic interference situations. "name" and "Constellation" is the first-level property of "commsystem". "transmitter", "receiver" and "interferer" is the second-level property of "CommSystem", which contains the name information.

Table 2. Model of interstellar electromagnetic interference.

Type	L1 Property	L2 Property	Details
CommSystem	name	/	/
	Constellation	transmitter	name
		receiver	name
		interferer	name

Artificial Electromagnetic Interference Due to the variety of technical means of generating artificial interference, its strong real-time nature, and its great impact on the quality of communication, this paper considers downlink interference in satellite-ground communication based on the use of space-based platforms such as unmanned aerial vehicle(uav) to simulate communication links affected by artificial electromagnetic interference. Accordingly, the model of artificial electromagnetic interference is designed in Table 3.

Table 3. Model of artificial electromagnetic interference.

Type	Property	Details
uav	name	/
	start_time	/
	waypoints	latitude, longitude, altitude, speed, turn radius
	transmitter	name, frequency, eirp, rate
	target	/

Among them, "uav" represents an unmanned space-based platform equipped with a interferer, which is the "Aircraft" object in STK and mainly contains properties of "name", "start_time", "waypoints", "transmitter" and "target". Properties of "waypoints" and "transmitter" can expand the detailed description.

4.2 Interference Links Analysis and Simulation

After the simulation scenario with electromagnetic interference is built, subsequent link simulation is required. The general process only needs to recalculate the link performance data of all links, and load the link performance data into the simulation scenario. However, not all links will be affected for interstellar electromagnetic interference, so an automatic extraction algorithm of interference links is proposed. It is necessary to further realize interference identification for artificial electromagnetic interference, so a link simulation technique combining performance data with interference signals.

Automatic Extraction Algorithm of Interference Links In STK, the calculation of link performance data on the direct link between a transmitter and a receiver does not consider the interference situation, and all interference links can only be analyzed and calculated by the "CommSystem" object. Meanwhile, the link analyzed by the "CommSystem" object comprises all communication links from the transmitter to the receiver. The automatic extraction algorithm of interference links is shown in Algorithm 1.

Algorithm 1 Automatic extraction algorithm of interference links.

```
Input: link_info
    //link_info contains the information of receivers, transmitters and links' detail.
Output: inter_link
    //inter_link contains the details of interference links.
recv_name ← link_info.receiver_info.receiver_name
trans_info ← link_info.transmitter_info
id_names ← empty dictionary
links ← link_info.link_detail
for link in links do
    link_name ← GetLinkNameByTransAndRecv(id_names[link.link_to_id],
    recv_name) //find the name of source and destination.
    if link_name not in inter_link then
        inter_link[link_name] ← empty dictionary
        inter_link[link_name]['time'] ← empty list
        inter_link[link_name]['ber'] ← empty list
    end if
    if link.(BER + I) == link.BER then
        //BER+I means the BER with interference
        Delete inter_link[link_name]
        Continue
    end if
    Append link.time To inter_link[link_name]['time']
    Append link.(BER + I) To inter_link[link_name]['ber']
end for
return inter_link
```

The algorithm takes the link information, which is the analysis result of the "CommSystem" object as input. First, we obtain the receiver name and transmitter information in the link information, where the transmitter information includes the transmitter name and ID, and the transmitter name corresponds to the ID. Second, based on the obtained link name, we obtain the specific link information data and decide whether to save the link data combined with the data before and after interference. Finally, the obtained information of interference links can be returned. Suppose that there are m transmitters, n receivers, j links, and k interference links; then, the time complexity of the algorithm is O(mn), and the spatial complexity is O(m) + O(k).

Link Simulation Combining Performance Data with Interference Signals. As a kind of interference that can be identified and countered, artificial electromagnetic interference can further improve the realism of simulation by reflecting the category of artificial electromagnetic interference through simulation, so the simulation process of interference link is expanded, and a link simulation technique combining performance data with interference signals is proposed. The main simulation process is shown in Fig. 2.

In the interference link simulation step, the simulation is divided into two parts, one part continues the original link performance simulation, and the other part is responsible for managing interference signals. The interference signal management part in the setting of artificial interference parameters, the parameters are loaded into the interference simulation node. The simulation node according to the received interference parameters uses the mathematical model to generate the corresponding interference signal and store the interference signal. When artificial electromagnetic interference is found in the simulation process, the simulation node will send the interference signal to the experimenter. Finally, the experimenter draws the spectrogram of the interference signal through the returned interference signal data.

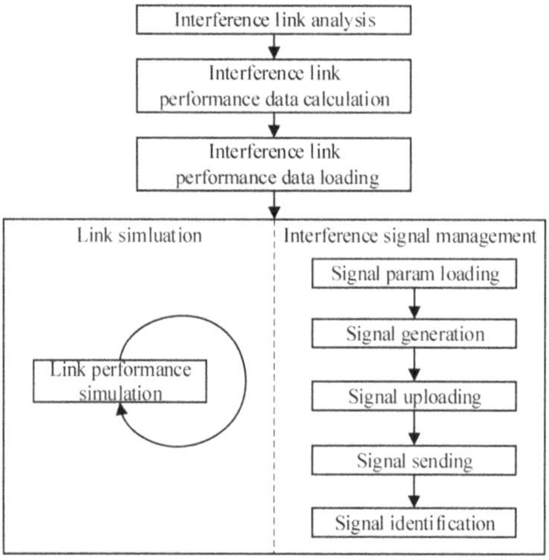

Fig. 2. The asynchronous link simulation process.

5 Experiments

5.1 Experimental Environment

The experimental environment discussed in this section consists of 1 Dell host and 2 Dell servers. The Windows 10 operating system is installed on the Dell server, and it is also equipped with the STK software, which is mainly used to generate STK simulation scenarios, calculate link data and load various electro-magnetic environments. The Centos 7.2 operating system is installed on 2 of the Dell servers. OpenStack Mitaka mainly used to build the simulation environment and create the simulation network is also deployed on the servers.

The experimental scenarios are depicted in Fig. 3. Figure 3(a) shows the first 3 satellite orbits of the Starlink satellite constellation with a total of 66 satellites and 3 ground stations (A, B, and C) that are deployed on the 3 orbital subsatellite trajectories for satellite-ground communication. Figure 3(b) additionally includes a new LEO satellite constellation (OneWeb) with a total of 108 satellites and a fourth ground station (D) on the basis of Fig. 3(a).

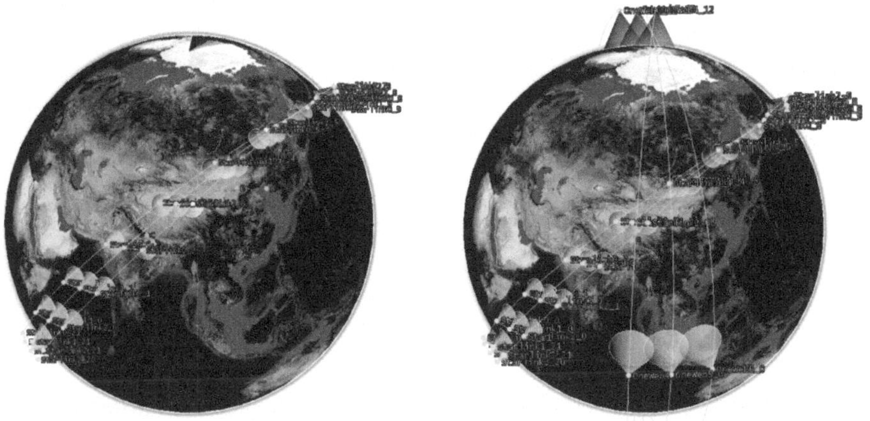

(a) Experimental scenario A (b) Experimental scenario B

Fig. 3. The experimental scenarios.

5.2 Results of Link Simulation in the Natural Electromagnetic Environment

We consider the scenario depicted in Fig. 3(a) in this analysis. Each satellite may be connected to a ground station, and the number of possible communication links between the 66 satellites and 3 ground stations is 198, too many to be convenient for analysis. Therefore, for this scenario, the simulation results for several specific links are selected for verification and analysis, namely Starlink1_20-A, Starlink0_11-B and Starlink1_12-C.

Due to limited space, only the model of cloud and fog attenuation is loaded into the scenario as shown in Table 4. The link performance data for satelliteground communication during the selected time period are extracted from the original scenario and the simulation scenario. The results are shown in Fig. 4.

Table 4. Model of cloud and fog attenuation.

Type	Model	Parameter
cloud and fog	ITU-R P840-3	cloud ceiling = 3 km cloud layer thickness = 0.5 km cloud temperature = 0 °C cloud liquid water density = 7.5 g/m^3

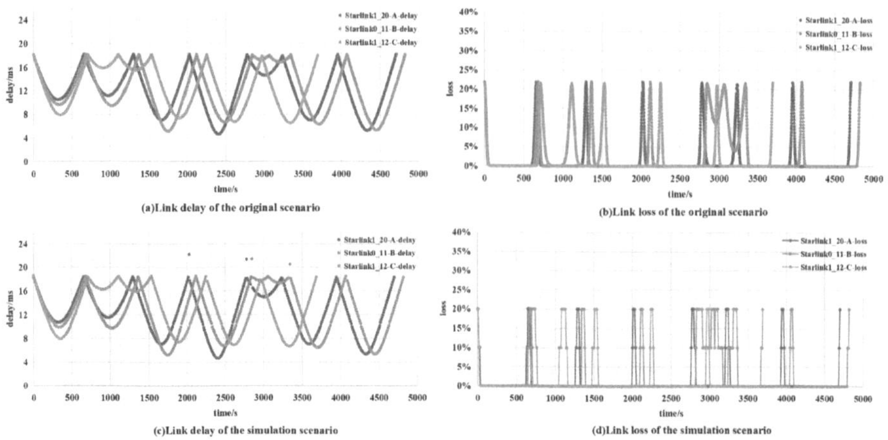

Fig. 4. Data comparison under the model of cloud and fog attenuation.

In the diagram of the experimental results, it can be calculated that the delay error of the simulation is between 0.12% and 6.53% and the error in the packet loss rate ranges from 0% to 13.45%, indicating high simulation fidelity.

5.3 Results of Link Simulation in the Interstellar Electromagnetic Environment

We consider the scenario depicted in Fig. 3(b) in this analysis. We select satellites 1, 3 and 15 in orbit 0, satellites 6, 14 and 15 in orbit 1, and satellites 3, 6 and 14 in orbit 2 on the Starlink constellation; On the Oneweb constellation, we select satellites 0, 12 and 24 in orbits 4, 5 and 6. For ground station D, the model of interstellar electromagnetic interference is shown in Table 5.

Table 5. Model of interstellar electromagnetic interference.

Type	L1 Property	L2 Property	Details
CommSystem	name	/	sat_fac
		transmitter	Starlink0_1, Starlink0_3, Starlink0_15, Starlink1_6, Starlink1_14, Starlink1_15, Starlink2_3, Starlink2_6, Starlink2_14
	Constellation	receiver	ground station D
		Interferer	OneWeb4_0, OneWeb4_12, OneWeb4_24, OneWeb5_0, OneWeb5_12, OneWeb5_24, OneWeb6_0, OneWeb6_12, OneWeb6_24

From the above table, We can extract 5 interference links: Starlink0_1-D, Starlink0_3-D, Starlink0_15-D, Starlink1_15-D and Starlink2_14-D. Comparative data on link performance are shown in Fig. 5.

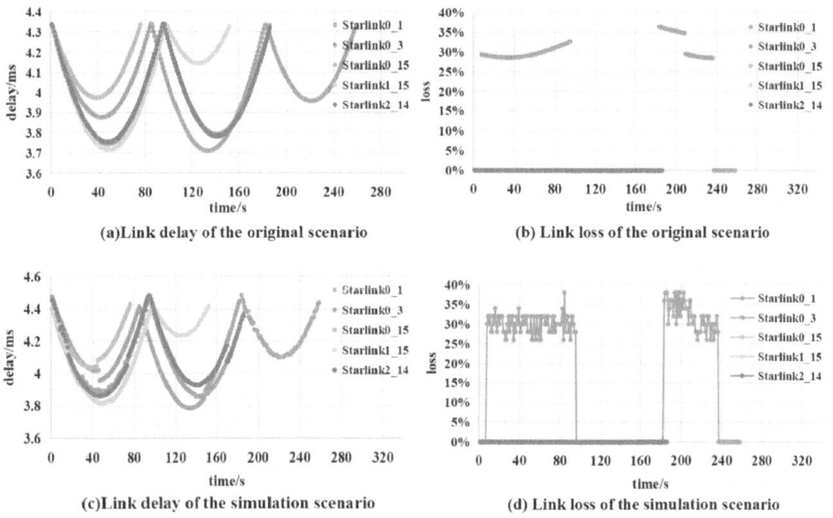

Fig. 5. Simulation results for the links connected to ground station D.

From the data comparison in Fig. 5, it can be calculated that the delay error of the simulation is between 0.01% and 4.39% and the error in terms of the packet loss is between 0% and 14.03%, indicating high simulation fidelity.

5.4 Results of Link Simulation in the Artificial Electromagnetic Environment

We again use Fig. 3(b) for this analysis. The model of the artificial electromagnetic environment is shown in Table 6. To verify the accuracy and fidelity of the link simulation, the link performance data in the original scenario are compared with the simulation scenario, and the results are shown in Fig. 6.

Table 6. Model of artificial electromagnetic interference.

Type	Property	Details
uav	name start_time waypoints transmitter target	UAV 1 Aug 2023 04:00 latitude = [48.13°,48.05°], longitude = [102.02°,99.63°] altitude = [10.66 km, 10.66 km], speed = [0.18 km/s, 0.18 km/s] turn radius = [11.68 km, 11.68 km] name = Manual, frequency = 12 GHz eirp = 30 dBW, rate = 20 Mbps ground station C

From the data comparison in Fig. 6, it can be calculated that the delay error of the simulation is between 0.5% and 3.17% and the error of in terms of the packet loss

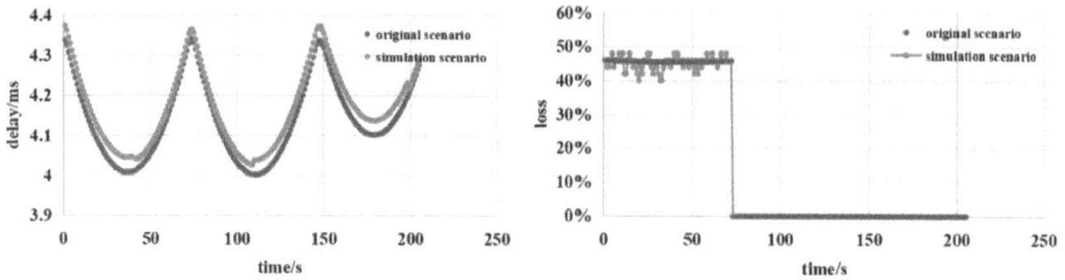

(a)Delay comparison of the scenario (b)Loss comparison of the scenario

Fig. 6. Link comparison between the original scenario and simulation scenario.

ranges from 0% to 12.22%, indicating high simulation fidelity. Next, for the analysis of the power spectrum characteristics of noise amplitude modulation interference and noise frequency modulation interference, it can be seen that the signal data in the figure conforms to the signal data of these two types of noise interference, which shows link simulation under artificial electromagnetic interference can be realized.

6 Conclusion

In this paper, a link simulation architecture named LEOLS-Stack based on the OpenStack cloud platform is proposed. The proposed architecture integrates three key techniques: (1) models construction of natural electromagnetic interference, interstellar electromagnetic interference and artificial electromagnetic interference; (2) automatic extraction of interference links; (3) link simulation combining performance data with interference signals. Experiments show that these techniques proposed in this paper can be used to can construct scenarios containing different electromagnetic environments and simulate links, improving the fidelity of link simulation under electromagnetic interference. Therefore, the work presented in this paper can provide strong support for the verification and evaluation of new techniques for LEO satellite constellations.

In future work, the integration of multiscale virtualization technology will be considered to carry out research on high-performance and high-fidelity simulation methods for LEO satellite constellation networks with larger scale and more complex electromagnetic environments.

Acknowledgements. This work was supported by the Major Key Project of PCL (Grant No. PCL2022A03) and the National Natural Science Foundation of China (Grant Nos. 62172191 and 61972182).

References

1. Alsafwani, N., Ali, M.A.M., Tahir, N.M.: Evaluation of the mobile ad hoc network (manet) for wormhole attacks using qualnet simulator. In: 2021 IEEE 11th International Conference on System Engineering and Technology (ICSET), pp. 46–49 (2021)
2. Asim, M., Wang, Y., Wang, K., Huang, P.: A review on computational intelligence techniques in cloud and edge computing. IEEE Trans. Emerg. Top. Comput. Intell. **4**(6), 742–763 (2020)

3. Cao, B., Gao, K., Xing, J., Jiang, T.: Emi analysis among leo satellites based on space conjunction theory. In: 2020 IEEE 3rd International Conference on Electronic Information and Communication Technology (ICEICT), pp. 375–378 (2020)
4. Dong, G., Li, Q., Jiang, X., Chen, S., Chen, X.: The analysis and suppression of electromagnetic interference on satellite automatic identification system. In: 2022 Asia-Pacific International Symposium on Electromagnetic Compatibility (APEMC), pp. 470–472 (2022)
5. Hou, L., Kang, S., Sun, S., Miao, D., Han, B., Zhang, M.: A load balancing routing method based on real time traffic in leo satellite constellation space networks. In: 2022 IEEE 95th Vehicular Technology Conference: (VTC2022-Spring), pp. 1–5 (2022)
6. Kim, C., Na, D.Y., Park, Y.: Electromagnetic wave propagation from low-earth orbit satellite to ground station considering interpolated atmospheric environments. IEEE Access **9**, 95853–95861 (2021)
7. Lai, J., Tian, J., Zhang, K., Yang, Z., Jiang, D.: Network emulation as a service (NEaaS): towards a cloud-based network emulation platform. Mobile Netw. Appl. **26**(2), 766–780 (2021)
8. Liang, Y., Min, M., Yu, Y., Xi, W., Xia, P.: Assessment on the diurnal cycle of cloud covers of fengyun-4a geostationary satellite based on the manual observation data in china. IEEE Trans. Geosci. Remote Sens. **61**, 1–18 (2023)
9. Lin, T., Chen, F., Zhao, K., Fang, Y., Li, W.: Mininete: a lightweight emulator for space information networks. In: Wu, Q., Zhao, K., Ding, X. (eds.) Wireless and Satellite Systems WiATS 2020. Lecture Notes of the Institute for Computer Sciences, Social Informatics and Telecommunications Engineering LNICSSITE, vol. 357, pp. 48–57. Springer, Cham (2021). https://doi.org/10.1007/978-3-030-69069-4_5
10. Liu, S., Lin, J., Xu, L., Gao, X., Liu, L., Jiang, L.: A dynamic beam shut off algorithm for leo multibeam satellite constellation network. IEEE Wirel. Commun. Lett. **9**(10), 1730–1733 (2020)
11. Nardini, G., Sabella, D., Stea, G., Thakkar, P., Virdis, A.: Simu5g–an omnet++ library for end-to-end performance evaluation of 5g networks. IEEE Access **8**, 181176–181191 (2020)
12. Valsamas, P., Mamatas, L., Contreras, L.: A comparative evaluation of edge cloud virtualization technologies. IEEE Trans. Netw. Serv. Manage. **19**(2), 1351–1365 (2022)
13. Wang, X., Chen, X., Ye, H., Liu, Y., Zhang, G.: Cloud-based experimental platform for the space-ground integrated network. Wirel. Commun. Mob. Comput. **2020**, 1–20 (2020)
14. Wang, Z., Cui, G., Li, P., Wang, W., Zhang, Y.: Design and implementation of ns3-based simulation system of LEO satellite constellation for IOTS. In: 2018 IEEE 4th international conference on computer and communications (ICCC), pp. 806–810. IEEE (2018)
15. Wu, C., Toosi, A.N., Buyya, R., Ramamohanarao, K.: Hedonic pricing of cloud computing services. IEEE Trans. Cloud Comput. **9**(1), 182–196 (2021)
16. Yong, D., Song, J., Dayang, W., Yang, B., Huichen, X., Jincheng, Z.: A stkbased constellation architecture implementation for 5g low-orbit satellites. In: 2022 IEEE 4th International Conference on Power, Intelligent Computing and Systems (ICPICS), pp. 602–606 (2022)
17. Zhang, L., Long, F., Zhao, G., Liang, Z.: A joint simulation framework based on opnet external system module for tactical data links. In: 2022 3rd International Conference on Information Science and Education (ICISE-IE), pp. 210–217 (2022)
18. Zhang, W.: Dynamic Link Emulation for Space-ground Integrated Network. Master's thesis, University of Electronic Science and Technology of China (2018)

A Cybersecurity Evaluation Oriented Attack Scheme Generation System

Mingsheng Yang, Jie Yang, Weihong Han(✉), and Jiawei Zhang(✉)

Peng Cheng Laboratory, Shenzhen 518000, China
{hanwh,zhangjw01}@pcl.ac.cn

Abstract. As the scale of network attacks continues to expand, cyber-security evaluation becomes crucial. The attack graph method provides a visual representation of attack paths, helping to identify vulnerabilities and optimize defense measures. However, the generation of attack graphs often faces the problem of exponential growth in size, and generating a complete graph is not always necessary. This paper presents a system that generates attack schemes in real-time based on targets and constraint conditions. This system helps improve the efficiency and effectiveness of cybersecurity evaluation.

Keyword: Cybersecurity evaluation · Attack graph · Attack scheme

1 Introduction

With the increasing scale of network attacks, cybersecurity evaluation has become increasingly important. In the field of cybersecurity evaluation, attack graph is an important method that provides a visual representation of potential attack paths that attackers may exploit. It helps defenders identify vulnerabilities in network systems, optimize defense measures, and enhance the security of network systems.

The attack graph method was initially proposed by Cunningham [1] in 1985 and quickly gained widespread recognition among experts and scholars. A considerable amount of research has been conducted domestically and internationally [2], proposing various methods for generating [3-7], optimizing [8, 9], and analyzing the application of attack graphs [10-17]. Based on the attack graph, the risk levels of various nodes in the network can be calculated [18], critical nodes in the network can be identified [19], and the benefits of attack-defense strategies can be quantified [20]. It's a commonly used method for cybersecurity evaluation. In a cyberrange oriented cybersecurity evaluation system, by simulating the target network in a cyber range, it is possible to use vulnerability scanning tools to scan for vulnerabilities and weaknesses within the target network. Based on the scan results, attack graphs can be generated. Then, by analyzing the attack graph and searching for attack paths for each node, the risk and exploitability of the attack paths can be quantitatively evaluated, and the risk levels of the nodes can be calculated. This information provides an overall risk evaluation of the target network. Currently, the analysis and processing of attack graphs mainly rely on manual intervention by evaluators, which is inefficient and may compromise the accuracy of the analysis

Z. Gu et al. (Eds.): NSE 2023, CCIS 2063, pp. 336–350, 2024.
https://doi.org/10.1007/978-981-97-4519-7_24

results. Furthermore, as the scale and complexity of the network increase, processing large-scale network clusters becomes increasingly challenging due to the exponential growth in the size and complexity of attack graphs. The presence of redundant information in attack graphs, such as multiple attack paths sharing the same components and vulnerabilities, as well as relatively long attack paths, does not significantly affect the evaluation results.

In summary, this paper proposes an attack scheme generation system for cybersecurity evaluation systems. By analyzing system vulnerability information, it identifies host nodes that can be attacked within the network and captures the attack state transition information between these host nodes. Through visualization, the system presents this information to the user, allowing them to select specific targets and automatically generate attack schemes. The system provides recommended attack steps and quantified evaluation values for different attack paths, assisting security assessors in conducting attack testing and system risk evaluation.

2 System Architecture

The system consists of several modules, including reasoning rule management, vulnerability information management, vulnerability information parsing, attack scheme generation, visualization, and database, as shown in Fig. 1.

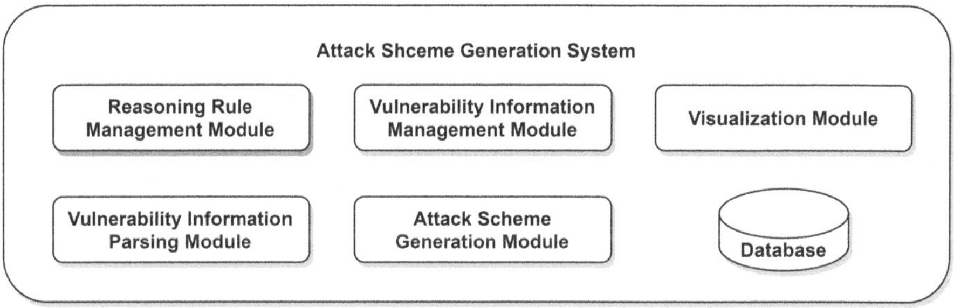

Fig. 1. Architecture diagram of attack scheme generation system.

Reasoning Rule Management Module: The reasoning rule management module manages the reasoning rules upon which the vulnerability information parsing relies, mainly include vulnerability exploitation, compromise propagation and multi-hop network access rules. The module provides interfaces for creating, querying and modifying rule data. The system defines a set of default reasoning rules that users can directly use. Users also have the option to create their own custom rules.

Vulnerability Information Management Module: The vulnerability information management module is responsible for managing vulnerability information and the attack steps information associated with exploiting those vulnerabilities. Vulnerability information is obtained from publicly available information provided by organizations

Table 1. Vulnerability information.

Name	Description	Example
vul_id	Vulnerability Id	"CVE-2023-27522", "CNNVD-202201-1572"
severity	Vulnerability severity level	Critical, High, Medium, Low
availability	The impact on the availability of the target system	None (N), Partial (P), Complete (C)
confidentiality	The impact on the confidentiality of data processed by the system	None (N), Partial (P), Complete (C)
integrity	The impact on the integrity of the exploited system	None (N), Partial (P), Complete (C)
access	Difficulty of exploiting the discovered vulnerability	High (H), Medium (M), Low (L)
exploitability	The exploitability scores of the vulnerability based on baseMetricV3	3.9
impact	The impact scores of the vulnerability based on baseMetricV3	3.6
base	The base scores of the vulnerability based on baseMetricV3.It represent characteristics of the vulnerability itself, have three subgroupings: Exploitability Metrics, Impact Metrics and Scope	7.5

such as CVE, includes key details such as vulnerability type, identifier, severity level, attack complexity, exploitability scores, impact scores, and so on, as shown in Table 1.

The vulnerability attack step information primarily includes detailed step-by-step information on how to exploit the vulnerability, such as recommended attack tools, operating steps, as shown in Table 2.

Users can query and modify vulnerability and vulnerability attack step information. Additionally, when new vulnerability information is released, users or administrators need to promptly import the new vulnerability information through the vulnerability management function and update the attack step information for exploiting that vulnerability.

Vulnerability Information Parsing Module: The vulnerability information parsing module is responsible for parsing and processing the input vulnerability scanning information. It identifies exploitable vulnerability information and, based on network configuration information, user account information and reasoning rules, identifies all attack state transition information along the attackable paths, as shown in Table 3.

Attack Scheme Generation Module: The attack scheme generation module is responsible for generating attack schemes specifically targeted towards a given target. Since there can be multiple attack paths for the same target, users can set the threshold of attack paths to be searched. Additional constraint conditions can also be set simultaneously,

Table 2. Attack step information.

Name	Description	Example
vul_id	Vulnerability Id associated with the attack step	"CVE-2023-27522", "CNNVD-202201-1572"
step_no	Attack step number	1
step_info	Detailed information about the attack step	Run the nmap tool to scan target ports

Table 3. Attack state transition information.

Name	Description	Example
source	Source host	192.168.1.1
destination	Destination host	192.168.1.2
vul_id	Vulnerability Id	"CVE-2023-27522", "CNNVD-202201-1572"
soft	Software with the vulnerability	http_server_11.1.1.9.0
protocol	Network protocols exploited in the attack	tcp
port	Port numbers exploited in the attack	80

such as including specified host nodes or the attack complexity is below a certain threshold, etc. After finding the attack path, for each host node along the path, recommended attack steps are provided based on the exploited vulnerability information, resulting in the generation of attack schemes.

Visualization Module: The visualization module is responsible for input and output operations, as well as graphically displaying the vulnerability parsing results and attack schemes. It provides a user-friendly interface for inputting relevant information and displaying the generated attack paths and attack schemes in a graphical format.

Database: The database is used to storing the reasoning rules, vulnerability information, parsing results, generated attack schemes, and other system configuration parameters. It manages a database or storage mechanism where these data elements are stored for easy retrieval and access by the system.

3 Processing Flow

First, a simulated system of the target network to be evaluated is created in the cyber range. Then, vulnerability scanning tools are used to get the vulnerability information. The scan results and network configuration information are inputted into the system for processing, mainly including vulnerability information parsing and attack scheme generation.

3.1 Parsing Vulnerability Information

The vulnerability information parsing process refers to the implementation of the MulVAL system [3]. It parsing vulnerability information based on reasoning rules, and the reasoning rules follow the Datalog format, where variables start with uppercase letters and constants start with lowercase letters. Datalog is an improved form of Prolog adapted for knowledge bases, and it is a restricted version of the Prolog language. A DataLog rule consists of three parts:

1. Rule head: A relational atom (e.g., student(x)).
2. Symbol: It can be read as "if", indicates a logical relationship, not an operator.
3. Rule body: One or more subgoals, denoted as L1, L2, ..., Ln. Subgoals can be relational atoms or arithmetic atoms (where the arithmetic result is a boolean value, e.g., a > 10). The subgoals are connected with AND.

The meaning of a rule is described as follows: Check all possible values of the variables in the rule. When these variables satisfy all the subgoals in the rule body, the rule head is true. A clause with an empty body is called a fact, and a clause with a non-empty body is called a rule.

The user's input of vulnerabilities, network configuration information, and reasoning rules are fed into the XSB engine for reasoning. Possible attack paths are determined based on the relationships between vulnerabilities, the attacker's privileges, and the paths of attack propagation.

After the reasoning process is completed, the output from the engine is processed to extract information about vulnerable hosts, vulnerability details, and the attack state transition information between the hosts. These information are presented to the user through a visualization system. The results are displayed using hosts as nodes, and clicking on a selected host node provides detailed information about the vulnerabilities exist on the host. Directed edges between nodes represent the attack state transition information. Selecting the corresponding edge allows viewing detailed information about the attack state transition information, such as the vulnerability, software, protocol types, and port numbers exploited, etc.

3.2 Attack Scheme Generation

Based on the completion of vulnerability information parsing, users can select a specific host node to generate attack schemes in real-time.

In scenarios with large-scale networks and numerous vulnerabilities, there is a potential for an explosion of attack paths, resulting in uncontrollable processing time and resource usage. Additionally, many redundant attack paths may not have an impact on the evaluation results. Therefore, the system support to set the threshold of attack paths. Once the number of discovered attack paths reaches the threshold, the search process stops. Users can adjust this threshold according to their needs.

The processing flow of attack scheme generation is illustrated in Fig. 2. Firstly, the attack paths that can reach the target host are searched in a shortest path-first manner. The processing flow starts by searching for the reachable paths from the internet to the target host, with the target host as the starting point. A temporary attack path is created,

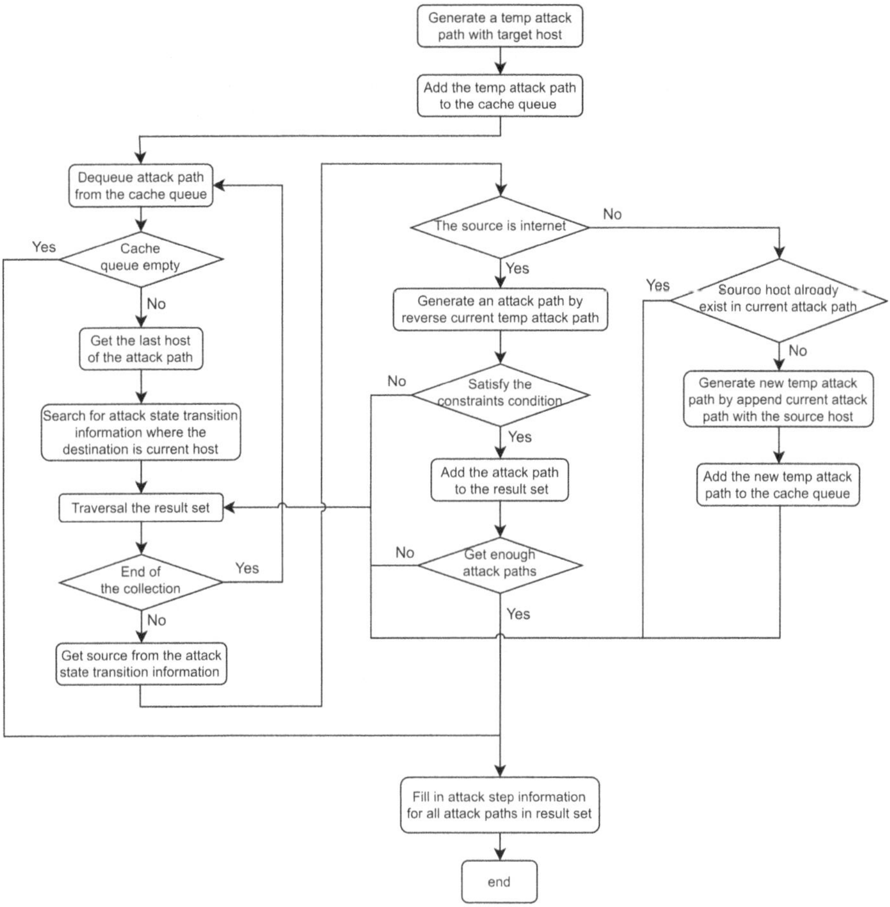

Fig. 2. Attack scheme generation process.

containing the information of the target host node, and add to the cache queue. Then, start the attack path search process: pop a temporary attack path from the cache queue and retrieving the information of the last host node in the path. The attack state transition information targeting the current host node is searched. If the search result is not empty, for each attack transition information, the source node information is extracted. If the source node is the internet, a valid attack path is found. A new attack path is generated based on the current temporary attack path, and if it satisfies the constraint conditions, the new attack path is added to the result set. Otherwise, if the source node is not in the list of host nodes in the current temporary attack path, a new temporary attack path is created and added to the cache queue by append the source node to the end of the path. The search process loops until the number of discovered attack paths reaches the threshold or the cache queue is empty. Then, attack schemes generated base on the attack paths in the result set, fill in vulnerability and attack step information.

The attack path search algorithm ensures prioritization of the shortest path, which is beneficial for larger scale network where the number of attack paths exceeds the threshold. This algorithm better meets the practical needs of cyber-security evaluation. The attack state transition information is stored in a hash table during the preprocessing stage, using the destination node information as the key. During the processing, searching for the attack state transition information of the current node can achieve a time complexity close to O(1), thereby improving processing efficiency. The attack schemes are presented to the user through a visualization system after generated. Users can sequentially view the source and destination nodes, vulnerability information exploited, and detailed attack step information for each step of the attack path.

4 Examples

We build a small-scale network in the cyber range for testing purposes, the network topology is show in Fig. 3, the subnet 192.168.8.0/24 represents the DMZ (Demilitarized Zone) area, where two servers with IP addresses 192.168.8.2 and 192.168.8.5 are deployed, offering email and website services, and ports 25 and 80 can be accessed separately via the TCP protocol from the Internet. The subnets 192.168.9.0/24 and 192.168.10. 0/24 represent the internal network area. Hosts within the same subnet can access all ports on each other. The servers with IP addresses 192.168.9.3 and 192.168.9.8 are respectively deployed with MySQL and HTTP services, and ports 3306 and 80 can be accessed separately via the TCP protocol from the DMZ and internal zone.

Scan all hosts in the simulated network using a vulnerability scanning tool, export the scan results, convert to a standard format, and supplemented with network configuration information. Table 4 shows the vulnerability information of each host discovered by scanning tool.

Input the vulnerability information into the attack scheme generation system for parsing, the parsing result is shown in Fig. 4. For comparison, the vulnerability information was also inputted into the MulVAL system to generate attack graph. It can be observed that the information about reachable attack hosts obtained from the attack scheme generation system is consistent with the results of the MulVAL-generated attack graph. Since the parsing results are displayed with hosts as nodes, it allows for easy identification of all reachable hosts and the attack state transition information between hosts. From the graph, it can be seen that there are numerous attack paths between host nodes within the same subnet and from 192.168.8.2 and 192.168.8.5 to 192.168.9.3. Due to the absence of externally accessible service ports, the subnet 192.168.10 .0/24 does not have any reachable hosts, so there are no hosts from that subnet in the results.

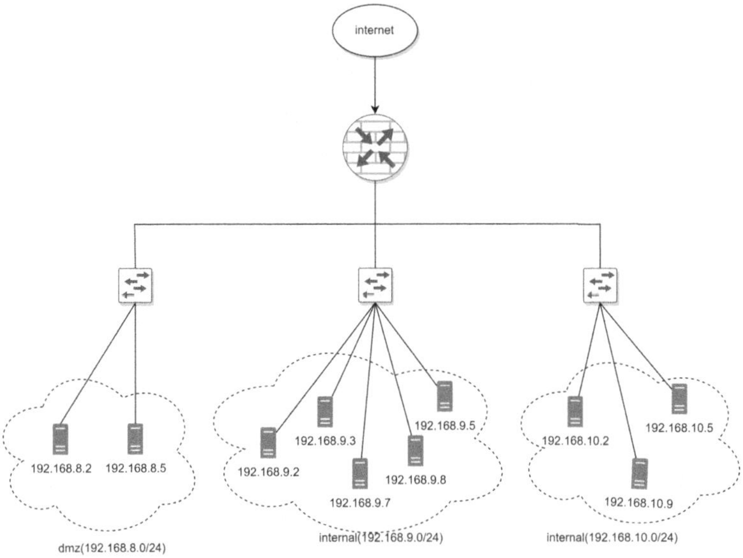

Fig. 3. Example.

Table 4. Vulnerability information.

Host	Vuln Id	Software	Protocol	Port
192.168.8.2	CVE-2013-2566	http_server_11.1.1.9.0	tcp	25
192.168.8.5	CVE-2023-25690	http_server_*	tcp	80
192.168.9.3	CVE-2016-6662	mysql_*	tcp	80
192.168.9.2	CVE-2016-2183	enterprise_linux_7.0	tcp	993
				995
192.168.9.5	CVE-2020-1350	windows_server_2008_r2	tcp	3389
				53
192.168.9.7	CVE-2013-2566	http_server_11.1.1.9.0	tcp	25
192.168.9.8	CVE-2023-25690	http_server_*	tcp	80
192.168.10.2	CVE-2016-2183	enterprise_linux_7.0	tcp	993
				995
192.168.10.5	CVE-2020-1350	windows_server_2008_r2	tcp	3389
				53
192.168.10.9	CVE-2015-2808	http_server_11.1.1.9.0	tcp	1433

Fig. 4. Vulnerability information parsing result.

Users can select target hosts to generate attack schemes in real-time. Table 5 presents the statistical information of attack schemes generated for each host. By selecting an attack scheme entry, users can view detailed information about the attack scheme. Table 6 presents the detailed information of a specific attack scheme generated for the target 192.168.9.2.

Table 5. Attack scheme generation statistics.

Destination	Attack Paths	Processing Time(ms)
192.168.8.2	2	0
192.168.8.5	2	0
192.168.9.2	432	2
192.168.9.3	188	1
192.168.9.5	432	1
192.168.9.7	368	2
192.168.9.8	188	1

The attack scheme includes the target name, number of hops, and detailed information for each hop in the attack path. It provides the names of the source and destination nodes, vulnerability ID, description, integrity impact score, exploitability score, access complexity, and recommended attack steps. This information is presented to assist users in evaluating the attack path.

Table 6. An attack scheme example.

Target	192.168.9.2		
Hop Count	3		
	Source	internet	
	Destination	192.168.8.5	
	Vuln Id	CVE-2023-25690	
	Software	http_server_*	
	Protocol	tcp	
	Port	80	
	Impact	5.9	
	Exp	3.9	
	Base	9.8	
	Access Com-plexity	Low	
	Description	Issue occurs when certain mod_proxy configurations on Apache HTTP server allow for HTTP request smuggling attacks	
	Attack Steps	1	Identifying the CRLF Injection:Determine if the httpd version is less than 2.4.55, more information: https://owasp.org/www-community/attacks/HTTP_Response_Splitting
			Internal HTTP Request Smuggling via Header Injection, more information: https://github.com/dhmosfunk/CVE-2023-25690-POC
2	Source	192.168.8.5	
	Destination	192.168.9.3	
	Vuln Id	CVE-2016-6662	
	Software	mysql_*	
	Protocol	tcp	
	Port	3306	
	Impact	5.9	
	Exp	3.9	
	Base	9.8	
	Access Com-plexity	Low	

(*continued*)

Table 6. (*continued*)

Target	192.168.9.2		
	Description	Some versions of MySQL allow local users to create arbitrary configurations and bypass certain protection mechanisms by setting general_log_file to a my.cnf configuration	
	Attack Steps	1	Inject malicious configurations into MySQL configuration files with weak or improper permissions
		2	Create new configuration files within the MySQL data directory on default MySQL installations without relying on improper configuration permissions
		3	Gain access to logging functions on all of the default MySQL installations and thus be in position to add or modify config files
3	Source	192.168.9.3	
	Destination	192.168.9.2	
	Vuln Id	CVE-2016-2183	
	Software	enterprise_linux_7.0	
	Protocol	tcp	
	Port	995	
	Impact	3.6	
	Exp	3.9	
	Base	7.5	
	Access Com-plexity	Low	
	Description	It easier for remote attackers to obtain cleartext data via a birthday attack against a long-duration encrypted session, as demonstrated by an HTTPS session using Triple DES in CBC mode, aka a "Sweet32" attack	
	Attack Steps	1	Download attack tool script: git clone https://github.com/drwetter/testssl.sh.git
		2	Run the script with the target IP address as a parameter, for example:./testssl.sh-W 192.168.1.1

As the network scale increases, the number of attack paths grows exponentially. We conducted a test by expanding the 192.168.9.0/24 subnet and compared the processing time of the Attack Scheme Generation System and MulVal, the results are shown in Table 7. After the number of hosts increased to 13, the processing time of MulVAL sharply increased to more than 45 min, additionally, the generated attack graph had numerous nodes, making it difficult to interpret and identify specific attack paths. The results from the Attack Scheme Generation System, as shown in Fig. 5, were easier to analyze since the nodes were displayed based on hosts, and the number of nodes did not explode. It provided a clear view of vulnerable hosts and their interrelated attack state transition information. As the number of hosts continued to increase, MulVAL has been running for over ten hours and still hasn't completed processing.

In the parsed results shown in Fig. 5, by selecting 192.168.9.5 as the target, we generated attack schemes by setting different thresholds for the number of paths and recorded the corresponding processing time. The results are shown in Fig. 6. The processing time increases approximately linearly with the increase of attack paths. When the number of attack paths is less than 10000, the processing time is less than 2 s, which can basically meet the real-time requirements. Due to the exponential growth of the number of paths with the scale of the network, without threshold limit, the processing time will be uncontrollable for large network.

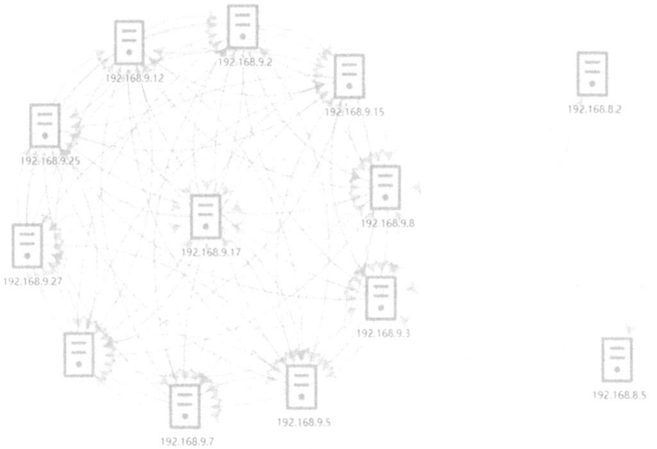

Fig. 5. Vulnerability information parsing result.

Table 7. Processing time.

Hosts	Attack Scheme Generation System	MulVal
7	187 ms	989 ms
10	276 ms	2177 ms
13	277 ms	45 min34 s
16	304 ms	-
19	312 ms	-

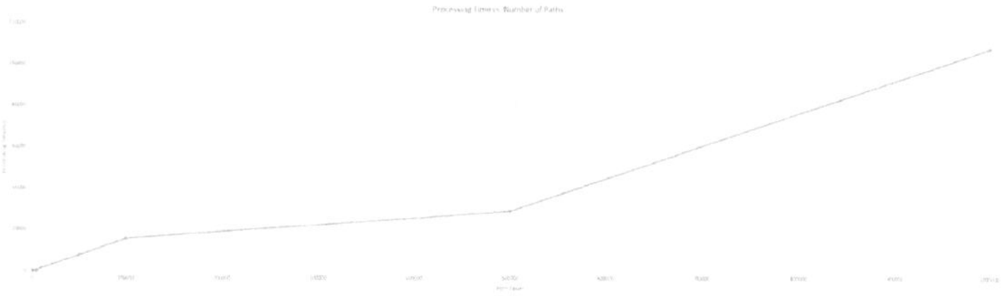

Fig. 6. Processing time vs. number of paths.

5 Conclusion

This paper presents a cybersecurity evaluation oriented attack scheme generation system. By inputting vulnerability information and network configuration, the system can dynamically analyze the at-risk hosts and the attack state transition information between hosts. The results are displayed through a graphical interface. Once the user selects a target host, the system can generate real-time attack schemes against that host, including detailed attack step information. When there are multiple attack paths, the system prioritizes providing attack schemes with shorter paths. This system addresses the issues of long processing times and lack of real-time requirements in traditional attack graph generation systems, as well as the difficulties in result analysis and comprehension. This system can be applied in cybersecurity evaluation to assist security assessors in generating system security testing plans and evaluating the security of the system. Additionally, the system can also support various applications, such as security personnel training, by generating attack schemes tailored to specific targets, guiding participants through simulated attacks step by step.

In future research, we can consider the following improvements and extensions:

1. Improve real-time performance: To address large-scale networks, further optimization of algorithms and data structures can be done to enhance the efficiency of vulnerability information parsing and attack scheme generation.
2. Consider additional attack vectors: The current system primarily focuses on vulnerability information. It can be expanded to consider other attack vectors such as malware and social engineering, to improve the comprehensiveness and accuracy of the system.

3. Support quantitative evaluation of attack schemes: Quantitatively assess attack schemes based on different metrics such as exploitability of vulnerabilities, attack complexity, damage potential, and other factors. This capability assists users in making informed choices among attack schemes based on different criteria and conditions.
4. Integrate automated evaluation capabilities: Integration of the attack scheme generation system with other security tools and systems can enable automated vulnerability scanning, automated attack testing and verification, and automated generation of evaluation reports, thereby improving overall efficiency.

Acknowledgements. This research was funded by the Major Key Project of PCL (Grant No. PCL2022A03-3), NSFC (No. 62072131, 61972106).

References

1. Cunningham, William H.: Optimal attack and reinforcement of a network. J. ACM **32**(3), 549–561 (1985). https://doi.org/10.1145/3828.3829
2. Lallie, H.S., Debattista, K., Bal, J.: A review of attack graph and attack tree visual syntax in cyber security. Comput. Sci. Rev. **35**, 100219 (2020)
3. Ou, X., Govindavajhala, S., Appel, A.W.: MulVAL: A logic-based network security analyzer. In 14th USENIX Security Symposium, Baltimore, MD, USA (2005)
4. Kuang, R.B.: Rule Based Security Checking. Technical Report, MIT Lab for Computer Science (1994)
5. Swiler, L.P., Phillips, C., Gaylor, T.: A Graph Based Network Vulnerability Analysis System, SAND97-3010/1. Sandia National Laboratories, Albuquerque, New Mexico and Livermore, California, 1998
6. Zhang, T., Hu, M.Z., Li, D., et al.: An effective method to generate. In: 2005 International Conference on Machine Learning and Cybernetics, Guangzhou, China, pp. 3926-3931 (2005)
7. Wang, S., Tang, G., Kou, G., et al.: An attack graph generation method based on heuristic searching strategy. In: 2016 2nd IEEE International Conference on Computer and Communications (ICCC), pp. 1180-1185. IEEE (2016)
8. Ammann, P., Wijesekera, D., Kaushik, S.: Scalable. graph-based network vulnerability analysis. In: Proceedings of the 9th ACM Conference on Computer and Communications Security (CCS.02), pp. 217-224. Washington DC, ACM (2002)
9. Ou, X., Boyer, W.F., Miles A.: McQueen. A scalable approach to attack graph generation. In: Proceedings of the 13th ACM Conference on Computer and Communications Security, Alexandria, VA, USA, October 30-November 3 (2006)
10. Durkota, K., etal.: Optimal network security hardening using attack graph games, pp. 526-532 (2015)
11. Nguyen, T.H., Wright, M., Wellman, M.P., et al.: Multi-stage attack graph security games: Heuristic strategies, with empirical game-theoretic analysis. In: Proceedings of the 2017 Workshop on Moving Target Defense. Dallas, TX, USA, pp. 87-97 (2017)
12. Cody, T., Rahman, A., Redino, C., et al.: Discovering Exfiltration Paths Using Reinforcement Learning with Attack Graphs. arXiv preprint arXiv:2201.12416 (2022)
13. Fadlallah, A., Sbeity, H., Malli, M., et al.: Application of attack graphs in intrusion detection systems: an implementation. Int. J. Comput. Netw. **8**(1), 1–12 (2016)

14. Durkota, K., Lisy, V., Bošansky, B., et al.: Optimal network security hardening using attack graph games. In: Proceedings of IJCAI, pp. 7-14 (2015)
15. Miehling, E., Rasouli, M., Teneketzis, D.: Optimal defense policies for partially observable spreading processes on Bayesian attack graphs. In: Proceedings of the Second ACM Workshop on Moving Target Defense, pp. 67-76 (2015)
16. Huang, J.H., Feng, D.Q., Wang, H.J.: A method for quantifying vulnerability of industrial control system based on attack graph. Acta Automatica Sinica **42**(5), 792–798 (2015)
17. Xu, L.J.: Industrial Control System Network's Potential Risk Analysis Based on Attack Graph. Beijing University of Posts and Telecommunications, Beijing (2015)
18. Polatidis, N., Pimenidis, E., Pavlidis, M., et al.: From product recommendation to cyber-attack prediction: generating attack graphs and predicting future attacks. Evol. Syst. **11**(3), 479–490 (2020)
19. Che, B., Liu, L., Zhang, H.: KNEMAG: key node estimation mechanism based on attack graph for IOT security. J. Internet of Things **2**(4), 145 (2020)
20. Liu, J., Zhang, Y., Hu, H., et al.: Efficient defense decision-making approach for multistep attacks based on the attack graph and game theory. Math. Probl. Eng. **2**, 1–12 (2020)

A Lightweight and Easily Scalable Tor Network Simulation Method Based on K8s

Wentao Huang[1], Han Wu[1], Zeyu Li[1], Qingfeng Zhang[2], Xuebin Wang[2], Jiapeng Zhao[1], and Jinqiao Shi[1](✉)

[1] Beijing University of Posts and Telecommunications, Beijing, China
{hyangwentao,shijinqiao}@bupt.edu.cn
[2] Institute of Information Engineering, Chinese Academy of Sciences, Beijing, China

Abstract. Criminals use Tor's powerful anonymity for illegal activities, which has attracted more and more researchers to study de-anonymization attacks on Tor. Researchers have difficulty conducting de-anonymization attack experiments in a real Tor network, they need a simulated environment. However, most of the existing Tor network simulation methods are difficult to meet the needs of de-anonymization attacks in terms of authenticity and flexibility, and some of them are not scalable, and other Tor network simulation methods that can meet the needs are costly. In this paper, we use K8s (Kubernetes) to build a lightweight and easily scalable Tor network simulation environment, using Docker as the basic unit for simulating the Tor relay. The results show that a thousand-scale Tor network can be simulated using four regular hosts. Meanwhile, we conducted two de-anonymization attack experiments to demonstrate that the simulated Tor network constructed in this paper has authenticity and flexibility.

Keyword: Tor · Network Simulation · De-anonymization Attacks

1 Introduction

The Second Generation Onion Router (Tor) [11] is currently the most popular low-latency anonymous communication system, with thousands of volunteer relays and millions of users worldwide [22]. Tor provides anonymity by forwarding traffic through multi-hop routing, hiding the IP addresses of clients and specific servers. Tor protects the integrity and confidentiality of message data through multi-level encryption. Tor's strong anonymity provides security for criminals to carry out their criminal activities.

To perform a de-anonymization attack on the Tor network and thus examine the criminal behavior within it, researchers need to conduct a large number of experiments in the Tor network. However, there are several problems with conducting experiments in a real Tor network: (i) Deploying a large number of relays in a real Tor network is not only costly but also subject to official scrutiny, leading to deployment failure. This is because a single organization controlling a large number of Tor relays can wreak havoc on the anonymity of the entire network. (ii) Relays in a real Tor network are widely distributed [15], and it is difficult for researchers to observe the complete anonymized

Z. Gu et al. (Eds.): NSE 2023, CCIS 2063, pp. 351–367, 2024.
https://doi.org/10.1007/978-981-97-4519-7_25

communication process of the Tor users. It makes it difficult for researchers to collect data and evaluate the effect in de-anonymization experiments. (iii) Conducting experiments in a real Tor network also raises ethical issues, as the experiments may undermine the anonymity of Tor users and affect the performance and functionality of the Tor network [41]. Therefore, we need Tor network simulation methods to help researchers conduct experiments in controlled environments to address the issues of cost, complexity, and ethics posed by the real Tor network.

Existing Tor network simulation methods can be classified into three categories: generic discrete event simulator-based, software-based, and hardwarebased Tor network simulation [45]. The generic discrete event simulator-based approach allows the use of a generic simulator to simulate discrete events in Tor. This type of approach does not run an actual Tor application, which lacks authenticity and has limited application scenarios. The software-based approach to Tor network simulation proposes a series of experimental tools dedicated to Tor research, capable of restoring realistic network topologies and simulating large-scale network nodes. However, this type of simulation approach currently replaces actual traffic transfers with memory reads and writes, and is unable to manipulate Tor relays in real-time during the simulation, and thus lacks a degree of authenticity and flexibility. Some of the software-based simulation methods for the Tor network are also limited in their scalability, as they are unable to distribute the simulation across multiple hosts. Hardware-based Tor network simulation methods use a cluster of virtual hosts or other hardware devices such as Raspberry Pi to build a private Tor network. This type of method is the closest to the real Tor network environment in terms of authenticity during simulation, but it is too costly for large-scale simulation.

Based on the above problems, this paper proposes a new method to simulate the Tor network and implements a Tor network experimental verification platform. We use container technology [2] to provide isolated environments such as network and file systems for Tor application processes and use K8s cluster technology [19] to manage multiple hosts during simulation, giving consideration to the authenticity, flexibility, scalability, and cost reduction of Tor network simulation. **This paper contributes as follows:**

- We present a new Tor network simulation approach that uses Docker as a simulation unit to provide lightweight, flexible, and authentic Tor network simulation.
- We use K8s to implement a scalable Tor network experimentation platform that can simulate a thousand-scale private Tor network using four regular hosts.
- We conducted a circuit association experiment and a hidden service traceability experiment in the Tor network experiment platform built in this paper, and these two de-anonymization attack experiments can verify the authenticity and flexibility of the Tor network simulated in this paper.

2 Background

In this section, we begin with a background on the Tor network. Then we describe de-anonymization attacks in the Tor network and their need for a Tor simulation environment. Finally, we present related work on Tor network simulation and summarize the research objectives of this paper.

2.1 Tor

The Tor network is an anonymous communication network consisting of a large number of relays, as shown in Fig. 1. Its design goal is to provide users with a more secure and privacy-protected Internet communication environment. Users can use the Tor Browser [38] to access websites anonymously or use Tor as an anonymous proxy to access application services such as SSH and FTP. The Tor network consists primarily of the following entities:

- **Directory Authority Servers:** Nine special relays in the Tor network, which are the central servers for the entire network, have their IP addresses hard-coded into the Tor code. A relay that wants to join the Tor network must first upload its descriptor information to Directory authority servers, and only after Directory authority servers have voted and written it into the consensus file can it be considered to have joined the Tor network. The consensus file records information about all relays in the Tor network, including the relays' IP addresses, fingerprints, bandwidths, ports, Tor versions, identity tags, etc.
- **Onion proxy:** A user-oriented Tor client that is built into the Tor browser or acts as a proxy for users to access the Tor network.
- **Guard:** It is the relay labeled "Guard" in the Tor consensus file and is the first hop relay in the circuit established by Tor. Relays can not be actively configured as a Guard relay, but need to meet a series of strict requirements to be given the "Guard" label, such as relay stability and long-term operation requirements. Therefore, the cost of deploying Guard relays in a real Tor network is high, and de-anonymization attack experiments often need to implant Guard relays and capture traffic in them.
- **Middle:** It is an ordinary routing relay in the Tor network and is in the position of a middle relay in the circuit constructed by Tor.
- **Exit:** It is the relay labeled "Exit" in the Tor consensus file and is the last hop relay in the circuit built by Tor.
- **RP (Rendezvous Point):** It is dedicated to the communication between the Onion proxy and the hidden service and is the last hop relay in a circuit randomly selected by the Onion proxy. The RP relay is responsible for forwarding the communication data between the Onion proxy and the hidden service.
- **Hidden service (Hidden server):** The hidden service is a special service that can only be accessed using the Tor network. It is the information publisher that can host various services, such as SSH, WEB, and FTP.

When users use the Onion proxy to access the Tor network, the Onion proxy first requests consensus information from Directory authority servers. With the consensus information, the Onion proxy can use the path selection algorithm to select the relays to build the circuit. In a scenario where a user accesses a common application service, the Onion proxy establishes a three-hop circuit to connect to the target service and negotiates a symmetric key for encrypting the data with each relay in the circuit. The data is first encrypted by the Onion proxy in three consecutive layers, and each of the three relays in the circuit decrypts one layer. Only the Exit relay can decrypt the innermost layer of data to get the real data content, but it only knows the address of the Middle relay and does not know who the original sender of the data is. The Guard relay only knows the

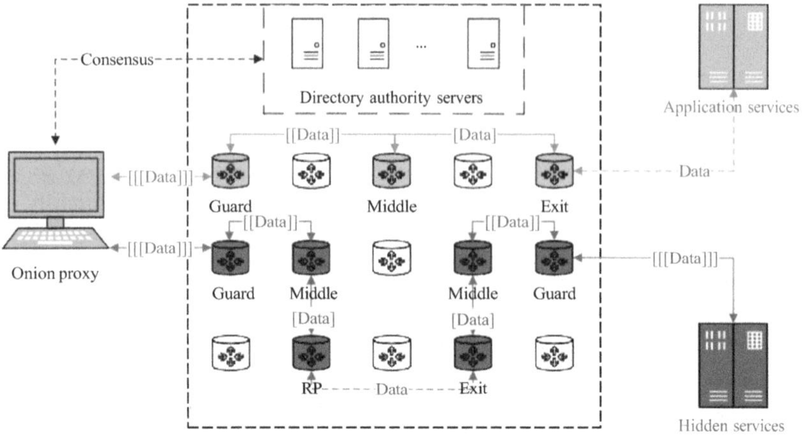

Fig. 1. Diagram of the Tor network structure, where blue indicates access to common application services and red indicates access to hidden services in Tor.

address of the sender of the data, but not the service it accesses, and the Middle relay neither knows the address of the sender of the data, nor the service it accesses, which guarantees the anonymity of the sender of the data.

In the scenario where a user accesses a hidden service, the Onion proxy randomly selects an existing circuit and uses its last hop relay as the RP relay. The Onion proxy informs the target hidden service about the address of its RP relay through other circuits and steps [31], and then the target hidden service establishes a circuit to the Onion proxy's RP relay, and the two parties start exchanging data for communication. Because the three-hop link and multi-layer encryption guarantee the anonymity of the data sender, both the user (Onion proxy) and the hidden service are anonymous in this scenario.

2.2 De-anonymization Attacks Against the Tor Network

Tor's anonymity is gradually being abused by criminals, such as building illegal hidden services, anonymous gun and drug deals, etc. [47]. Researchers trace back to locate the criminals by performing de-anonymization attacks on Tor, which can be classified into user-side de-anonymization attacks and hidden-service-side de-anonymization attacks according to the target of tracing [27].

The goal of the user-side de-anonymization attack is to trace the correspondence between the user and the target service, which mostly uses flow association methods [18, 20, 28, 34, 35]. In this attack scenario, the researcher first deploys some controlled relays in the Tor network, and when malicious traffic is detected at the target service or controlled Exit relay, it will be correlated and matched with the traffic at the controlled Guard relay, and if the correlation is successful, it can be traced back to the IP address of the user (Onion proxy). The target of hidden service side de-anonymization attacks is to trace back the hidden service, which requires the deployment of controlled Guard relays and can be classified into passive and active attacks according to the type of attack. Passive attacks mostly use the WFP (Website Fingerprinting) method [3, 9, 43,

44], where the traffic of the target hidden service is collected for training the model. In the detection phase, controlled Guard relays use the model to detect the traffic to the target hidden service, and then the IP address of the hidden service can be traced. Active attacks mostly use watermarking methods [8, 29, 30, 46], researchers inject, modify, and delete traffic and other ways to make watermarks, using Onion proxy to send watermarks to the target hidden service, if the water-mark is detected at controlled Guard relays can be traced back to the IP address of the hidden service.

There are several reasons why de-anonymization attack experiments are difficult to carry out in a real Tor network. First, the de-anonymization attack requires the target user or hidden service to use the controlled relays deployed by the researcher. Because of the large number of relays in the Tor network, achieving this goal requires the researcher to deploy as many controlled relays as possible, especially Guard relays. This is costly and difficult in real Tor networks. Second, it is difficult to guarantee that the target user or hidden service has been using the researcher's controlled relays throughout the de-anonymization attack, as Tor periodically changes circuits [36]. This makes it difficult for researchers to observe the complete user behavior. In addition, such attacks may undermine the anonymity of normal Tor users, raising ethical issues.

Providing a private and controllable network environment through the Tor network simulation method can solve the above problems. However, the deanonymization attack also has the following requirements for the simulation environment: (i) The simulated environment is authentic, involving real Tor applications and generating real Tor traffic. Because the de-anonymization attack process requires analyzing real Tor protocol packets and has multiple complex phases. (ii) The simulation environment needs to be sizable and scalable. On the one hand, a small network size cannot help researchers assess the probability of a target user selecting a controlled relay in a de-anonymization attack. On the other hand, the scale of the real Tor network is increasing from more than 6,000 relays in 2022 to nearly 8,000 relays in 2023 [22], so the simulation environment should be scalable to better approximate the real effect. (iii) The simulation environment has the flexibility to be able to operate Tor relays in real time during the simulation. Because the de-anonymization attack has multiple phases, it is necessary to formulate different traffic processing and script execution strategies at different phases, and the ability to operate the Tor relay in real time can greatly facilitate the researchers.

2.3 Related Work and Motivation

Existing Tor network simulation methods can be categorized into generic discrete event simulator-based, software-based, and hardware-based Tor network simulation. Approaches based on generic discrete event simulators often use components such as OMNET++ and NS2 to build network simulators. In 2013, Doswell et al. [14] simulated a mobile user at various mobility speeds (e.g. walking) alongside a range of Tor circuit build times. This type of simulation can simulate discrete anonymous communication events, but without the involvement of real Tor applications or the generation of real Tor traffic, and tends to be used only in the preliminary research phase.

In the software-based approach to Tor network simulation, the researchers designed a series of experimental tools. In 2011, Jansen et al. introduced Shadow [24], a Tor network simulator that encapsulates real Tor applications in a plug-in wrapper and loads them once into memory. Shadow intercepts the application's sockets, event libraries, and other function calls and redirects them to the simulated environment. It can simulate large, diverse Tor-specific networks, but can only use a single host to simulate discrete events and is not scalable. In the same year, Bauer et al. proposed ExperimenTor [5], a Tor experimentation tool based on the ModelNet network simulation testbed. It requires at least two hosts (which can be scaled up using a cluster of machines), where one host is responsible for simulating the network, e.g., bandwidth, topology, etc., and the other host is responsible for running the actual Tor application. In 2012, Elahi et al. proposed COGS [16] for evaluating the impact of Tor Guard relay selection on user privacy. It utilizes a consensus file to recreate the state of the Tor network at a given time, runs extensive path selection simulations, and generates logs for analysis. In 2014, Singh proposed SNEAC [42], which is similar to ExpeimenTor, which is also based on ModelNet. However, it replaces the FreeBSD-based emulator with a lightweight Linux container that can emulate larger Tor networks with limited resources. In 2014, 2021, and 2022, Jansen et al. made several improvements to Shadow [23, 25, 26], including more accurate modeling of the Tor network from a geographic perspective, runtime optimizations, more realistic TCP connection limits and network stack improvements, and improved simulation speeds, but still did not address the scalability issue. These software-based approaches to Tor network simulation suffer from authenticity versus flexibility issues because they use memory reads and writes instead of actual traffic transfers and simulate discrete events that can't adapt to highly dynamic environments.

Hardware-based approaches to Tor network emulation typically use highperformance servers running virtual machine clusters or other hardware devices such as Raspberry Pi to build private Tor networks. PlanetLab [10] is a globally distributed overlay network that allows researchers to run experiments on specific shares of the network known as slices. PlanetLab consists of 1,335 nodes in 645 locations and is intended to serve as a microcosm of the Internet. Bauer et al. [4] and Akhoondi et al. [1] have used PlanetLab to simulate and analyze Tor's routing algorithm. DeterLab [6, 33] is a cybersecurity simulation testbed. In contrast to PlanetLab, researchers can provide network topologies and run arbitrary applications. Chakravarty et al. [7] demonstrated a traffic analysis attack against Tor users by using Deter to inject bandwidth fluctuations on the server side and observing these fluctuations along the circuit. In 2020, Wang et al. [45] built a private Tor network with 20 nodes via Raspberry Pi and conducted experiments to validate it. Compared with other platforms of the same type, this platform has the advantages of low cost, easy scalability, and a high degree of network protocol recovery.

Hardware-based Tor network simulation methods can only simulate relays of a size equal to the number of virtual machines or other hardware devices. This approach is expensive to simulate and is not suitable for lightweight simulations with limited resources. Generic discrete event simulator-based approaches do not meet the needs of de-anonymization attack experiments because they do not involve real Tor applications. Existing software-based approaches to Tor network simulation also fail to meet the needs of de-anonymization attacks in terms of authenticity and flexibility, and some of them are not scalable. Therefore, a Tor network simulation method with lightweight, flexibility, authenticity, and scalability is needed to satisfy the de-anonymization attack experiment scenarios, which is the research goal of this paper.

3 Method

In this section, we present the design idea and deployment process of the K8s-based Tor network simulation method.

3.1 Design

In this paper, we use Docker container technology to address the issues of lightweight, authenticity, and flexibility in the Tor network simulation, and use K8s clustering technology to address scalability issues. The overall structure of the private Tor network simulation environment in this paper is shown in Fig. 2. The Docker image repository has images for the Directory Authority Server, relay, and Exit relay, which are used to create the corresponding containers during simulation. Users can also make various custom Docker images related to Tor experiments, such as FTP server images, web server images, etc. Users operate the entire private Tor network through the Kube-apiserver component. In the private Tor network, Pod is the basic unit of K8s scheduling, which runs a Docker container to simulate a Tor relay. The Master node host schedules Pods to different physical node hosts (with real IP addresses) and can assign simulated IP addresses to each Pod through a network plugin to ensure that simulated Tor relays from different hosts are connected.

To provide lightweight, authenticity, and flexibility, this paper uses Docker to simulate each Tor relay with the help of the isolated environment it provides for application processes. The virtualization technology [21] virtualizes separate hardware devices, such as CPU, memory, etc., for each virtual machine through Hypervisor, and then installs the operating system (i.e., Guest OS) on the virtualized hardware devices. Unlike virtualization technology, Docker does not virtualize hardware devices and operating systems. Docker provides an isolated environment by setting various NameSpace parameters for application processes through the Docker Engine, and the application processes share the host's CPU, memory, and other hardware resources. Therefore, the overhead of Docker is less than that of the virtual machine and other hardware devices, and it belongs to the software-based Tor network simulation in terms of architecture, which can meet the lightweight demand. Docker can isolate a separate network, file system, etc. for application processes. In the network, each application process has its own separate virtual NIC (Network Interface Card), IP stack, IP routing table, and other network state information, and thus normal traffic transmissions can be carried out to meet the authenticity requirements. Meanwhile, researchers can always log in to manipulate application processes in the container created by Docker, which provides flexibility. To provide scalability, this paper uses K8s clustering technology to ensure that Docker containers on different hosts can connect, making it easy to scale hosts to increase network size.

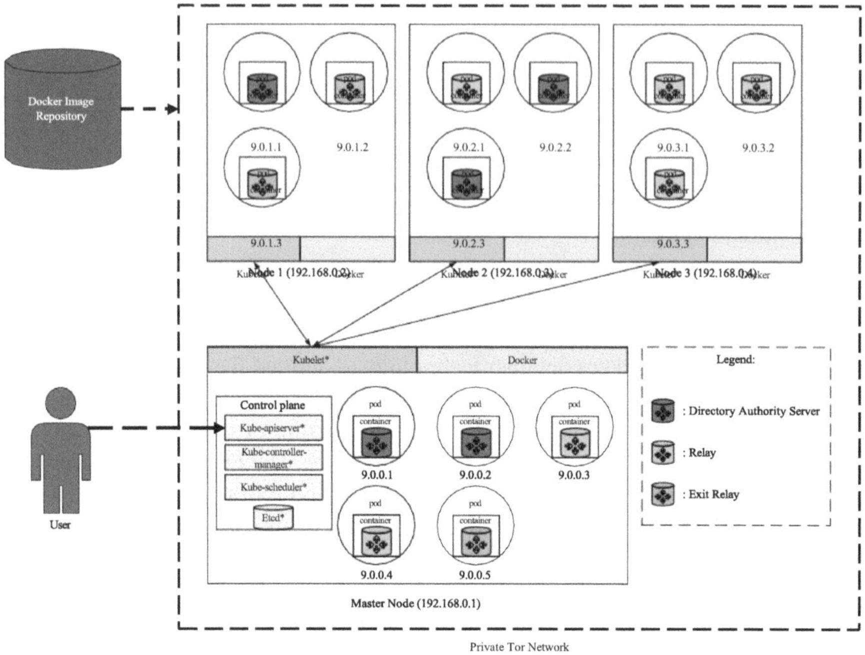

Fig. 2. Tor network simulation environment based on K8s. *represents the components of K8s, of which etcd is the database that stores the state of the entire cluster; apiserver provides the only entry point for users to operate resources; controller-manager is responsible for maintaining the state of the cluster; scheduler is responsible for scheduling Pods to the appropriate machines according to a predefined scheduling policy; kubelet is responsible for maintaining the container lifecycle, as well as Volume (CVI) and network (CNI) management.

3.2 Deployment Process

Creating Docker Images: We first need to make Tor relay images to create the appropriate Tor relays for use in the simulation. In this paper, we use *Dockerfile* [12] to make images. According to the role of Tor relays in the network, they can be classified as Directory Authority Servers, common relays, Exit relays, etc., which run the same application program, but with a different *torrc* configuration file at startup. To make a Directory Authority Server, you need to add "AuthoritativeDirectory 1" and "V3AuthoritativeDirectory 1" to the *torrc* configuration file, and to make an Exit relay, you need to add "ExitRelay 1" in the *torrc* configuration file, and other configurations can be found in the official Tor documentation [37].

K8s Cluster Build: To build a simulated private Tor network using K8s, we need to deploy the Master node and other nodes and the network plugin to provide a clustered environment for the private Tor network. First, we use the "kubeadm init" command to deploy the Master node. *Kubeadm* [17] images the relevant components in the Master node's control panel and deploys them as containers (the kubelet component needs to be deployed separately on the host). The Master node is mainly used for managing and scheduling cluster resources as well as running user commands. Then we use the "kubeadm join" command to deploy other nodes for the Pod to run on (the Pod can also run on the Master node). Finally, we deploy the *flannel*[39] network plugin, which maintains network IP address assignments with the help of the etcd component and assigns each node a different IP address segment, e.g., the network of the three nodes in the cluster is divided into the *9.0.1.0/24*, *9.0.2.0/24*, and *9.0.3.0/24* subnets in that order.

Build a Private Tor Network: Finally, we simulate a private Tor network, the core idea of which is to have Tor relays use private Directory Authority Servers. The process is shown in Fig. 3. First, the private Directory Authority Servers are run to generate their IPs, fingerprints, etc., which are then written to a public document; other types of Tor relays read the Directory Authority Servers in the public document and use the private Directory Authority Servers by adding "DirAuthority" configuration items[37] to their *torrc* configuration file before starting; after starting, the relays upload their descriptors to the private Directory Authority Servers; the private Directory Authority Servers write the information of these relays into the consensus file to add these relays into the private Tor network; all Tor relays download and use the latest consensus file information, and then the private Tor network has been built.

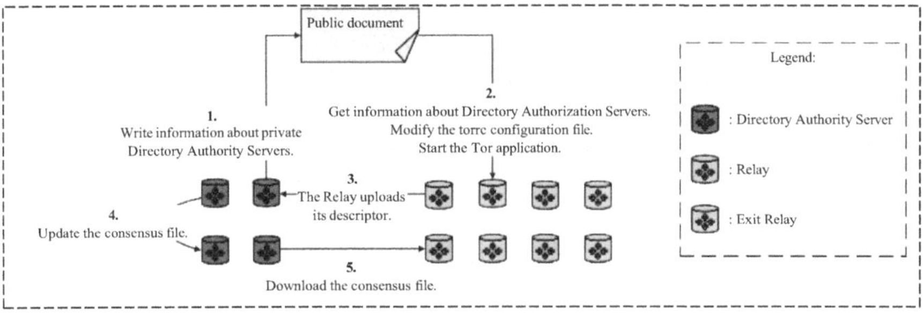

Fig. 3. Private Tor network organization process.

4 Experimentation and Evaluation

In this section, we design experiments to verify that the Tor network simulation approach in this paper is lightweight and scalable while being able to meet the needs of de-anonymization attacks in terms of authenticity and flexibility.

4.1 Design of Experiments

In this paper, the following three sets of experiments were designed:

- **Performance Experiment:** The performance of the proposed method is investigated and analyzed by recording the parameters related to the simulation of private Tor networks of different sizes. In this group of experiments, the parameters are recorded at simulated relay sizes of 400, 600, 800, and 1000, which include startup time and memory consumption.

- **Circuit Association Experiment:** This is a user-side de-anonymization attack designed to associate the IP address of a user (Onion proxy) with the IP address of the application service it accesses. Unlike existing flow association methods [18, 20, 28, 34, 35] that use flows for association, we use the number of cells in Tor circuits to associate relays at both ends (Tor uses the cell as the communication unit [32]) to check that the Tor network simulation method in this paper creates circuits properly. The experimental scenario is shown in the blue box in Fig. 4. Firstly, we deploy a controlled relay and a controlled Exit relay using a modified Tor program that records information about circuits and cells. Secondly, we deploy an Onion proxy and then manipulate it using the *stem* library [13] to use the two controlled relays as Guard and Exit relays, respectively. We then use the Onion proxy to upload a large file to the FTP server to simulate the behavior of a single access by a user using the Tor network. During one access behavior of a user, we capture Onion proxy to Guard circuits in real-time at the controlled Guard relay (denoted as "OP->Guard") and Exit to FTP server circuits in real-time at the controlled Exit relay (marked as "Exit->S"). At the end of a user's access behavior, we count the number of cells per second for each "OP->Guard" circuit to form the vector $\mathbf{X}([X_1, X_2, \ldots, X_i, X_{i+1}, \ldots])$, where each X_i is shaped like $[x_1, x_2, \ldots, x_j, x_{j+1}, \ldots]$ and represents a vector of the number of cells per second for the ith circuit. Similarly, we count the number of cells per second for each "Exit->S" circuit to form the vector $\mathbf{Y}$ $([Y_1, Y_2, \ldots, Y_i, Y_{i+1}, \ldots])$. Finally, we use Eq. 1 to calculate the correlation between $\mathbf{X}$ and $\mathbf{Y}$, where 1 means correlation and 0 means no correlation, and $Person(X_i, Y_j)$ represents the Pearson's correlation coefficient [40] between X_i and Y_j. The meaning of this formula is that $\mathbf{X}$ and $\mathbf{Y}$ are considered to be associated whenever there is a circuit in "OP->Guard" and a circuit in "Exit->S" and the Pearson correlation coefficient between them is greater than or equal to 95%. The correlation between $\mathbf{X}$ and $\mathbf{Y}$ means that the same Tor user access behavior was observed on both the controlled Guard and the controlled Exit, and then the user (Onion proxy) and its corresponding FTP server in this access behavior can be traced back to obtain their IP addresses.

$$f(\mathbf{X}, \mathbf{Y}) = \begin{cases} 1, \text{ if } \exists X_i \in \mathbf{X}, \exists Y_j \in \mathbf{Y}, Person(X_i, Y_j) \geq 95\% \\ 0, \text{ if } \forall X_i \in \mathbf{X}, \forall Y_j \in \mathbf{Y}, Person(X_i, Y_j) < 95\% \end{cases} \tag{1}$$

- **Hidden Service Traceability Experiment:** This is a hidden-service-side de-anonymization attack using the WFP approach [3, 9, 43, 44], which uses traffic fingerprinting to trace the IP address of the target hidden service since traffic fingerprinting is different for different hidden services. We use this experiment to verify that the Tor network emulation method in this paper can generate traffic correctly. This experimental scenario is shown in the red box in Fig. 4 and is similar to the WFP experiment of Sirinam et al. [43]. First, we deploy ten controlled relays. Then we build ten hidden services and make each hidden service use a controlled relay as its Guard relay. The content of the hidden service is taken from the homepage of the top 10 Alexa sites. For each hidden service, we use an Onion proxy to access it 512 times and capture the traffic between the Guard and the hidden service in real time on the corresponding controlled Guard relay, where the first 500 access behaviors are used to train the DF model [43], and the last 12 access behaviors are used for actual testing. In the training phase, the packet direction sequences of each flow are extracted as features from the traffic of the ten hidden services, and these features are fed into the DF model to set up a ten-class task to train the model to recognize which hidden service the flow belongs to. In the testing phase, the trained DF model is deployed on controlled Guard relays and classifies the behavior of users accessing the hidden service to find out which hidden service they access, which in turn is traced back to the IP address of the corresponding hidden service.

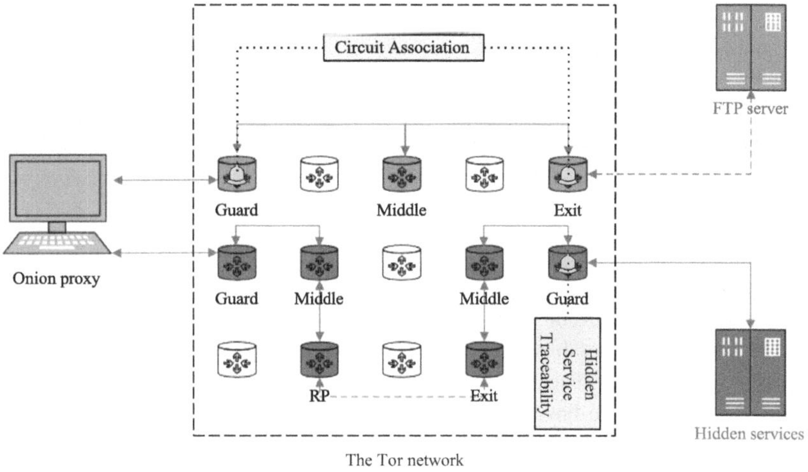

Fig. 4. Flow Association and Hidden Service Traceability Scenario Diagram.

The Performance Experiment verifies the performance of the Tor network simulation method in this paper in terms of lightweight. The Circuit Association Experiment and the Hidden Service Traceability Experiment belong to the de-anonymization attacks on Tor and are used to verify that the Tor network simulation method in this paper meets the need for authenticity and flexibility of the de-anonymization attack experiments, as they require a Tor network simulation environment that can generate real traffic data and manipulate Tor relays in real-time. In addition, we use different numbers of physical hosts for the Performance Experiment and the de-anonymization attack experiments, respectively, to demonstrate that the Tor network simulation approach in this paper is scalable. The fact that these experiments can be completed will also indicate that the private Tor network simulated in this paper is functioning properly, i.e., circuits can be created, users can access services anonymously, hidden services can be constructed, etc.

4.2 Experimental Results and Analysis

The experimental environment is shown in Table 1. Four hosts were used for the Performance Experiment and two hosts for the de-anonymization attack experiments to demonstrate that the Tor network simulation method in this paper is scalable by changing the number of hosts to vary the size of the simulation. Performance Experiment has a 2: 3 ratio of Exit relays to regular relays and also includes three Directory Authority Servers with relay sizes ranging from 400 to 1000. De-anonymization attack experiments include 10 Directory Authority Servers, 120 regular relays, 70 Exit relays, 11 Onion proxies, 10 hidden services, and 1 FTP server, with a relay size of 222.

Table 1. Table of the experimental environment.

Experimental category	K8s node type	Operating system	Memory sizes	CPU core count	Hard disk storage size	Number of hosts
Performance Experiment	Master	Centos 7.5	20G	8	100G	1
	Other	Centos 7.5	20G	8	100G	3
De-anonymization Attacks	Master	Centos 7.9	11G	8	189G	1
	Other	Centos 7.9	11G	8	189G	1

Performance Experiment. The startup time and memory consumption when simulating Tor networks of different sizes using four hosts(the relays are evenly distributed among the four hosts) are shown in Table 2. The resource consumption in simulating one thousand relays is also acceptable, which reduces the simulation cost significantly compared to the hardware-based Tor network simulation approach.

Circuit Association Experiment. We conducted 10 Circuit Association Experiments, and the results are shown in Table 3, all experiments are associated successfully. This demonstrates that the Tor network simulation method in this paper satisfies both the authenticity of building circuits and the flexibility of operating Tor relays in real-time, as well as ensuring that users have normal access to the private Tor network. It is worth noting that since this experiment is run in a simulated environment where the Onion proxy is controllable, the researcher can observe the complete behavior of the user and the collection of circuit information at the controlled relay will not be disturbed by circuits from other Onion proxies. This phenomenon leads to a high association success rate. In real scenarios, the Onion proxy is not controllable, and even if the Onion proxy is made controllable during the experiment, there is no guarantee that the controlled relay used will not be interfered with by the circuits of other Onion proxies. Therefore user-side de-anonymization attacks in real scenarios still face many problems, but the Tor network simulation approach in this paper can help researchers replicate various complex problems to aid in their research.

Hidden Service Traceability Experiment In the testing phase, we concurrently traced 10 target hidden services for a total of 120 times, so the number of accesses to each target hidden service is not strictly 12, which deviates from the experimental design, but does not affect the significance of the experiment and analysis of the results. The results are shown in Table 4, totaling 113 successful traceability attempts with a traceability accuracy of 94.17%. This shows that the Tor network simulation method in this paper can satisfy the authenticity of generating traffic with the flexibility of operating Tor relays in real time, and can build hidden services properly. It is worth noting that the hidden

Table 2. Table of Performance Experimental results.

Number of simulated relays	Startup time	Average memory consumption	K8s node type	Initial memory	Memory after startup completion
400	2 min 15 s	4.2G	Master	18.4G	13.6G
			Other Other	18.4G 18.4G	14.3G 14.5G
			Other	18.4G	14.3G
600	3 min	6.4G	Master	18.4G	11G
			Other Other	18.4G 18.4G	12.4G 12.2G
			Other	18.4G	12.5G
800	3 min 50 s	8.2G	Master	18.4G	8.9G
			Other Other	18.4G 18.4G	10.7G 10.7G
			Other	18.4G	10.8G
1000	4 min 30 s	11.9G	Master	18.4G	4.4G
			Other Other	18.4G 18.4G	7.1G 7.3G
			Other	18.4G	7.2G

Table 3. Table of Circuit Association Experimental results.

Index	The maximum value of $Person(X\text{i}, Y\text{j})$	$f\,(\mathbf{X}, \mathbf{Y})$
1	0.9981741464736523	1
2	0.9950434560291763	1
3	1.0	1
4	0.9973259696600676	1
5	0.9965850090502247	1
6	0.9958361479001489	1
7	0.9980901342096359	1
8	0.9952349663525573	1
9	0.9969461005607609	1
10	0.9985147913732642	1

service in this experiment is controlled, so it can be made to use the controlled relay as its Guard, which in turn can capture the traffic of the target hidden service on the controlled Guard relay. However, hidden services in real scenarios are uncontrollable and anonymous, making it difficult to collect data in real scenarios. The Tor network simulation method in this paper can replicate this scenario to help researchers in their studies.

Table 4. Table of Hidden Service Traceability Experiment results.

Index	Number of access	Number of successful traceability	Individual hidden service traceability accuracy
1	12	11	0.9167
2	12	11	0.9167
3	11	11	1
4	12	12	1
5	15	13	0.8667
6	12	12	1
7	13	13	1
8	10	9	0.9
9	11	10	0.9091
10	12	11	0.9167

5 Discussion

In this paper, we propose a Tor network simulation method based on K8s and experimentally verify that the method is lightweight, authentic, flexible, and scalable, and can meet the needs of de-anonymization attack experiments. The simulation environment in this paper can simulate not only Tor relays but also other types of services (e.g., FTP servers), which have a certain degree of universality. However, the simulation environment in this paper does not consider the differences in network latency of relays due to different geographic locations, different bandwidths, etc., which can be further investigated in the future to enhance the authenticity of the simulated private Tor network.

Acknowledgements. This work was supported by the 2023QY0202.

References

1. Akhoondi, M., Yu, C., Madhyastha, H.V.: Lastor: A low-latency as-aware tor client. In: 2012 IEEE Symposium on Security and Privacy, pp. 476–490. IEEE (2012)
2. Anderson, C.: Docker [software engineering]. IEEE Softw. **32**(3), 102-c3 (2015)
3. Bang, J., Jeong, J., Lee, J.: Fedfingerprinting: a federated learning approach to website fingerprinting attacks in tor networks. IEEE Access **11**, 78431–78444 (2023). https://doi.org/10.1109/ACCESS.2023.3299174

4. Bauer, K., McCoy, D., Grunwald, D., Kohno, T., Sicker, D.: Low-resource routing attacks against tor. In: Proceedings of the 2007 ACM workshop on Privacy in electronic society, pp. 11–20 (2007)
5. Bauer, K., Sherr, M., Grunwald, D.: {ExperimenTor}: a testbed for safe and realistic tor experimentation. In: 4th Workshop on Cyber Security Experimentation and Test (CSET 11) (2011)
6. Benzel, T.: The science of cyber security experimentation: the deter project. In: Proceedings of the 27th Annual Computer Security Applications Conference, pp. 137–148 (2011)
7. Chakravarty, S., Stavrou, A., Keromytis, A.D.: Traffic analysis against low-latency anonymity networks using available bandwidth estimation. In: Gritzalis, D., Preneel, B., Theoharidou, M. (eds.) Computer Security – ESORICS 2010: 15th European Symposium on Research in Computer Security, Athens, Greece, September 20-22, 2010. Proceedings, pp. 249–267. Springer Berlin Heidelberg, Berlin, Heidelberg (2010). https://doi.org/10.1007/978-3-642-15497-3_16
8. Chen, M., Wang, X., Shi, J., Zhao, C., Wang, M., Fang, B.: Napping guard: deanonymizing tor hidden service in a stealthy way. In: 2020 IEEE 19th International Conference on Trust, Security and Privacy in Computing and Communications (TrustCom), pp. 699–706. IEEE (2020)
9. Cherubin, G., Jansen, R., Troncoso, C.: Online website fingerprinting: Evaluating website fingerprinting attacks on tor in the real world. In: 31st USENIX Security Symposium (USENIX Security 22), pp. 753–770 (2022)
10. Chun, B., Culler, D., Roscoe, T., Bavier, A., Peterson, L., Wawrzoniak, M., Bowman, M.: Planetlab: an overlay testbed for broad-coverage services. ACM SIGCOMM Comput. Commun. Rev. **33**(3), 3–12 (2003)
11. Dingledine, R., Mathewson, N., Syverson, P.F., et al.: Tor: the second-generation onion router. In: USENIX Security Symposium, vol. 4, pp. 303–320 (2004)
12. dockercon: Dockerfile reference. https://docs.docker.com/engine/reference/builder/. Last accessed 15 Sep 2023
13. Docs, S.: Api. https://stem.torproject.org/api.html, last accessed 18 Sep 2023
14. Doswell, S., Aslam, N., Kendall, D., Sexton, G.: The novel use of bridge relays to provide persistent tor connections for mobile devices. In: 2013 IEEE 24th Annual International Symposium on Personal, Indoor, and Mobile Radio Communications (PIMRC), pp. 3371–3375. IEEE (2013)
15. Edman, M., Syverson, P.: As-awareness in tor path selection. In: Proceedings of the 16th ACM conference on Computer and communications security, pp. 380–389 (2009)
16. Elahi, T., Bauer, K., AlSabah, M., Dingledine, R., Goldberg, I.: Changing of the guards: a framework for understanding and improving entry guard selection in tor. In: Proceedings of the 2012 ACM Workshop on Privacy in the Electronic Society, pp. 43–54 (2012)
17. feiskyer: kubeadm. https://kubernetes.feisky.xyz/setup/cluster/kubeadm. Last accessed 18 Sep 2023
18. Galteland, H., Gjøsteen, K.: Adversaries monitoring tor traffic crossing their jurisdictional border and reconstructing tor circuits. arXiv preprint arXiv:1808.09237 (2018)
19. Google: K8s: Production-grade container orchestration. https://kubernetes.io/. Last accessed 4 Aug 2023
20. Guan, Z., Liu, C., Xiong, G., Li, Z., Gou, G.: Flowtracker: improved flow correlation attacks with denoising and contrastive learning. Comput. Secur. **125**, 103018 (2023)
21. Hat, R.: Understanding virtualization. https://www.redhat.com/en/topics/virtualization. Last accessed 15 July 2023
22. Inc., T.T.P.: Tor metrics portal. https://metrics.torproject.org/. Last accessed 4 Aug 2023

23. Jansen, R., Geddes, J., Wacek, C., Sherr, M., Syverson, P.: Never been {KIST}:{Tor's} congestion management blossoms with {Kernel-Informed} socket transport. In: 23rd USENIX Security Symposium (USENIX Security 14), pp. 127–142 (2014)
24. Jansen, R., Hopper, N.J.: Shadow: Running tor in a box for accurate and efficient experimentation (2011)
25. Jansen, R., Newsome, J., Wails, R.: Co-opting linux processes for {High-Performance} network simulation. In: 2022 USENIX Annual Technical Conference (USENIX ATC 22), pp. 327–350 (2022)
26. Jansen, R., Tracey, J., Goldberg, I.: Once is never enough: foundations for sound statistical inference in tor network experimentation. In: 30th USENIX Security Symposium (USENIX Security 21), pp. 3415–3432 (2021)
27. Karunanayake, I., Ahmed, N., Malaney, R., Islam, R., Jha, S.K.: De-anonymisation attacks on tor: A survey. IEEE Commun. Surv. Tutorials **23**(4), 2324–2350 (2021)
28. Li, J., Gu, C., Zhang, X., Chen, X., Liu, W.: Attcorr: A novel deep learning model for flow correlation attacks on TOR. In: 2021 IEEE International Conference on Consumer Electronics and Computer Engineering (ICCECE), pp. 427–430. IEEE (2021)
29. Li, T., Liu, K., Feng, W., Yang, C., Luo, X.: Heterotic: a robust network flow watermarking based on heterogeneous time channels. Comput. Netw. **219**, 109424 (2022)
30. Ling, Z., Luo, J., Xu, D., Yang, M., Fu, X.: Novel and practical sdn-based traceback technique for malicious traffic over anonymous networks. In: IEEE INFOCOM 2019-IEEE Conference on Computer Communications, pp. 1180–1188. IEEE (2019)
31. Mathewson, N.: Tor rendezvous specification. https://gitlab.torproject.org/tpo/core/torspec/-/blob/HEAD/rend-spec-v3.txt. Last accessed 12 June 2023
32. Mathewson, N.: Tor's protocol specifications. https://gitweb.torproject.org/torspec.git/tree/tor-spec.txt. Last accessed 12 June 2023
33. Mirkovic, J., Benzel, T.V., Faber, T., Braden, R., Wroclawski, J.T., Schwab, S.: The deter project: Advancing the science of cyber security experimentation and test. In: 2010 IEEE International Conference on Technologies for Homeland Security (HST). pp. 1–7. IEEE (2010)
34. Nasr, M., Bahramali, A., Houmansadr, A.: Deepcorr: Strong flow correlation at-tacks on tor using deep learning. In: Proceedings of the 2018 ACM SIGSAC Conference on Computer and Communications Security. pp. 1962–1976 (2018)
35. Oh, S.E., Yang, T., Mathews, N., Holland, J.K., Rahman, M.S., Hopper, N., Wright, M.: Deepcoffea: Improved flow correlation attacks on tor via metric learning and amplification. In: 2022 IEEE Symposium on Security and Privacy (SP), pp. 1915–1932. IEEE (2022)
36. project, T.: How often does tor change its paths? https://support.torproject.org/about/change-paths/. Last accessed 15 June 2023
37. project, T.: Tor -stable manual. https://2019.www.torproject.org/docs/tor-manual.html.en. Last accessed 17 Sep 2023
38. project, T.: Tor browser. https://www.torproject.org/download/. Last accessed 10 June 2023
39. rbrtbnfgl: flannel. https://github.com/flannel-io/flannel. Last accessed 19 Sep 2023
40. Sedgwick, P.: Pearson's correlation coefficient. BMJ **345**(jul04 1), e4483–e4483 (2012)
41. Shirazi, F., Goehring, M., Diaz, C.: Tor experimentation tools. In: 2015 IEEE Security and Privacy Workshops, pp. 206–213. IEEE (2015)
42. Singh, S.: Large-scale emulation of anonymous communication networks. Master's thesis, University of Waterloo (2014)
43. Sirinam, P., Imani, M., Juarez, M., Wright, M.: Deep fingerprinting: Undermining website fingerprinting defenses with deep learning. In: Proceedings of the 2018 ACM SIGSAC Conference on Computer and Communications Security, pp. 1928–1943 (2018)

44. Wang, M., Li, Y., Wang, X., Liu, T., Shi, J., Chen, M.: 2ch-tcn: A website fingerprinting attack over tor using 2-channel temporal convolutional networks. In: 2020 IEEE Symposium on Computers and Communications (ISCC), pp. 1–7. IEEE (2020)
45. Wang, Q., Cao, W.: A tor anonymity attack experiment platform driven by raspberry pi. In: 2020 11th International Conference on Prognostics and System Health Management (PHM-2020 Jinan), pp. 569–574. IEEE (2020)
46. Wang, X., Chen, S., Jajodia, S.: Network flow watermarking attack on low-latency anonymous communication systems. In: 2007 IEEE Symposium on Security and Privacy (SP'07), pp. 116–130. IEEE (2007)
47. William, P., Jawale, M.A., Pawar, A.B., Bibave, Rahul R., Narode, Priyanka: Systematic approach for detection and assessment of dark web threat evolution. In: Rawat, R., Kaur, U., Khan, S.P., Ranjana Sikarwar, K., Sankaran, S. (eds.) Using Computational Intelligence for the Dark Web and Illicit Behavior Detection, pp. 230–256. IGI Global (2022). https://doi.org/10.4018/978-1-6684-6444-1.ch013

Enhancing Internal Networks Security and Limiting Lateral Movement Through Hidden Services in Anonymous Communication

Yu Xiang, Runnan Tan, Silin Zhao, and Qingfeng Tan(✉)

Cyberspace Institute of Advanced Technology, Guangzhou University, Guangzhou 511363, China
tqf528@gzhu.edu.cn

Abstract. In the ever-evolving landscape of the digital domain, the frequency and complexity of network attacks continue to increase, creating significant challenges for organizations seeking to strengthen internal network security. Among the various threats, lateral movement attacks are of particular concern. Traditional defenses such as network segmentation, access control, and firewalls have become increasingly inadequate to counter these intrusions. Therefore, this paper proposes an innovative strategy using hidden services in anonymous communication. This strategy consists of two essential components: the intranet service hiding module and the hidden service publishing module. When in operation, it hides regular services by listening on the loopback address and uses the isolation capabilities provided by anonymous networks to confine intranet access to a dedicated anonymous channel. To evaluate the effectiveness of our approach, experiments were conducted in two different internal network test environments-one that implemented our strategy and one that did not. Data analysis, including insights into attack paths and success rates, indicates that implementing our strategy significantly reduces an attacker's ability to detect and exploit internal network resources. While we are confident in the practicality and potential value of our strategy, we recognize its limitations and potential challenges. As a result, we are committed to continually refining our strategy to further improve its security, robustness, and ease of use.

Keywords: Intranet security · lateral movement · anonymous communication system

1 Introduction

The Internet, as a constantly evolving technology, has fundamentally transformed our personal and professional lives. Organizations are increasingly building their own internal networks to facilitate information sharing, improve collaboration, and optimize resource utilization. This move also reduces reliance on external networks, thereby enhancing network security and reliability.

Z. Gu et al. (Eds.): NSE 2023, CCIS 2063, pp. 368–383, 2024.
https://doi.org/10.1007/978-981-97-4519-7_26

However, corporate intranets face a growing threat from sophisticated and covert network intrusion attacks. These attackers employ stealthy techniques to breach network systems undetected, with the intent of pilfering sensitive information or causing damage.

Once a device on the corporate intranet is compromised, attackers use it as a stepping stone to laterally traverse the network by various means, seeking additional targets to attack, thereby expanding the scope of their influence. These intranet lateral movement attacks pose a significant threat to data security, business stability, and result in substantial economic losses and damage to an organization's reputation. Thus, preventing intranet lateral movement has become an urgent concern.

One common method to defend against intranet attacks is network segmentation [14, 21], which divides the network into multiple security domains and restricts communication between them to reduce the likelihood of intranet penetration. However, this method does not completely halt attackers from moving laterally within the intranet. Attackers can still employ methods like malicious scanning to discover host vulnerabilities or clues in known subnets for further attacks.

Another widely-used method is access control [16], involving user and device permission verification and management. Access control effectively limits attackers' access to intranet resources but cannot entirely prevent intranet entry through other means, such as malicious scanning. Additionally, configuring and maintaining access control demands substantial time and effort, and poorly set rules can disrupt normal business operations and reduce intranet availability.

Weak passwords represent a common vulnerability in intranet security, with numerous instances of breaches attributable to weak passwords [2, 12]. Strengthening password complexity and enforcing strong password policies is an effective defense against intranet attacks [17]. However, this policy requires users to adopt stronger, harder-to-remember passwords, potentially affecting the user experience. Users may resort to writing down or sharing their passwords, compromising password security. While strong password policies mitigate password guessing and weak password attacks, they are less effective against other forms of attacks, such as social engineering.

In addition to these methods, technologies like firewalls [5, 20, 23], intrusion detection systems [3, 4, 6, 11], security patches, monitoring, and auditing enhance intranet security. However, these measures are not foolproof and do not completely thwart intranet lateral movement, whereby attackers can still traverse the intranet, gain control of additional machines, or exploit sensitive files once access is gained.

Hence, this paper introduces an effective method based on hidden services to secure internal networks and restrict lateral movement. This approach conceals service ports, preventing attackers from discovering exploitable vulnerabilities through scanning. Simultaneously, it unlocks anonymous services and utilizes the isolation of an anonymous network to restrict intranet access to a dedicated anonymous channel. This prevents external attackers from accessing intranet services using ordinary network tools or protocols. The method ensures anonymity and obfuscation of intranet nodes and services without disrupting normal services. This paper designs, implements, and evaluates the method in an experimental environment. The results indicate that this method effectively thwarts intranet traversal attacks.

The primary contributions of this paper include:

- **Automated Intranet Service Concealment:** We introduce an automated configuration modification method for concealing intranet services, significantly bolstering the security of internal networks.
- **Lateral Movement Restriction:** Our method is designed to effectively limit lateral movement within internal networks, further fortifying their defense.
- **Method Design, Implementation, and Evaluation:** We meticulously detail the design and implementation of our proposed method, presenting comprehensive testing and evaluation results obtained in an experimental environment.
- **Comprehensive Analysis:** We conduct an in-depth analysis of our method, elucidating its advantages and disadvantages, and we offer constructive suggestions for potential improvements.

2 Background

2.1 The Invisible Internet Project

Invisible Internet Project (I2P) is a distributed and anonymous communication system, which is unique in that it does not rely on any trusted third party, but adopts garlic routing [8] technology similar to onion routing [15] to realize low-latency anonymous communication among users. The goal of I2P is to build a highly reliable, scalable, and censorship-resistant network platform to protect users' communications from interference and monitoring by third parties such as Internet Service Providers (ISPs) [24].

Figure 1 shows the system architecture of I2P, which consists of about 55,000 computers [1] spread around the world, which are called I2P routers. The network of I2P contains three basic components: the I2P routers, the tunnels, and the network database (netDb).

I2P router: A node in an I2P network that is responsible for forwarding traffic and providing services. Anyone can join an I2P network by running an I2P router, and the communication between any two routing nodes is encrypted using the I2P secure transport protocol.

Tunneling: I2P uses end-to-end encryption to protect user traffic and sends packets through paths consisting of multiple randomly selected nodes called "tunnels". In an I2P network, all communication takes place through tunnels, which are categorized as inbound tunnels and outbound tunnels. An inbound tunnel is a path used only to send messages to the tunnel creator and an out-bound tunnel is a path used only to send messages away from the tunnel creator [7].

netDb: A distributed database based on Kademlia Distributed Hash Table (DHT) [13], I2P selects the nodes with higher bandwidth, which account for about 6% of the total number of nodes, as the seed nodes (floodfill router) from all routing nodes, and these nodes constitute the I2P Kad network. netDb plays a vital role in the network, which not only stores the information of all the nodes in the I2P network, but also provides query function for the nodes. netDb is the most important database in the I2P network.

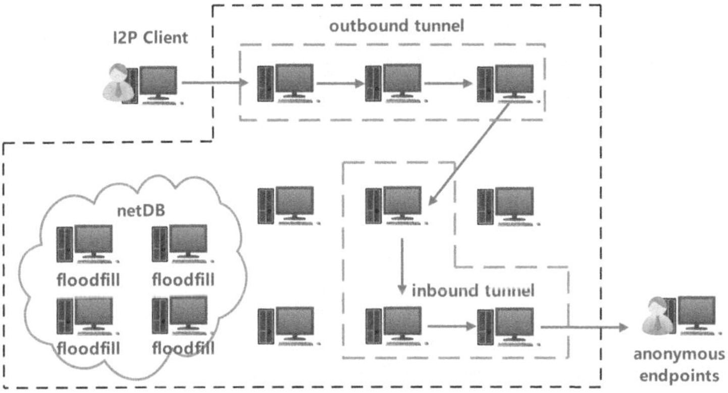

Fig. 1. To access the I2P website, the user uses their own I2P router to select three nodes from the network as the outbound tunnel and sends the encrypted message along these nodes to the inbound tunnel of the target website.

The communication process of I2P is shown in Fig. 2, when Alice wants to communicate with Bob, firstly, Alice will query Bob's inbound gateway address through NetDb and then send a message to Bob's inbound tunnel through his own outbound tunnel. Similarly, if Bob wants to reply to Alice's message, he needs to first query Alice's inbound tunnel gateway address and then reply to Alice's message through his own outbound tunnel. During the whole communication process, both Alice and Bob only know each other's inbound tunnel gateway address, but not each other's real identity and IP address, which effectively protects the privacy of both parties.

Within an I2P network, users can not only communicate anonymously but also access websites known as I2P EepSites. These sites are hosted anonymously within the I2P network via HTTP tunnels and are only accessible via I2P, not via the Internet [10]. To resolve the mapping relationship between domain names and keys of I2P EepSites, a distributed and decentralized domain name system called SusiDNS (Susi Domain Name System [9]) is used. Users can manage and maintain SusiDNS by themselves without relying on any authoritative server or registry.

The URLs of I2P EepSites always end in.i2p, such as idk.i2p or xxxxx.b32.i2p. These names are actually aliases for the destination identifier of the target server. With these destination identifiers, users can protect their real IP addresses and identities and securely access various services and resources.

2.2 Intranet Penetration Testing

An intranet penetration attack is an act in which hackers take advantage of intranet vulnerabilities or weaknesses to attack other hosts from the inside or outside to steal or control sensitive resources. This kind of attack may cause data leakage, system paralysis, business interruption and other serious consequences. In order to test the security of the intranet system, we usually conduct penetration testing.

Penetration testing is the process of assessing the security of a target system and identifying and repairing potential risks by simulating real attacks. In the process of

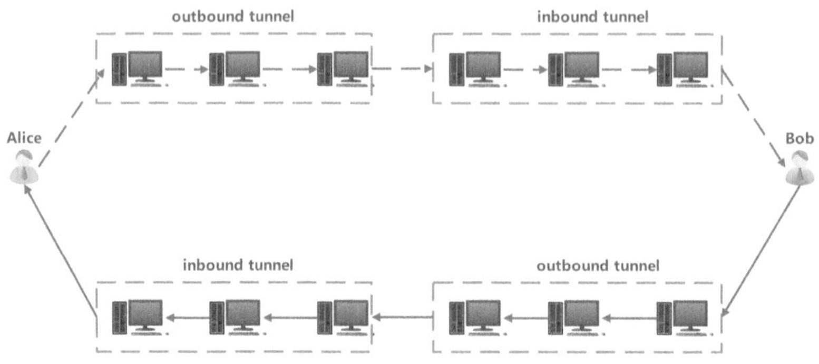

Fig. 2. This is a schematic illustration of i2p communication

penetration testing, the thing that mimics the hacker to carry out the attack is Penetration Testing Red Team. The Red Team's attack methods are usually not limited to the technical level, but use a variety of innovative and covert ways, including social engineering, physical penetration, supply chain attacks, etc., to achieve the intended goal. Typically, the work of a red team can be divided into three phases:intelligence gathering, establishing a stronghold and lateral movement [25].

Intelligence gathering Intelligence gathering in the context of penetration testing involves the activities of gathering relevant information about the target organization. This information includes details about the target's network topology, system configurations, security measures, business relationships, personnel data, and more. The primary goal of intelligence gathering is to uncover potential vulnerabilities and attack vectors within the target, thereby increasing the efficiency and success rate of the penetration testing process.

By gathering information, the red team gains an in-depth understanding of the target organization's network environment, service landscape, vulnerabilities, and personnel information. This knowledge helps identify available attack surfaces and increases the efficiency and success rate of penetration testing. In addition, intelligence gathering helps the red team develop well-informed attack strategies and methodologies, reducing the likelihood of blind or ineffective attacks and minimizing the risk of detection or blocking by the target organization. Intelligence gathering also enables the red team to assess the target organization's security posture, uncover security weaknesses and vulnerabilities, and provide a basis for subsequent security recommendations and improvements.

Establishment of Strongholds. Once sufficient information has been gathered, the red team will proceed to establish a controlled entry point into the target network and then launch further attacks. One of the most common techniques used is remote code execution through vulnerability exploitation. To gain control of the target system, this technique involves executing malicious code or commands by exploiting vulnerabilities in the target system or services, such as buffer overflow, command injection, or file upload exploits.

In addition to technical exploits, social engineering tactics may also be used. These tactics include activities such as the creation of phishing emails or web pages designed to trick users into clicking on malicious links or attachments. Such techniques are designed to trick target users into unknowingly executing malicious code or downloading malicious files, thereby giving the attacker control of the user's system.

Lateral Movement. Once a foothold has been successfully established, the red team continues to gather information and intelligence on the machines or servers already under their control. This includes gathering data such as current network connections, process lists, command execution history, database information, user details, administrator credentials, password policies, and patch update frequencies. They also perform reconnaissance on IPs, host names, open ports, services, and applications running on other machines or servers within the intranet.

To expand the scope of their attacks, red teams exploit weaknesses within the intranet, including vulnerabilities, security holes, or shared passwords. This allows them to perform lateral movement, jumping from one machine or server to another, with the target system or others becoming potential targets for their attacks.

2.3 Enterprise Intranet Architecture

Enterprise intranet architecture refers to the network structure and design within an enterprise. In order to meet different business needs and security requirements, the current mainstream enterprise intranet is generally divided into three core functional areas, namely, the DMZ zone, the internal core server zone and the office network zone, as shown in the Fig. 3 [22].

Demilitarized Zone (DMZ): The DMZ is a buffer zone between the corporate intranet and the external network. It is used to provide some services that are open to the public, such as web servers, mail servers, etc. The role of the DMZ zone is to isolate and protect the corporate intranet, preventing external attackers from directly accessing internal resources. At the same time, it can also reduce the load on internal servers and improve performance.

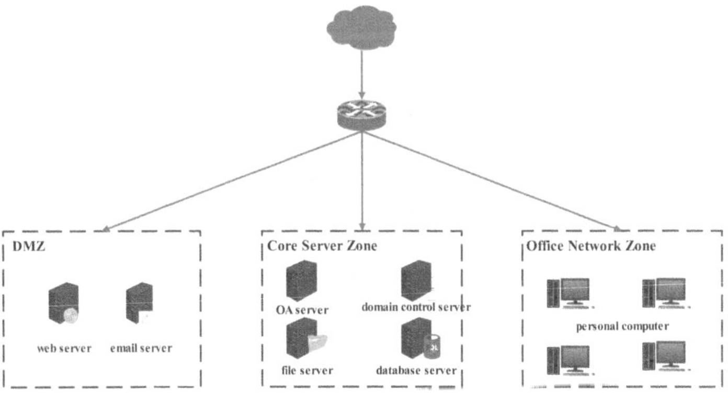

Fig. 3. Intranet structure of common enterprises

Core Server Zone (CSZ): This zone contains the core servers of the enterprise intranet, such as domain controllers, database servers, application servers, and so on. These servers are responsible for core business functions and require high reliability and security. The core server area is usually open only to authorized personnel and requires strict authentication and access control. The core server area also requires reasonable firewall rules to restrict unnecessary traffic.

Office Network Zone (ONZ): This is the main area of an organization's intranet and is used to connect employees' work devices, such as computers and cell phones. The Office Network usually allows access to the Internet and internal resources. Since a lot of sensitive information and data is involved here, security is very important. Office networks require a number of security measures such as firewalls, antivirus software, encrypted communications, etc. It also requires security education and supervision of employees to avoid leaking or destroying data.

3 Methodology Design and Implementation

This paper aims to address the security and availability issues of intranet services and proposes a method to ensure internal network security and limit lateral movement based on hidden services. The design goal of the method is to prevent attackers from discovering and attacking intranet services through scanning, probing, and lateral movement, while ensuring that intranet users can access and use intranet services normally. The architecture of this system is shown in Fig. 4 and consists of the following two main modules.

3.1 Intranet Service Hiding Module

Hiding network assets is an effective strategy to defend against intranet penetration attacks. An intranet penetration attack is a network attack technique that aims to take advantage of vulnerabilities or weaknesses that exist in an intranet to break through firewalls or other security measures, enter the target network and take control of the target host. When conducting an intranet penetration attack, attackers usually first scan the intranet segment for active hosts.

Therefore, to defend against potential scanning attacks, measures must be taken to hide network assets, such as server IP addresses and open ports, to make them difficult for attackers to detect or exploit. The easiest way to do this is to take advantage of the properties of a loopback address, which cannot be detected or forwarded by other hosts and can only be accessed locally. This can be accomplished by modifying the service configuration file so that the service only listens on the loopback address.

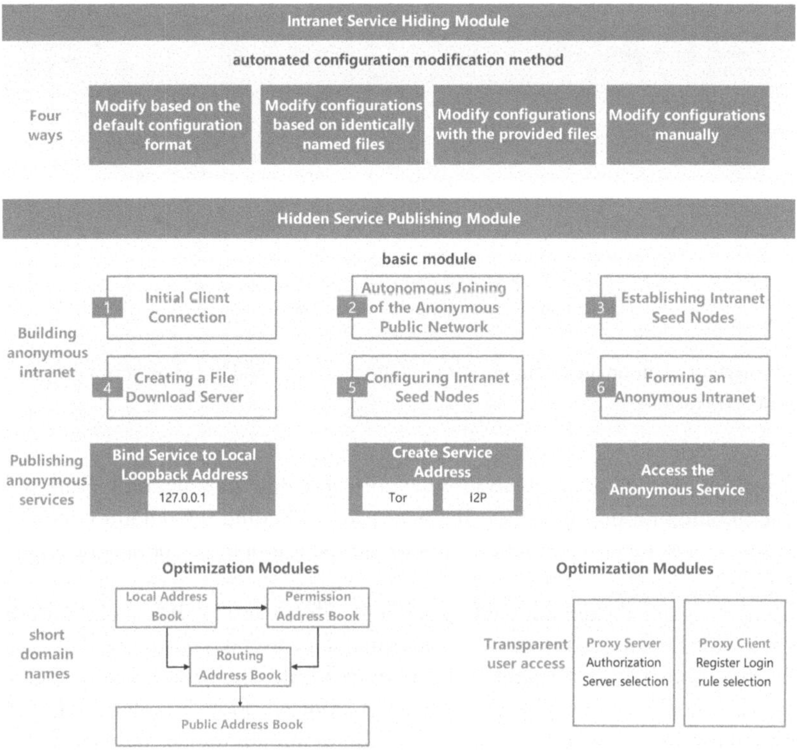

Fig. 4. Our method consists of two modules, the Intranet Service Hiding Module and the Hidden Service Publishing Module

This approach effectively hides open ports inside the server, reducing the risk of detection and abuse by attackers. However, servers usually have as many as 65,535 available ports, and manually modifying the configuration file of each service is a huge workload and also prone to introducing errors. Therefore, this paper proposes an automatic configuration modification method, based on script programming and regular expressions. This approach automatically recognizes and modifies the parameters related to the listening address in the configuration file, and also generates corresponding log files and backup files. Such an approach significantly improves the efficiency and accuracy of configuration modification, and facilitates subsequent management and recovery operations. The flowchart of the method is shown in Fig. 5:

Firstly, the method builds a default configuration database, which stores the names, versions, configuration file names, default paths, modification formats, and other information of common intranet services, as well as common configuration file formats with corresponding configuration commands. For example, the default configuration file location for mysql 5.7 in ubuntu system is /etc./mysql/mysql.conf.d/mysqld.cnf, and the configuration command is bind-address = 127.0.0.1.

The method then runs the "netstat -tulpn" command to discover locally running services and their corresponding ports. Once the service name and version information

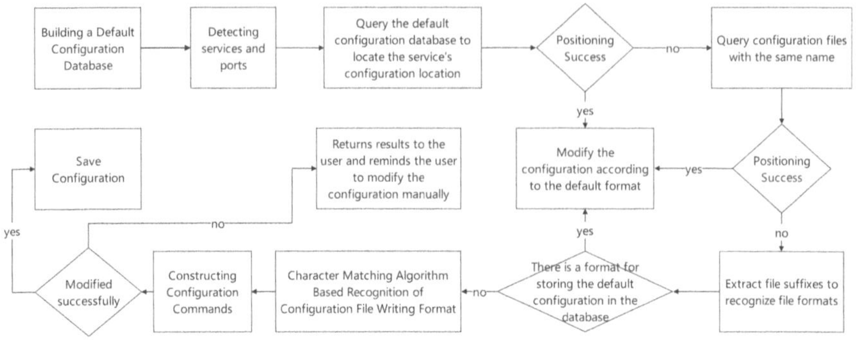

Fig. 5. Flowchart of automated configuration modification method based on script programming and regular expressions

is obtained, the method attempts to find the corresponding default configuration file in the default configuration repository. If it succeeds in finding the default configuration file, it changes the local listening address to the local loopback address in the default format.

If it fails to locate the default configuration file, the method will first check to see if a file with the same name exists locally to determine if the user has customized the location of the configuration file. If the method successfully locates the custom configuration file, it changes the local listening address to the local loopback address in the default format. If it still cannot locate the configuration file, the method prompts the user, and if the user provides a path to the configuration file, the method retrieves the configuration file at that path and modifies it accordingly. This method ensures that configuration files are handled correctly whether they are stored in the default path or in a user-specified location.

For internal development of user units and unusual service configurations, the method extracts the configuration file based on the configuration file path provided by the user and detects its file suffix to determine whether the configuration file is in a common file format, such as ini, json, yaml, and so on. If it can determine that the configuration file is in a common file format, it modifies the service's listening address according to the appropriate configuration rules. If the file format cannot be recognized, the method uses character-matching techniques (e.g., regular expressions) to analyze the writing format of the configuration file's contents and generates the appropriate configuration commands based on the format, then writes these commands to the configuration file and saves them. This ensures that the service's listening address can be effectively modified even when dealing with unusual configuration files.

If none of the above can successfully modify the service configuration, the result is returned to the user and the user is reminded to modify the configuration manually.

3.2 Hidden Service Publishing Module

Based on the anonymous communication network, this part proposes a method to externally publish an intranet service that prohibits remote access as described in Sect. 3.1

for normal use and access by other users, whose detailed steps are shown in the Fig. 4. The method has four modules, which are building an anonymous intranet, publishing anonymous services, user transparent access and publishing short domain names. Among them, building an anonymous intranet and publishing an anonymous service are core functional modules, and user transparent access and publishing short domain names are functional optimization modules.

Building an Anonymous Intranet Anonymity network is a network system that realizes anonymous communication by using special software and special configuration. It hides the identities and relationships of the communicating parties through layered encryption and multiple forwarding of communication data, and is characterized by a high degree of anonymity, tracking resistance and dynamism. Anonymity networks not only allow users to access the Internet anonymously, but also allow service providers to deploy hidden network services. When an anonymous network client initially connects to the anonymous public network, it retrieves seed node information from the anonymous public network via the built-in nodes provided officially. It then propagates through these seed nodes to autonomously join the anonymous public network. To ensure that newly connected nodes enter the anonymous intranet through our nodes rather than the official built-in nodes, we establish seed nodes within the intranet. Additionally, we create a file download server to store and distribute the information about these intranet seed nodes.

Upon client startup, the original client downloads information about the official seed nodes from the provided link on the official built-in nodes. However, our modified client downloads information about the intranet seed nodes from the file download server link. The configuration of intranet seed nodes can vary depending on the specific type of anonymity network we are using. For example, in the Tor network, we establish both a hidden service directory server node and a directory server node, whereas in the I2P network, we set up a Floodfill node. Since the seed nodes are all intranet nodes, the new nodes will no longer be connected to the anonymous public network, but will constitute an anonymous intranet. Also, it should be noted that in order to be able to distribute hidden services to the outside world based on the anonymous network, we need to install and run an anonymous intranet client on the intranet server host.

Publishing Anonymous Services. To publish an anonymous service, we begin by running the anonymous intranet client on the intranet server host. We then open the corresponding anonymous service based on the specific type of anonymous network. Subsequently, we bind this anonymous service to the local loopback address and the intranet service's port.

This setup enables the anonymous network client to create a service address using the anonymous network's service publishing mechanism and upload this service information to designated nodes. Other anonymous intranet nodes can access and use this anonymous service via the provided service address. This achieves the dual purpose of enabling normal intranet service use within the intranet while concealing the IP address and port number of the intranet server. For example, if we wish to publish a web service to the outside world using the I2P anonymous communication software, we start the modified I2P client and create a web service through the I2P client's hidden service manager page. Next, we bind the web service to 127.0.0.1:8080. The I2P client generates a service address for the anonymous web service based on its service publishing mechanism.

When the client uploads the details to the Floodfill node, we can access intranet services normally within the intranet using this.i2p address, all while keeping the intranet server's IP address and port number concealed from the outside world.

Publishing Short Domain Names. In the context of anonymous services, accessing an I2P service typically requires knowledge of its lengthy and irregular.i2p address, making it inconvenient for users to remember. To address this issue, we aim to assign easily memorable short domain names for these intranet services.

To implement this feature, we draw inspiration from the I2P address book design and create a similar address book structure on the intranet server, as depicted in Fig. 4. This address book comprises four main components: the public address book, local address book, permission address book, and routing address book.

The public address book holds mappings between.i2p addresses and user-defined, easy-to-remember domain names for all intranet services. For instance, if the intranet web service's address is <x> .i2p, the public address book entry might be: <x>.i2p: <z>.i2p, where <x> represents the base32-encoded string and <z> is the user's chosen, memorable domain name.

The local address book is user-managed, private, and resides on the user's device. Users can selectively share parts of their local address book with others by publishing them to the routing address book or permission address book. When publishing to the permission address book, the publisher must define access rights. In contrast, publishing directly to the routing address book defaults to granting access rights to all users. Once published, the address book entries migrate to the routing address book. Periodically, the Routing Address Book updates the Public Address Book with the addresses contained within, enabling sharing with other users.

Transparent User Access. As we introduced before, to access anonymous services, we need to use special software and special configuration, which will affect the original business of the intranet and increase the technical difficulty and operational burden of unrelated business personnel. In order to solve this problem, we build a proxy server and develop a proxy client, so that intranet users can connect to the proxy server through the proxy client, thus realizing access to and use of hidden services in the intranet. In this way, intranet users do not need to install any specialized software or dependent environment, but only a simple proxy client.

4 Experimental Verification

In order to verify the effectiveness and security of our proposed method, we designed two experiments for each of the two modules. One is to test the ability of Intranet Service Hiding Module to prevent external scanning and probing, and the other is to evaluate the attack resistance of our approach in penetration testing.

4.1 Protection from External Scanning and Probing

In this section, we use an open source network scanning tool, Nmap, to simulate an external attacker scanning and probing our system. We have deployed our system on an intranet server and enabled some common network services such as SSH, HTTP, and MySQL services on this server. We scanned this server with the following command:

$nmap - Pnip_you_want_to_scan

This command means to disable host discovery and perform port scanning and service probing directly on the target host. We ran this command with our module turned on and off at the same time and recorded the scan results as the Fig. 7.

From the Fig. 6, we can clearly see that before using Intranet Service Hiding Module, we were able to scan 5 ports, 4 of which were in the open state. After using Intranet Service Hiding Module, all the open ports disappeared because their listening address changed to 127.0.0.1, so their status changed to closed. In this way, whether you try to access the web service or ssh service, you'll find that you can't connect to it, which shows that our method can effectively block port scanning.

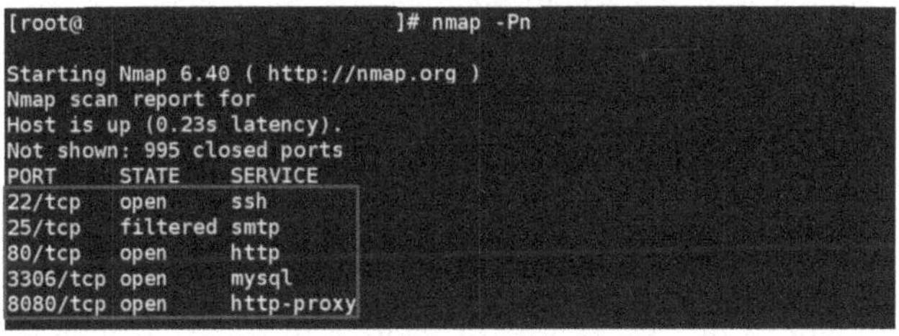

Before using Intranet Service Hiding Module

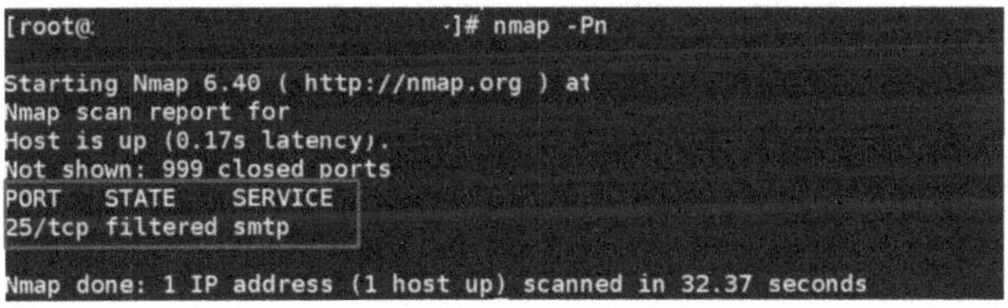

After using Intranet Service Hiding Module

Fig. 6 Before and after using Intranet Service Hiding Module

To rule out chance in our scanning results, we applied our method on 25 servers and counted the number of ports successfully closed and the success rate, as shown in Fig. 7.

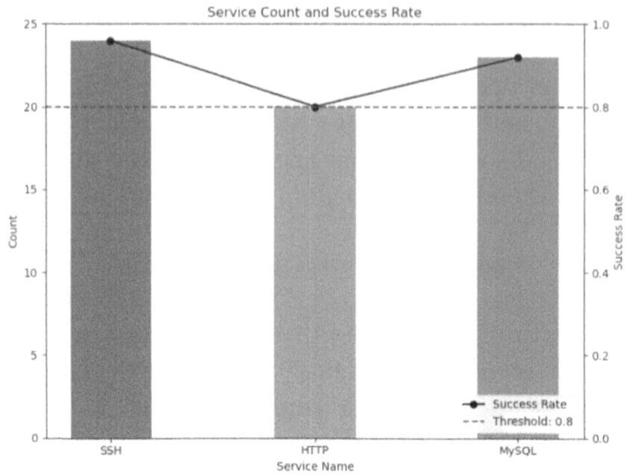

Fig. 7. Before and after using Intranet Service Hiding Module

4.2 Practical Penetration Testing

In this section, we validate the effectiveness and security of our approach through a real-world penetration test. We established two similar intranet systems: one secured using our method and the other without it. We ensured that both systems had identical configurations and network environments, as detailed in Table 1, for comparative testing. To maintain consistency between the two penetration tests, we adhered to the standard procedure for conducting and assessing a penetration test.

According to our penetration test report, we observed that all machines lacking our security method fell prey to attacks, primarily due to weak password vulnerabilities. In contrast, the machines with our security method, despite having some vulnerabilities, remained resilient to attackers. There are three key reasons for this, with the exception of the DMZ zone web service, which is designed to provide services to the outside world and cannot be concealed.

- **Hide the real IP address and service port of the intranet system**: The method leverages the anonymity provided by the anonymous network to conceal the actual IP address of the intranet system behind a randomly generated anonymous network domain name. This effectively prevents external attackers from directly accessing the intranet system or discovering its existence through scanning or locating tools.
- **Encrypted Intranet communications**: The method capitalizes on the strong encryption capabilities of anonymity networks to encrypt intranet communications at multiple layers. This ensures that external attackers are unable to eavesdrop, manipulate intranet communications, or gain access to intranet information through the analysis of traffic or data packets.
- **Restricted Intranet Access**: The method leverages the isolation provided by an anonymous network to confine intranet access to a dedicated anonymous channel. This effectively prevents external attackers from using standard network tools or protocols to access intranet services or exploiting common vulnerabilities and weak passwords for lateral penetration.

In summary, the hidden service-based intranet security protection system can effectively prevent lateral movement and improve the security of the intranet system.

Table 1. Network Information

Zone	Server	Services Tested	Open Ports	Vulnerabilities
DMZ	Website	Web, SSH	8080, 2233	Weak Password
DMZ	Email System	Web, SMTP, SSH	8010, 25, 2233	Weak Password
CSZ	OA	SSH	2233	Weak Password
CSZ	MySQL	MySQL, SSH	3306, 2233	Weak Password
ONZ	Computers (Linux)	SSH	2233	Weak Password
ONZ	Computers (Windows)	RDP	3389	Weak Password, MS17-010

5 Conclusion

In this paper, we propose an intranet security protection system based on hidden services, which utilizes the anonymity and covert nature of hidden services to provide an effective security protection mechanism for intranets. The system aims to prevent lateral movement attacks and protect sensitive information and resources in the intranet. This paper makes some theoretical and practical contributions to the field of intranet security protection by designing the architecture of the system and experimentally verifying the security and feasibility of the system.

However, the system proposed in this paper has not yet been deployed and tested on a large scale, and thus may have some potential security risks and performance issues. Therefore, future research can improve and extend the system in the following aspects:

First, the selection and configuration of the intranet service hiding module can be optimized to improve the reliability and efficiency of this module.

Second, Hidden services are susceptible to a wide range of potential attack threats [18, 19], which underscores the paramount need to increase the robustness and fault tolerance of the system. This can be achieved by increasing the diversity and redundancy of hidden services, thereby ensuring that the system can maintain proper functionality even if some hidden services are attacked or fail.

In addition, other security technologies, such as encryption, authentication and auditing, can be combined to enhance the security and trustworthiness of the system. Encryption techniques can be used to encrypt and protect intranet communications, authentication techniques can be used to verify the identity of a user or service, and auditing techniques can be used to monitor and record system operations so that potential security problems can be detected and dealt with in a timely manner.

In summary, the hidden service-based intranet security protection system proposed in this paper has some significance in both theory and practice. Despite some potential problems, the reliability, robustness and security of the system can be further improved by optimizing the selection and configuration of hidden services, increasing the diversity and redundancy of hidden services, and combining other security technologies. Future research can explore and improve on these aspects in depth to further improve the system.

References

1. I2p anonymous network. https://geti2p.net/. Last accessed on 25 Oct 2023
2. Phisher steals $500,000 due to weak password (2023). https://www.itro.com.au/phisher-steals-500000-weak-password. Last accessed on 25 Oct 2023
3. Abdulganiyu, O.H., Ait Tchakoucht, T., Saheed, Y.K.: A systematic literature review for network intrusion detection system (ids). Int. J. Inf. Secur. **22**(5), 1125–1162 (2023)
4. Aljanabi, M., Ismail, M.A., Ali, A.H.: Intrusion detection systems, issues, challenges, and needs. Int. J. Comput. Intell. Syst. **14**(1), 560–571 (2021)
5. Bellovin, S.M., Cheswick, W.R.: Network firewalls. IEEE Commun. Mag. **32**(9), 50–57 (1994)
6. Borkar, A., Donode, A., Kumari, A.: A survey on intrusion detection system (ids) and internal intrusion detection and protection system (iidps). In: 2017 International conference on inventive computing and informatics (ICICI), pp. 949–953. IEEE (2017)
7. De Boer, T., Breider, V.: Invisible internet project (i2p). In: System and Network Engineering, pp. 1–16 (2019)
8. Dingledine, R., Mathewson, N., Syverson, P.F., et al.: Tor: The second-generation onion router. In: USENIX security symposium. vol. 4, pp. 303–320 (2004)
9. i2pd Documentation: I2pd user guide: Susidns (No Date). https://i2pd.readthedocs.io/en/latest/user-guide/susidns/. Last accessed on 25 Oct 2023
10. Gao, Y., Tan, Q., Shi, J., Wang, X., Chen, M.: Large-scale discovery and empirical analysis for i2p eepsites. In: 2017 IEEE Symposium on Computers and Communications (ISCC), pp. 444–449. IEEE (2017)
11. Khraisat, A., Gondal, I., Vamplew, P., Kamruzzaman, J.: Survey of intrusion detection systems: techniques, datasets and challenges. Cybersecurity **2**(1), 1–22 (2019)
12. Mathews, L.: A password set to "password" leads to a $15 million ransom demand (2022). https://www.forbes.com/sites/leemathews/2022/03/20/a-password-set-to-password-leads-to-a-15-million-ransom-demand/. Last accessed on 25 Oct 2023
13. Maymounkov, P., Mazières, D.: Kademlia: a peer-to-peer information system based on the xor metric. In: Druschel, P., Kaashoek, F., Rowstron, A. (eds.) Peer-to-Peer Systems, pp. 53–65. Springer Berlin Heidelberg, Berlin, Heidelberg (2002). https://doi.org/10.1007/3-540-45748-8_5
14. Mhaskar, N., Alabbad, M., Khedri, R.: A formal approach to network segmentation. Comput. Secur. **103**, 102162 (2021)
15. Reed, M.G., Syverson, P.F., Goldschlag, D.M.: Anonymous connections and onion routing. IEEE J. Select. Areas Commun. **16**(4), 482–494 (1998)
16. Sandhu, R.S., Samarati, P.: Access control: principle and practice. IEEE Commun. Mag. **32**(9), 40–48 (1994)
17. Siponen, M., Puhakainen, P., Vance, A.: Can individuals' neutralization techniques be overcome? a field experiment on password policy. Comput. Secur. **88**, 101617 (2020)
18. Tan, Q., Gao, Y., Shi, J., Wang, X., Fang, B., Tian, Z.: Toward a comprehensive insight into the eclipse attacks of tor hidden services. IEEE Internet of Things J. **6**(2), 1584–1593 (2018)
19. Tan, Q., Wang, X., Shi, W., Tang, J., Tian, Z.: An anonymity vulnerability in tor. IEEE/ACM Trans. Netw. **30**(6), 2574–2587 (2022)

20. Wack, J., Cutler, K., Pole, J.: Guidelines on firewalls and firewall policy. NIST Spec. Publ. **800**, 41 (2002)
21. Wagner, N., et al..: Towards automated cyber decision support: a case study on network segmentation for security. In: 2016 IEEE Symposium Series on Computational Intelligence (SSCI), pp. 1–10. IEEE (2016)
22. Xu, Y.: Analysis of attack and defense technologies under the new architecture of enterprise intranet. Netw. Secur. Informatization 108–113 (2023)
23. Yue, X., Chen, W., Wang, Y.: The research of firewall technology in computer network security. In: 2009 Asia-Pacific Conference on Computational Intelligence and Industrial Applications (PACIIA), vol. 2, pp. 421–424. IEEE (2009)
24. Zantout, B., Haraty, R., et al.: I2p data communication system. In: Proceedings of ICN. pp. 401–409. ICN, Springer Singapore (2011)
25. Zhang, G., Zhou, R.: Research on the breakthrough of defense system from the perspective of red team. In: Xu, Z., Parizi, R.M., Loyola-González, O., Zhang, X. (eds.) Cyber Security Intelligence and Analytics: 2021 International Conference on Cyber Security Intelligence and Analytics (CSIA2021), Volume 1, pp. 213–218. Springer International Publishing, Cham (2021). https://doi.org/10.1007/978-3-030-70042-3_31

A Cyber Range Based Method for Digital Twin Construction of a Blockchain Network

Lin Chen[1], Qunhong Sun[1], Shen Su[1,4(✉)], Yu Cui[2], Man Zhang[3], and Zhihong Tian[1]

[1] Guangzhou University, Guangzhou, China
{2112233047,2112233066}@e.gzhu.edu.cn, {sushen, tianzhihong}@gzhu.edu.cn
[2] Rangesoft Network Technology Co., Ltd., Beijing, China
[3] PENG CHENG LABORATORY, Shenzhen, China
zhangm@pcl.ac.cn
[4] Pazhou Lab, Guangzhou, China

Abstract. In recent years, with the rapid development of blockchain technology, researchers have placed increasing emphasis on the performance and security of blockchain networks. However, conducting experiments on existing blockchain networks always consumes significant hardware resources and time costs. Therefore, the mainstream approaches involve either developing custom blockchain network simulation platforms or deploying blockchain networks through automated systems. Nevertheless, existing blockchain network simulation platforms tend to over-look the underlying structural characteristics of blockchain networks, and blockchain automation deployment systems suffer from issues such as distorted blockchain network traffic data. In this paper, we propose a cyber range based method for digital twin construction of a blockchain network. It utilizes cyber range virtualization to address the issues of physical resource consumption and the simulation of underlying blockchain network structural features. It involves the addition of probe nodes in the original blockchain network for traffic data collection and real-time analysis, thereby resolving the issue of traffic data distortion. Furthermore, we explore the application of this method in three original blockchain network scenarios: small-scale, large-scale, and borderless. To overcome the limitations of probe nodes in large-scale original blockchain network scenario, we introduce a reverse tracing strategy to optimize the selection of probe node locations. finally, we conduct experimental evaluations to assess the effectiveness of the reverse tracing strategy and the accuracy of method for digital twin. We anticipate that the technological approach proposed in this paper will assist scholars, institutions, and developers in their research endeavors within the blockchain field.

Keywords: Blockchain · Cyber Range · Digital Twin · Reverse Tracing Strategy

Z. Gu et al. (Eds.): NSE 2023, CCIS 2063, pp. 384–399, 2024.
https://doi.org/10.1007/978-981-97-4519-7_27

1 Introduction

Blockchain is a tamper-proof distributed ledger system built upon P2P networks. It was originally proposed by Satoshi Nakamoto in the Bitcoin whitepaper [1] in 2008, with the primary objective of solving the trust problem in transactions. The subsequent year marked a significant milestone when the Bitcoin blockchain network, featuring a decentralized digital currency and payment platform, was officially launched, signifying the blockchain's introduction to the public and the commencement of the blockchain 1.0 era. In 2014, Vitalik Buterin developed Ethereum [2], a decentralized programmable platform with Turing completeness, building upon the foundation laid by Bitcoin blockchain. This development unequivocally propelled blockchain technology to a new zenith, heralding the advent of the blockchain 2.0 era. Over the course of more than a decade, blockchain technology has demonstrated robust vitality, giving rise to notable applications such as Decentralized Autonomous Organization (DAO), Decentralized finance (Defi), and Non-Fungible Token (NFT). Concurrently, alliance chains have emerged, denoting blockchains that only permit participation following authorization from designated entities. Prominent examples of alliance chains projects include Hyperledger Fabric [3] and the Enterprise Ethereum Alliance (EEA) [4].

However, the performance of blockchain network has gradually become the bottleneck of its development [5], and the security situation of blockchain network is becoming increasingly severe. By the date of submission, the security accidents of blockchain reached 1187, resulting in property losses of about $30,453,196,652.24 [6]. These incidents have triggered heightened consideration and research by a growing number of scholars, institutions, and developers regarding the performance and security of blockchain networks. However, due to the distributed nature of blockchain technology, conducting comprehensive experiments on existing blockchain networks has proven to be exceedingly challenging. Building mirror blockchain networks using non-virtualized techniques, mirroring the original blockchain networks at a one-to-one scale, leads to unforeseeable consumption of physical resources and imposes significant human resource costs.

Within existing research, two primary methods have been proposed to over-come these difficulties: blockchain network simulation platform technology and blockchain automated deployment technology. Blockchain network simulation platform technology aims to emulate the behavioral characteristics of blockchain networks through predetermined algorithms. It provides an assessment frame-work for analyzing and evaluating these behavioral characteristics. However, it tends to overlook the underlying structural features of blockchain networks, including network topology, propagation mechanisms, transaction traffic, and other crucial elements. On the other hand, blockchain automated deployment technology focuses on automating the deployment process of blockchain networks. Yet, it tends to neglect the real-time nature of business data, such as transaction traffic and ledgers, within blockchain networks, as well as the authenticity of data sources.

To address these challenges, we introduce a cyber range based method for digital twin construction of a blockchain network. This method entails the realtime acquisition of blockchain network topology, transaction traffic, ledger data, and other pertinent information by embedding probe nodes within the original blockchain network. Subsequently, the core module analyzes and restructures this data, generating usable blockchain network topology and transaction traffic data for integration into cyber range, thus achieving a faithful digital twin of the original blockchain network.

In our work, the simulation challenge associated with the underlying structural characteristics of blockchain networks is effectively addressed through the utilization of cyber range techniques. The introduction of probe nodes into the original blockchain network enables data collection, real-time analysis of topology and traffic data, and synchronized updates of traffic and ledger data. This resolves issues related to the distortion of blockchain network simulation data. To overcome the challenge of probe nodes operating ineffectively in large-scale blockchain network scenarios, we introduce a novel reverse tracing strategy to optimize the selection of probe node locations, consequently enhancing the efficiency of measurement and monitoring. Additionally, the establishment of a mirror library further bolsters the scalability of our method. This paper uses probe nodes to collect data in three scenarios, where the original blockchain network is small-scale, large-scale and borderless. In each scenario we use the core module to realize network topology splicing by analyzing and modeling data, then deploy it in cyber range. After that, the transaction traffic and ledger data are imported into the deployed blockchain network in real time to achieve business data synchronization. In conclusion, our work offers researchers the most viable solution for deploying highly realistic and independently controllable blockchain network experimental environments. There are three main contributions of our work:

1. We propose an innovative method for digital twin based on cyber range, which realizes the digital twin of blockchain network. Furthermore, our work can solve the resource consumption and distortion of underlying network structural features of blockchain network deployment in reality.
2. For blockchain networks of different sizes, we propose the integration of probe nodes within the original blockchain network for real-time data collection. To overcome the challenges of probe nodes operating effectively within large-scale original blockchain network scenarios, this paper further introduces a reverse tracing strategy to optimize probe node placement. This strategy significantly enhances the efficiency of measurement and monitoring.
3. We introduce a mirror library based on real-world network infrastructure and node resources. The library enhances system scalability while offering flexible support for the digital twin of various blockchain networks within the cyber range, using authentic and usable data.

2 Related Work

In recent years, substantial scholarly and institutional efforts have been dedicated to blockchain network simulation. Numerous blockchain network simulation tools have emerged. In 2016, Arthur Gervais et al. introduced the Bitcoin Simulator framework [7] for the quantitative analysis of the performance of the Bitcoin blockchain network and the security of consensus algorithms. They proposed optimal adversarial strategies related to double-spending and selfish mining. In 2017, Lyubomir Stoykov et al. proposed VIBES [8], a configurable blockchain simulator for large-scale peer-to-peer networks. VIBES enables the exploration of crucial blockchain network features and metrics, simulating networks like Bitcoin, Ethereum, and Hyperledger with the capability to simulate large-scale blockchain networks comprising thousands of nodes. Tien Tuan Anh Dinh et al. introduced BLOCKBENCH [9], a tool for benchmarking and evaluating the performance of blockchain networks such as Ethereum and Hyperledger. It aims to assist developers, researchers, and system administrators in conducting performance testing and comparisons of different blockchain networks and protocols. In 2019, Yusuke Aoki et al. presented SimBlock [10], a tool for performance modeling and simulation of blockchain networks, including Bitcoin and Ethereum. It helps researchers and developers assess the impact of various blockchain protocols, algorithms, and parameters on system performance by simulating how blockchain networks operate.

However, these tools share a significant limitationthey have a restricted scope when it comes to simulating blockchain network feature data. These studies primarily focus on simulating macro-level characteristics of existing blockchain networks, emphasizing research on consensus algorithms, network behavior, and network performance. They overlook the underlying structural features of existing blockchain networks and do not consider the occurrence of business data within the network, such as transactions and ledgers. Moreover, a considerable number of security issues in existing blockchains are closely related to transactions and ledgers. Destructive transactions can be employed as tools by attackers, while ledger inconsistencies can lead to forks in blockchain networks. These issues are critical to the survival of existing blockchain networks and the viability of blockchain applications.

In the realm of research on the automated construction of blockchain networks, tools like PlaTIBART [11], Hyperledger Bevel [12], MixBytes Tank [13], and Blockchain as a Service (BaaS) platforms such as uBaaS [14] and Nut-BaaS [15] have emerged one after another. These tools and service platforms offer customized frameworks for blockchain construction. However, it is observed that their applicability is limited in that they support only specific blockchain networks. To address this limitation, NVAL [16] proposed an automated blockchain network deployment software framework, implementing an architecture-driven and community-supported approach to accomplish blockchain deployment and evaluation. This work suggested involving a technical community to provide more blockchain network models to enhance the framework's scalability. However, it is evident that the introduction of a community brings uncertainties in community management, as it cannot guarantee the enthusiasm and security of community contributions regarding blockchain network resources. Moreover, the framework did not propose a solution concerning the authenticity and real-time nature of deployed blockchain network. Consequently, generated blockchain networks, lacking real-time and authentic data, may lose their experimental value for scholars, institutions, and developers researching blockchain network performance and security.

3 Design

3.1 Architecture

As shown in Fig. 1, the overall architecture of the method can be divided into three main components: measurement and monitoring of the original blockchain network, blockchain network digital twin core framework and virtualized cyber range.

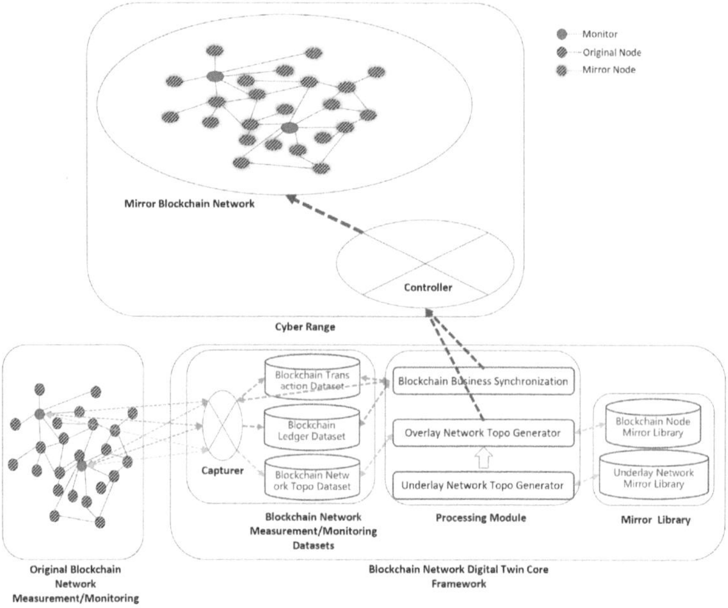

Fig. 1. The architecture of the digital twin construction of blockchain networks.

Measurement and Monitoring of the Original Blockchain Network. We use probe nodes to establish data connections with the original blockchain network, and synchronize transaction traffics, ledgers, and other data across the entire blockchain network. They also act as topology measurement nodes in borderless blockchain network scenarios. Additionally, their clocks are synchronized to ensure relative order in receiving transaction data. They continuously send the monitored transaction data, ledger data, and topology data in real-time to the blockchain transaction dataset, blockchain ledger dataset, and blockchain network topo dataset.

Blockchain Network Digital Twin Core Framework. The underlay network topo generator within the blockchain network digital twin core framework retrieves real underlay network mirror data from the mirror library to generate the underlay network topology profile for the blockchain network digital twin. We use the overlay network topo generator to create the overlay network topology profile, referred to as the blockchain network topology profile. Specifically, we fetch actual blockchain node mirror data from the library and combines it with the original blockchain network's topology data obtained from the blockchain network topo dataset. This profile is then deployed into the virtualized cyber range environment by the controller.

Cyber Range. The blockchain business synchronization module obtains transaction and ledger data from the blockchain transaction dataset and blockchain ledger dataset. It analyzes this data and then, through the controller, imports it into the deployed mirror blockchain network in cyber range, achieving synchronization of transaction and ledger data.

3.2 Module

Measurement and Monitoring Module. In order to ensure that the data of the mirror blockchain network originates from the existing blockchain network as much as possible, this paper proposes a dynamic measurement and monitoring method for the original blockchain network. The monitoring methods are categorized into ledger data monitoring and transaction traffic monitoring. For original blockchain networks of different sizes, the main difference between the monitoring methods is the difference in the number of probe nodes.

Ledger data monitoring ensures the synchronization of ledger data among nodes in the blockchain network. By introducing probe nodes into the original blockchain network, consistency is maintained. In the original blockchain network, probe nodes have the capacity to receive and relay transactions. When a probe node receives a transaction, a capturer immediately captures the transaction data in real-time using packet sniffing. This real-time data capture process facilitates transaction traffic monitoring.

For the measurement method, it can be divided into two specific cases:

Original Blockchain Network with Known Structure. In this scenario, the structural information of the original blockchain network, including node locations, degrees, and interconnections, is already available, eliminating the need for network measurement. However, monitoring requirements vary depending on the scale of the original blockchain network. We further divide the original blockchain networks into small-scale networks and large-scale networks.

For relatively small-scale blockchain networks, it is possible for probe nodes to monitor only the original blockchain nodes directly connected to them. This means that transaction traffic originating from original blockchain nodes is directly forwarded to probe nodes, without the need for multiple intermediate relays. This not only enhances the efficiency of synchronizing transaction and ledger data but also minimizes the resource consumption of probe nodes (Fig. 2).

In large-scale blockchain networks, due to the large scale of nodes in the original blockchain network, probe nodes cannot achieve rapid monitoring by directly connecting to each original blockchain node. Moreover, in practice, the monitoring scale of each probe node is uniform, and all probe nodes will eventually receive transaction traffic and other data from the entire original blockchain network. This makes it a key challenge to ensure that a limited number of probe nodes can still rapidly synchronize transaction traffic and other data.

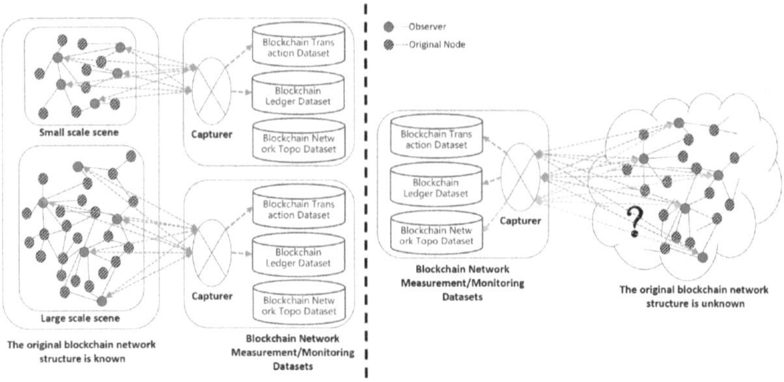

Fig. 2. The process of measurement and monitoring.

Given that the number and degree of probe nodes are limited, addressing the challenges mentioned above requires selecting a monitoring radius for probe nodes and minimizing the intersection between their monitoring ranges. To address this, we introduce a heuristic strategy for selecting probe node positions called reverse tracing. To provide a clearer understanding of the reverse tracing strategy, an example is illustrated in Fig. 3.

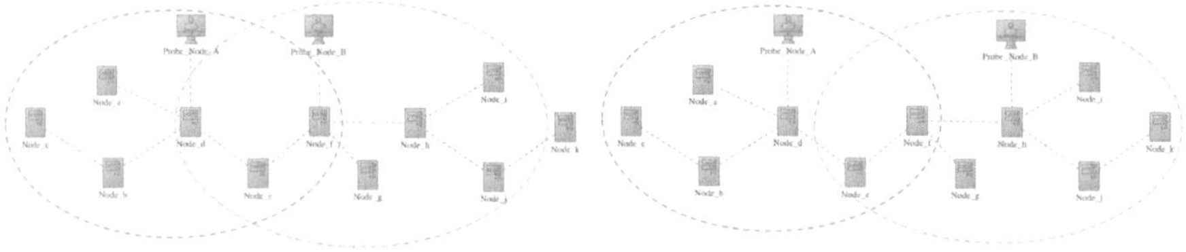

(a) Initial monitoring range of the probe nodes.

(b) The monitoring range of the probe nodes after reverse tracing.

Fig. 3. The monitoring range of the probe nodes.

Assuming that each probe node has a monitoring radius of 3 hops. Probe node A is directly connected to node d, and its monitoring range is $P = \{a, b, c, d, e, f\}$. Probe node B is directly connected to node f, and its monitoring range is $Q = \{d, e, f, g, h, i, j\}$. The intersection of monitoring ranges for probe node A and B, denoted as W, can be expressed as follows:

$$W = P \cap Q = \{d, e, f\} \tag{1}$$

Therefore, we need to backtrack probe node B in the opposite direction of probe node A to narrow down the intersection of their monitoring ranges. At this point, there are two choices for backtracking.

Option 1: Probe node B is directly connected to node g, resulting in a new monitoring range for B, denoted as $Q1 = \{e, f, g, h\}$.

$$W_1 = P \cap Q_1 = \{e, f\} \tag{2}$$

It can be observed that the intersection of monitoring ranges for probe nodes A and B has narrowed down. However, because probe node B is directly connected to the boundary node g, the monitoring range has been significantly reduced. This outcome is undesirable as it greatly reduces the monitoring efficiency of probe node B and wastes probe node resources.

Option 2: Probe node B is directly connected to node h, resulting in a new monitoring range for B, denoted as $Q2 = \{e, f, g, h, i, j, k\}$, as shown in Fig. 3b.

$$W_2 = P \cap Q_2 = \{e, f\} \tag{3}$$

Similarly, the intersection of monitoring ranges for probe nodes A and B has narrowed down, while the monitoring range for probe node B has not decreased. In other words, the monitoring efficiency of probe node B remains high, which is a favorable outcome.

Therefore, compared to Option 1, the reverse tracing strategy is more inclined toward Option 2. When probe nodes backtrack in reverse, we summarize three criteria for probe node position selection during reverse tracing:

1. They should refrain from establishing direct connections with boundary nodes to avoid narrowing their monitoring range, diminishing monitoring efficiency, and squandering probe node resources.
2. Prioritizing direct connections with high-degree nodes within the blockchain network is recommended. This approach expands the monitoring range of probe nodes and enhances monitoring efficiency compared to connecting with nodes of lower degrees.
3. When determining the backtracking scale, it should encompass the boundary nodes slated for monitoring. In the provided example, probe node B's back-tracking scale is limited to boundary node g, significantly improving probe node monitoring efficiency while ensuring synchronization with boundary node g.

To validate the effectiveness of the reverse tracing strategy, we compared the relationships between probe node quantity, monitoring radius, and transaction traffic monitoring rate before and after probe node position optimization. Experimental evaluations are presented in the Sect. 4.

Original Blockchain Network with Unknown Structure. This means that the location and degree information of the nodes in the blockchain network and the connection information between nodes are unknown. This requires measuring the original blockchain network first, and then monitoring the ledger data and transaction traffic. This paper primarily focuses on the measurement and monitoring methods of borderless blockchain networks, such as Ethereum and Bitcoin blockchains, as the measurement methods for borderless blockchain networks are applicable to previously defined scales of blockchain networks. In this context, monitoring ledger data and transaction traffic methods are no different from those in previously defined scales of blockchain networks. Therefore, discussing measurement methods for borderless blockchain networks when the blockchain network structure is unknown is more representative.

Numerous measurement methods and techniques have emerged in existing research on blockchain network topology analysis and measurement [17–26]. In this paper, we utilized existing measurement methods to measure the blockchain network, obtaining valuable topology feature parameters such as node degrees, broadcast hop counts, and average delays. These measurements provide real and valuable topology feature data for our digital twin of the original blockchain network.

Core Framework. The core framework proposed in this paper for the blockchain network digital twin primarily consists of three components: blockchain network measurement and onitoring datasets, mirror library, and processing module. The following will provide specific explanations for these three components.

Blockchain Network Measurement and Monitoring Datasets. The datasets consists of capturer, transaction dataset, ledger dataset and topology dataset. Capturer is primarily responsible for capturing data from the original blockchain network's probe nodes and storing it in the corresponding dataset. For instance, it employs dynamic packet capture techniques to obtain transaction traffic data from the original blockchain network, which is stored in the blockchain transaction dataset. Additionally, to ensure that transactions are promptly injected into the mirror blockchain network within the cyber range, capturer also forwards the real-time captured transaction traffic to the processing module's blockchain business synchronization for parsing. In summary, the blockchain network measurement and monitoring datasets play a crucial role in supporting the deployment and data synchronization of the mirror blockchain network within the cyber range.

Mirror Library Synchronization Module. As shown in Fig. 4, this paper introduces a mirror library based on real-world network data and node resources, enhancing the system's scalability and providing flexible support for various blockchain networks on cyber range. We retrieved underlay network instance data from the internet open network database CAIDA [27], extracted underlay network node datasets and underlay network AS datasets that meet the core framework's requirements. For server, router, and other resource mirror data, we utilized open-source mirrors such as CentOS and Vyos. Ultimately, the data, along with BGP and OSPF protocol rule data, were consolidated into an underlay network mirror library. In practice, blockchain mirror data is open-source on the internet.

We can download various types of blockchain mirror data from the internet to local storage. Additionally, we support custom modifications of downloaded blockchain node client mirror data to create improved versions that meet specific requirements. These are then gathered into a blockchain node mirror library stored on a local server for access.

The design of this mirror library significantly enhances the scalability ofour method, while also reducing complexity and costs compared to other methods [16].

Processing Module. In Fig. 5, the processing module is divided into three layers: the underlay network topology generation layer, the overlay network topology generation layer, and the blockchain business synchronization layer.

The underlay network topology generation layer generates the underlay network topology profile, providing foundational network topology data for the overlay network topology generation layer. The overlay network topology generation layer builds the overlay network topology, known as the blockchain network topology profile, based on

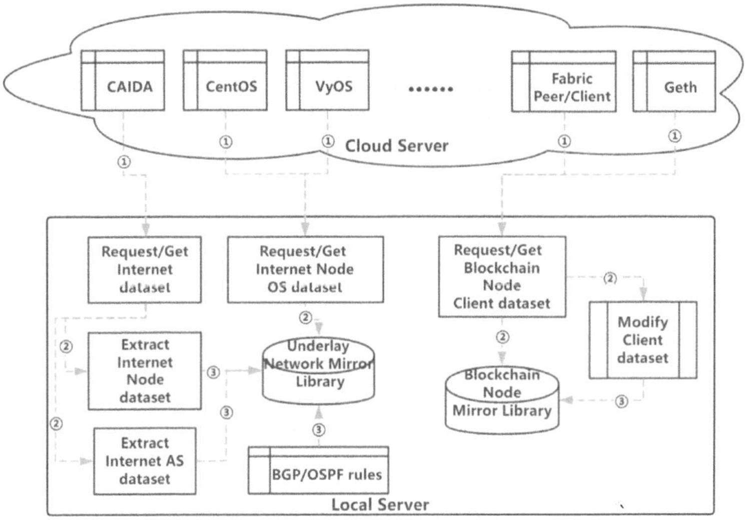

Fig. 4. The mirror library synchronization process.

this underlay topology profile. finally, after deployment in cyber range, the blockchain business synchronization layer sends transaction data from the original blockchain network to the mirror blockchain network.

It executes transactions and synchronizes ledger data to maintain consistency between the original blockchain network and the mirror blockchain network.

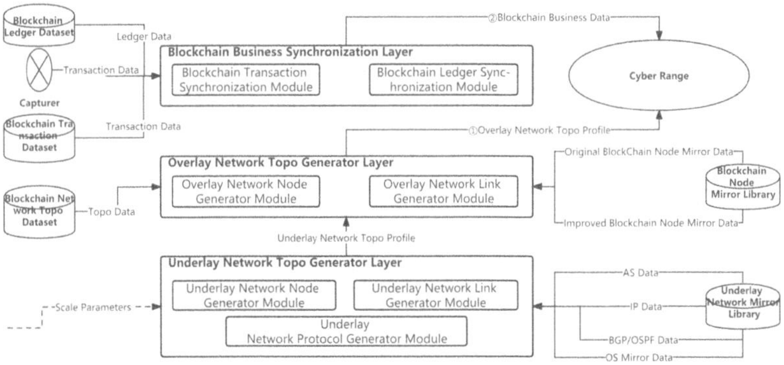

Fig. 5. The flow of the processing module.

Specifically, the underlay network topology generation layer consists of three functional modules: the underlay network node generation module, the underlay network connection generation module, and the underlay network protocol generation module. These modules are responsible for generating underlay network nodes, connections, and network protocols, and they are invoked sequentially to create the underlay network topology profile.

Here is a concrete process of the underlay network topology generation layer. first, the underlay network node generation module is called according to the network scale parameter to obtain the IP data and OS image data of the underlay network node from the underlay network mirror library to generate the underlay network node profile. In the second step, the underlay network connection generation module obtains AS data from the mirror library and combines it with the underlay network node profile generated in the previous step to create subnet and connection data for the underlay network. These data are then integrated according to rules to form the underlay network topology structure data. finally, the underlay network protocol generation module retrieves routing protocol rule data, such as BGP and OSPF protocols, from the underlay network mirror library. It combines this data with the IP and subnet data of routing nodes to generate routing profiles, which serve as the communication foundation for the underlay network. Through these three steps, the underlay network topology profile is generated and forwarded to the overlay network topology generation layer. This layer can be formally expressed as follows:

$$f(n) = H(v, e) \tag{4}$$

n: Underlay network scale parameter. $f(n)$: Function to retrieve underlay network mirror data based on the underlay network scale. $H(v, e)$: The result set of $f(n)$, including node data(v) and connection data(e).

$$G(H, p) = F \tag{5}$$

p: Routing protocol rule data. $G(H, p)$: Function to generate the underlay network topology profile based on the underlay network node, connection data, and routing node protocol data. F: The final result of $G(H, p)$, i.e., the underlay network topology profile.

The overlay network topology generation layer is responsible for creating the topology of the mirror blockchain network. It consists of two main functional modules: the overlay network node generation module and the overlay network connection generation module.

The process of overlay network topology generation layer is a bit shorter. firstly, it obtains the underlay network topology profile generated by the underlay network topology generation layer. The overlay network node generation module retrieves blockchain node mirror data from the mirror library and matches it with the nodes in the underlay network topology profile to create overlay network node profiles. Additionally, the overlay network connection generation module periodically synchronizes the topology data from the original blockchain network to establish connection relationships among overlay network nodes. This process ultimately leads to the generation of the overlay network topology profile, which is then deployed in cyber range by controller. This completes the basic construction of the mirror blockchain network.

The blockchain business synchronization layer is primarily responsible for the analysis and synchronization of business data from the original blockchain network. It includes modules such as transaction data synchronization module and ledger data synchronization module.

Blockchain ledger data synchronization involves obtaining ledger data from the blockchain ledger dataset and importing it into the deployed mirror blockchain network. During the initialization of the mirror blockchain network, the blockchain transaction synchronization module parses a large batch of transaction data obtained from the blockchain transaction dataset. Since the clock of the probe nodes is synchronized, the probe nodes add timestamps to the received transaction data according to their local clocks. The blockchain transaction synchronization module first sorts the transactions based on their timestamps to maintain their relative order. It then extracts data such as account addresses and content from the transactions. finally, these transactions are mirrored and executed in the mirror blockchain network. Once the mirror blockchain network reaches a consistent state with the original blockchain network, the blockchain transaction synchronization module will continuously and directly synchronize the transaction data captured by the capturer in real-time. After that this module will parse the data, and import it into the mirror blockchain network for execution, ensuring the accuracy of the digital twin is maintained.

Cyber Range. Cyber range is a network system simulation platform that simulates a target network based on a set of software and hardware resources. It serves various purposes such as network attack and defense exercises, security tool evaluations, and validation of new network technologies. It is considered a scientific instrument for research in cybersecurity. The three main functions of cyber range are as follows:

1. *Simulation of target network.* It utilizes computer and network hardware and software resources to simulate a target network.
2. *Deployment of network applications.* The controller deploys specified network resources and initiates the operation of the entire network.
3. *Simulation of network behavior.* It simulates the behavior of users and systems within the network environment.

Since the overlay network topology profile generated by the core framework is only logical data and lacks the corresponding resources, it cannot be directly used. To generate usable blockchain network scenarios, the controller will allocate available resources in cyber range based on the overlay network topology profile for instantiated deployment. Then cyber range will subsequently synchronize the transaction and ledger data from the blockchain business synchronization layer.

4 Evaluation

In this section, we focus on evaluating the effectiveness of the reverse tracing strategy and the accuracy of our method for digital twin. The effectiveness of the reverse tracing strategy is measured through relevant metrics of probe nodes, such as synchronization time, monitoring radius, and the number of probe nodes. Additionally, we assess the accuracy of the method for digital twin by comparing the differences in blocks between the original blockchain network and the mirror blockchain network. The experimental environment for our cyber range consists of 12 servers, each configured with 32 GB of RAM and 4 TB of storage.

4.1 Evaluation of the Reverse Tracing Strategy

When evaluating the reverse tracing strategy, we compared the synchronization time, monitoring radius, and the number of probe nodes before and after applying the strategy in the same original blockchain network scenario. Synchronization time refers to the time it takes for a probe node to receive a transaction after it was sent, and the monitoring radius represents the maximum number of hops a probe node can monitor. These metrics help gauge the optimization effects of this strategy, and the results are depicted in the Fig. 6.

We can find that in the original blockchain network scenario, as the monitoring radius of probe nodes decreases, the synchronization speed increases, but it requires a larger number of probe nodes to achieve this. Furthermore, when probe nodes have the same monitoring radius, the scenario utilizing the reverse tracing strategy exhibits a reduction of approximately 35% in the required number of probe nodes compared to the scenario without this strategy.

It is evident that the reverse tracing strategy we proposed significantly optimizes the selection of probe node positions. While maintaining the same synchronization speed, it reduces the consumption of probe node resources, providing substantial assistance to the entire digital twin construction of blockchain networks.

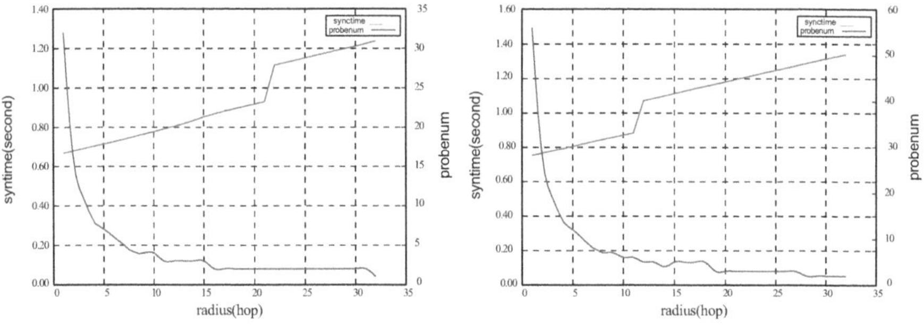

(a) Before applying reverse tracing strategy. (b) Before applying reverse tracing strategy.

Fig. 6. The synchronization time, monitoring radius, and the number of probe nodes.

4.2 Evaluation of the Digital Twin Construction of Blockchain Network

In the existing evaluation framework for digital twin [28], researchers commonly focus on the accuracy of digital twin. In our evaluation of the method for digital twin construction of blockchain networks, we also emphasize accuracy as a crucial assessment metric.

When evaluating the accuracy of our method, we primarily analyze the degree of difference between blocks in the original blockchain and blocks in the mirror blockchain. Furthermore, we introduce Levenshtein Distance as a specific representation of the degree of difference.

In Fig. 7, we can see that the blocks in the original and mirror blockchains continue to align as the transactions begin to synchronize. It is not until after about the 7th block of synchronization that the levenshtein distance of the blocks starts spiking and remains so for some time. We further analyze this to get that there is no change in the order of transactions in the blocks, but it appears that the transactions are not in the same block. The reason could be that the transaction synchronization speed cannot keep up with the block out speed of the mirror blockchain. After synchronizing about 15 blocks, a significant fluctuation occurs because the order of transactions in the blocks of the mirror blockchain is out of order. At this moment it is necessary to synchronize the ledger data of the original blockchain to maintain consistency.

It can be seen that the digital twin technique is able to ensure that there is no need to synchronize the ledger data within a range of about 15 blocks, and the digital twin technique has considerable accuracy.

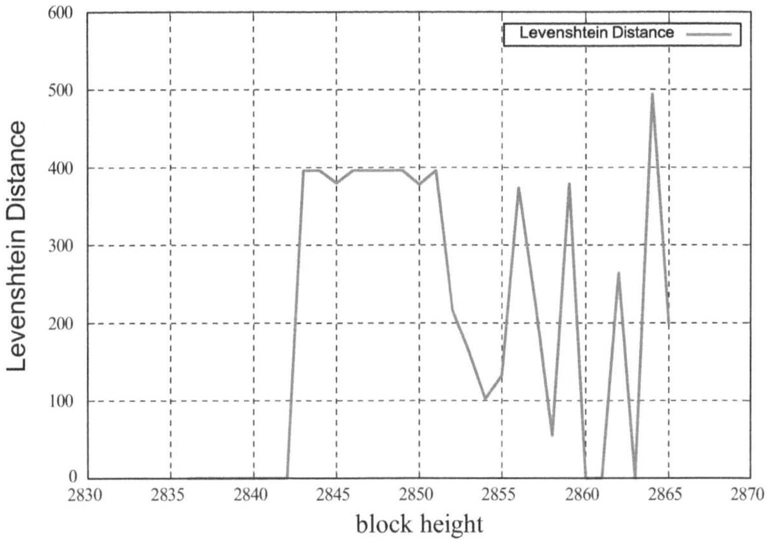

Fig. 7. The Levenshtein Distance of the blocks.

5 Conclusion

We propose a cyber range based method for digital twin construction of a blockchain network, encompassing the entire process from original blockchain network data collection, network topology generation, and business data synchronization to the instantiation and deployment in cyber range. This process achieves a highly realistic mirror blockchain network. Furthermore, the construction of an underlying network mirror library enhances the scalability of the proposed approach. In addition, we present a novel reverse tracing strategy to optimize the selection of probe node positions, improving measurement and monitoring efficiency. The effectiveness of this strategy and the accuracy of the digital twin method are assessed through extensive evaluations.

The research anticipates that the proposed methodology will provide valuable support to scholars, institutions, and developers in the field of blockchain research. By addressing challenges in data measurement, monitoring, and automation, this method contributes significantly to advancing frontier research and solving practical issues in the blockchain field. In summary, this paper holds great significance in driving cutting-edge research directions and addressing real-world problems in the field of blockchain.

Acknowledgements. This work was supported by the National Key Research and Development Program (2022YFB2702300) and Cultivation Project of PZL (No. PZL2022KF0013).

References

1. Nakamoto, S.: Bitcoin: A peer-to-peer electronic cash system. Decentralized business review (2008)
2. Buterin, V., et al.: A next-generation smart contract and decentralized application platform. White Paper **3**(37), 2–1 (2014)
3. Androulaki, E., et al.: Hyperledger fabric: a distributed operating system for permissioned blockchains. In: Proceedings of the thirteenth EuroSys conference, pp. 1–15 (2018)
4. EEA Homepage, https://entethalliance.org/. Last Accessed 21 Sep 2023
5. Zheng, Z., Xie, S., Dai, H.N., Chen, X., Wang, H.: Blockchain challenges and opportunities: a survey. Int. J. Web Grid Serv. **14**(4), 352–375 (2018)
6. Slowmist Homepage: https://hacked.slowmist.io/. Last Accessed 15 Sep 2023
7. Gervais, A., Karame, G.O., Wüst, K., Glykantzis, V., Ritzdorf, H., Capkun, S.: On the security and performance of proof of work blockchains. In: Proceedings of the 2016 ACM SIGSAC conference on computer and communications security, pp. 3–16 (2016)
8. Stoykov, L., Zhang, K., Jacobsen, H.A.: Vibes: fast blockchain simulations for large-scale peer-to-peer networks. In: Proceedings of the 18th ACM/IfiP/USENIX Middleware Conference: Posters and Demos, pp. 19–20 (2017)
9. Dinh, T.T.A., Wang, J., Chen, G., Liu, R., Ooi, B.C., Tan, K.L.: Blockbench: A framework for analyzing private blockchains. In: Proceedings of the 2017 ACM international conference on management of data, pp. 1085–1100 (2017)
10. Aoki, Y., Otsuki, K., Kaneko, T., Banno, R., Shudo, K.: Simblock: A blockchain network simulator. In: IEEE INFOCOM 2019-IEEE Conference on Computer Communications Workshops (INFOCOM WKSHPS), pp. 325–329. IEEE (2019)
11. Walker, M.A., Dubey, A., Laszka, A., Schmidt, D.C.: Platibart: a platform for transactive iot blockchain applications with repeatable testing. In: Proceedings of the 4th Workshop on Middleware and Applications for the Internet of Things, pp. 17–22 (2017)
12. Hyperledger Bevel: https://github.com/hyperledger/bevel. Last Accessed 13 Sep 2023
13. Mixbytes Tank: https://github.com/mixbytes/tank. Last Accessed 11 Sep 2023
14. Lu, Q., Xu, X., Liu, Y., Weber, I., Zhu, L., Zhang, W.: UBaaS: a unified blockchain as a service platform. Futur. Gener. Comput. Syst. **101**, 564–575 (2019). https://doi.org/10.1016/j.future.2019.05.051
15. Zheng, W., Zheng, Z., Chen, X., Dai, K., Li, P., Chen, R.: Nutbaas: a blockchain as-a-service platform. IEEE Access **7**, 134422–134433 (2019)
16. Tran, N.K., Babar, M.A., Walters, A.: A framework for automating deployment and evaluation of blockchain networks. J. Netw. Comput. Appl. **206**, 103460 (2022)
17. Wang, T., Zhao, C., Yang, Q., Zhang, S., Liew, S.C.: Ethna: Analyzing the underlying peer-to-peer network of ethereum blockchain. IEEE Trans. Netw. Sci. Eng. **8**(3), 2131–2146 (2021)

18. Franzoni, F., Salleras, X., Daza, V.: Atom: Active topology monitoring for the bitcoin peer-to-peer network. Peer-to-Peer Netw. Appl. **15**(1), 408–425 (2022). https://doi.org/10.1007/s12083-021-01201-7
19. Li, R., Zhu, J., Xu, D., Wu, F., Gao, J., Zhu, L.: Bitcoin network measurement and a new approach to infer the topology. China Commun. **19**(10), 169–179 (2022)
20. Gebraselase, B.G., Helvik, B.E., Jiang, Y.: Bitcoin P2P network measurements: a testbed study of the effect of peer selection on transaction propagation and confirmation times. IEEE Trans. Netw. Serv. Manag. **19**(4), 3975–3987 (2022). https://doi.org/10.1109/TNSM.2022.3216955
21. Grundmann, M., Baumstark, M., Hartenstein, H.: On the peer degree distribution of the bitcoin p2p network. In: 2022 IEEE International Conference on Blockchain and Cryptocurrency (ICBC), pp. 1–5. IEEE (2022)
22. Li, K., Tang, Y., Chen, J., Wang, Y., Liu, X.: Toposhot: uncovering ethereum's network topology leveraging replacement transactions. In: Proceedings of the 21st ACM Internet Measurement Conference, pp. 302–319 (2021)
23. Cao, T., Yu, J., Decouchant, J., Luo, X., Verissimo, P.: Exploring the monero peer-to-peer network. In: Bonneau, J., Heninger, N. (eds.) FC 2020. LNCS, vol. 12059, pp. 578–594. Springer, Cham (2020). https://doi.org/10.1007/978-3-030-51280-4_31
24. Su, M., Zhang, H., Du, X., Fang, B., Guizani, M.: A measurement study on the topologies of bittorrent networks. IEEE J. Sel. Areas Commun. **31**(9), 338–347 (2013)
25. Cortes-Goicoechea, M., Bautista-Gomez, L.: Discovering the ethereum2 p2p network. In: 2021 Third International Conference on Blockchain Computing and Applications (BCCA), pp. 81–88. IEEE (2021)
26. Huang, Z.B., Yang, G., Shen, Z.K.: Design of a network storage system based on peer-to-peer network with supernode. In: CIBDA 2022; 3rd International Conference on Computer Information and Big Data Applications, pp. 1–7. VDE (2022)
27. CAIDA Data Server: https://publicdata.caida.org/datasets/topology/ark/ipv4/itdk/. Last Accessed 6 Sep 2023
28. Liu, S., Sun, Y., Zheng, P., Lu, Y., Bao, J.: Establishing a reliable mechanism model of the digital twin machining system: an adaptive evaluation network approach. J. Manuf. Syst. **62**, 390–401 (2022)

Author Index

Z. Gu et al. (Eds.): NSE 2023, CCIS 2063, pp. 401–403, 2024.
https://doi.org/10.1007/978-981-97-4519-7

SPRINGER NATURE

GPSR Compliance

The European Union's (EU) General Product Safety Regulation (GPSR) is a set of rules that requires consumer products to be safe and our obligations to ensure this.

If you have any concerns about our products, you can contact us on ProductSafety@springernature.com

In case Publisher is established outside the EU, the EU authorized representative is:

Springer Nature Customer Service Center GmbH
Europaplatz 3
69115 Heidelberg, Germany

The manufacturer's authorised representative in the EU is Springer Nature Customer Service Centre GmbH, Europaplatz 3, 69115 Heidelberg, Germany. If you have any concerns regarding our products, please contact ProductSafety@springernature.com

Printed and bound by CPI Group (UK) Ltd, Croydon, CR0 4YY
01/10/2026
02225834-0013